Günter Helmchen
with J. Dibo, D. Flubacher, B. Wiese (Eds.)

Organic Synthesis *via* Organometallics
OSM 5

Günter Helmchen
with
Jörg Dibo
Dietmar Flubacher
Burkhard Wiese (Eds.)

Organic Synthesis *via* Organometallics

OSM 5

Proceedings of the Fifth Symposium in Heidelberg, September 26 to 28, 1996

Produced by Lengericher Handelsdruckerei, Lengerich
Printed on acid-free paper

ISBN 978-3-642-49350-8   ISBN 978-3-642-49348-5 (eBook)
DOI 10.1007/978-3-642-49348-5

Contents

Preface

Interdisciplinarity is an important aspect of modern science. A convincing example is provided by organometallic chemistry, a discipline at the interface of inorganic and organic chemistry. Over the last 20 years, organometallic chemistry has served as an inexhaustible source of new methods of ever increasing value to organic synthesis both in the laboratory and in large scale industrial plants using transition metal catalysts.

In order to strengthen collaboration between inorganic and organic chemists the "Vokswagen-Stiftung" started a new program "Organic Synthesis *via* Organometallics" in 1986. This program grew up to more than 60 projects over a period of six years. In addition to its main topic organic synthesis it covered mechanistic and structural aspects of organometallic chemistry. Scientists participating in the program met with other experts from academia and industry in a series of five symposia held in Hamburg (February 1986), Würzburg (October 1988), Marburg (July 1990) and Aachen (November 1992).

The fifth and final symposium took place in Heidelberg from September 26 to 28, 1996. More than 20 well-known experts from Germany and abroad presented recent developments in a broad range of timely areas: polymer synthesis *via* organometallics, synthesis of new organometallics, mechanism of catalytic processes, asymmetric catalysis of carbon carbon bond forming reactions and stereoselective synthesis of biologically active compounds. We are confident that the results collected in this volume present a useful survey over the fascinating and expanding field of organometallic chemistry and catalysis, and we hope that this book will help to stimulate further progress in this important area of modern chemistry.

Heidelberg, March 15, 1997

Günter Helmchen
Jörg Dibo
Dietmar Flubacher
Burkhard Wiese

Carbon-rich Organoruthenium and Selective Catalytic Transformations of Alkynes

Pierre H. Dixneuf and Christian Bruneau

URA CNRS 415, Université de Rennes, Campus de Beaulieu, 35042 Rennes Cedex
(France), "dixneuf@univ rennes1.fr"

The activation of terminal alkynes under mild conditions is a topic of current interest for both the design of new organometallics containing a conjugated chain and the selective catalytic transformations of alkynes into functional organic molecules.

Organometallics containing π-conjugated, carbon-rich, rigid chains have a potential for the design of geometrically arranged molecules with specific properties for material science or as precursors for conjugated oligomers and metal containing polymers[1,2]. Organometallics with carbon-rich, unsaturated chains have already shown liquid crystals[3] or non linear optical properties[4]. Their new intrinsic properties have motivated the search for the access to new unsaturated, rigid organometallic molecules or bimetallic complexes containing carbon-rich bridges.

Unsaturated alkynes and polyynes constitute the key-stones for the building of carbon-rich ligands, bridges or organometallics. They have already been used to produce $LnM(C)_xML_n$ bimetallic systems containing bridges made of carbon wire[5] or to generate bimetallic complexes with carbon-rich bridges[6].

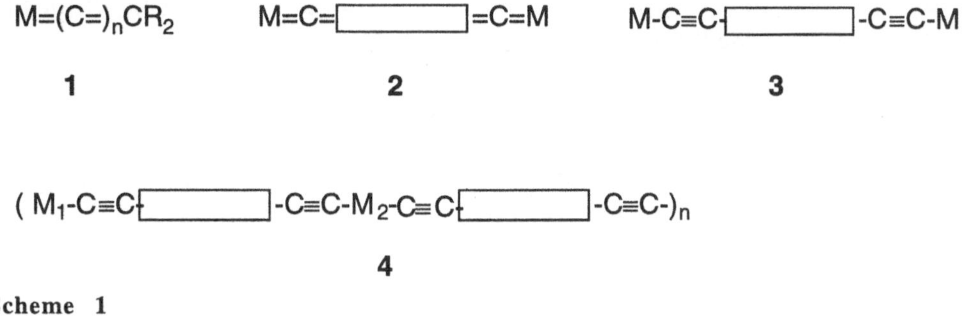

Scheme 1

The search for novel carbon-rich organometallics by the Rennes team has been orientated toward the generation of new metallacumulenes of type **1**, bimetallic complexes containing bis cumulene or bis alkynyl bridges of type **2** and **3**, and their use for the access to metal containing oligomers of type **4** (Scheme 1).

To reach these target molecules the activation of unsaturated alkynes and functional diynes by selected ruthenium complexes has been studied. The activation of terminal alkynes, has shown that *electron-rich ruthenium(II)* precursors favour the stoechiometric activation of alkynes and stabilize the resulting unsaturated complexes. By contrast *electron-deficient ruthenium(II)* complexes favour *electrophilic* activation of terminal alkynes, allow regioselective additions to the carbon-carbon triple bond and offer selective *catalytic* transformations of alkynes into a variety of useful functional molecules.

It is the aim of this review to present both aspects of the ruthenium activation of alkynes:

    1. the access to carbon-rich organoruthenium derivatives and

    2. the ruthenium-catalysed transformations of alkynes.

1 Carbon-rich Organoruthenium Derivatives

1.1 Preparation of Metallacumulenes $M=(C=)_nCR_2$

Metal-vinylidene complexes constitute the simplest metallacumulenes $LnM=C=CR_2$. Although the first complex was reported in 1972,[7] metal-vinylidene derivatives have gained interest since they have been produced by simple activation of terminal alkynes[8]. Initial studies with ruthenium complexes showed that the activation of terminal alkynes in methanol depended on the electron-richness of the ruthenium(II) precursors. Electrophilic $RuCl_2(PR_3)(arene)$[9] complexes afforded carbene complexes, *via* the vinylidene intermediate (Eq. 1), whereas electron-rich ruthenium complexes $RuCl(PR_3)_2(C5R_5)$[10,11] and $RuCl_2(Ph_2P(CH_2)_nPPh_2)_2$[12,13] led to stable vinylidene ruthenium(II) complexes (Eq. 2).

Eq. 1

    (Ru-Cl + HC≡C-R + MeOH + NaPF₆ $\longrightarrow$ (Ru=C(OMe)CH$_2$R]PF$_6$

    (Ru-Cl : $(C_6Me_6)(Me_3P)RuCl_2$

Eq. 2

    [(Ru-Cl + HC≡C-R + MeOH + NaPF$_6$ $\longrightarrow$ [Ru=C=CHR]PF$_6$

    [Ru-Cl : $RuCl(PR_3)_2(C_5R_5)$, $RuCl_2(diphosphine)_2$

Although the first allenylidene-metal complexes $M=C=C=CR_2$ were reported in 1976 by Fischer[14] and Berke[15], this is only after the discovery by Selegue[16], showing that

$RuCl(PMe_3)_2(C_5H_5)$ promoted the dehydration of a propargyl alcohol derivative to directly generate the allenylidene $[(C_5H_5)(PMe_3)_2Ru=C=C=CPh_2]PF_6$, that the chemistry of allenylidene derivatives has known new developments.

The study of the activation of prop-2-yn-1-ols by ruthenium(II) complexes has shown that, depending on the electron-richness of the ruthenium(II) precursors, several complexes could be formed and isolated in the presence of a non coordinating salt (Scheme 2). The electron-richness of the ruthenium precursor is evaluated by its oxidation potential measured by cyclic voltammetry. With very electron-rich complexes such as $RuCl(PMe_2Ph)_2(C_5Me_5)$[11] ($E° = +0.15$ V_{SCE}) the vinylidene complex **B** is isolated and can not be dehydrated. With $RuCl_2(Ph_2P(CH_2)_nPPh_2)_2$[17,18] ($n = 1, 2$; $E_o = +0.4 - 0.5$ V_{SCE}), the formed allenylidene complex **C** is stable and inert toward the addition of methanol. By contrast, with the electron-deficient $RuCl_2(PMe_3)(C_6Me_6)$[19] precursor ($E° = +0.77$ V_{SCE}) the allenylidene intermediate **C** is readily formed and readily adds methanol to afford, in a one pot reaction, the alkenyl carbene complexes of type **D** (Scheme 2).

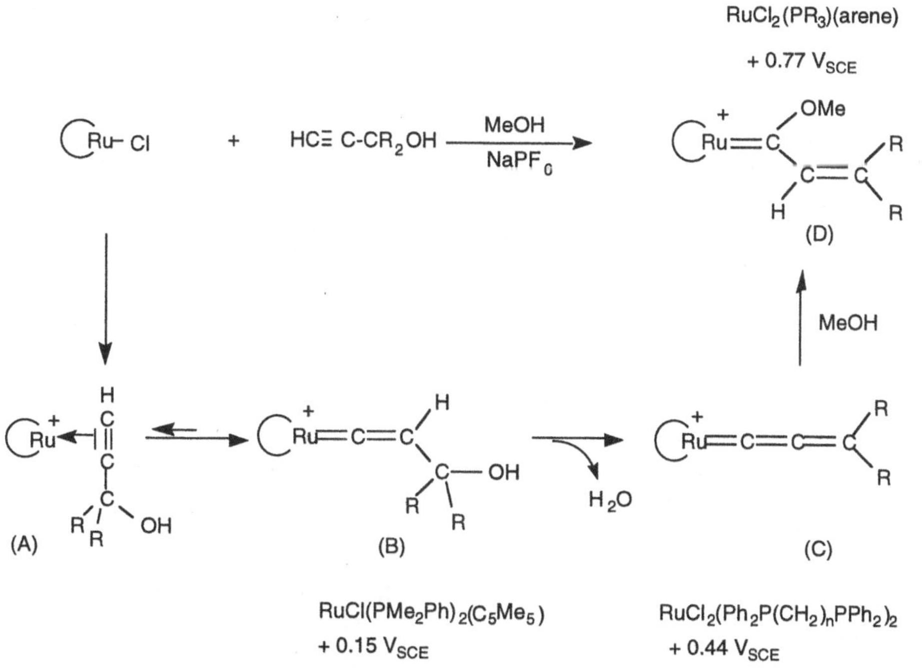

Scheme 2

These studies suggested that in order to produce allenylidene complexes $Ru=(C=)_nCR_2$ ($n = 2$) or higher metallacumulenes ($n \geq 2$) electron-rich precursors $RuCl_2(diphosphine)_2$ should be selected. Indeed, a variety of allenylidene ruthenium complexes have been prepared by activation of prop-2-yn-1-ols by suitable ruthenium derivatives as shown in (Eq. 3), the activation allows to produce stable secondary and alkenyl allenylidene complexes (Scheme 3)[17,20].

Eq. 3

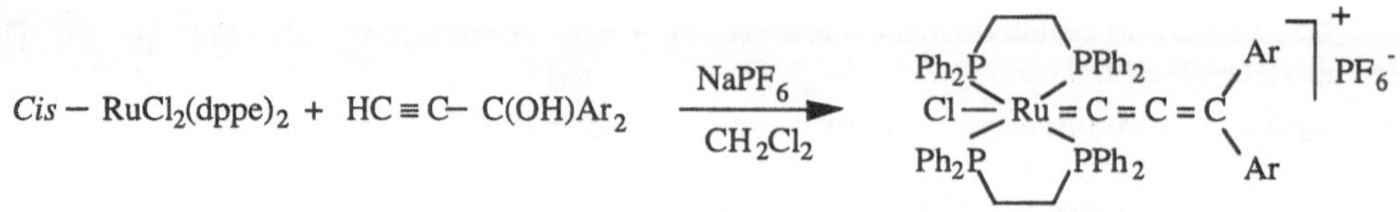
Cis – RuCl2(dppe)2 + HC ≡ C– C(OH)Ar2
NaPF6
CH2Cl2
PF6-

Ar = C_6H_5, p-C_6H_4-Cl, p-C_6H_4-F

This method has been recently applied to the formation of binuclear bis allenylidene complexes from $RuCl_2(PPh_3)_3$[20] (Scheme 4). Several other allenylidene-ruthenium complexes have been reported recently.

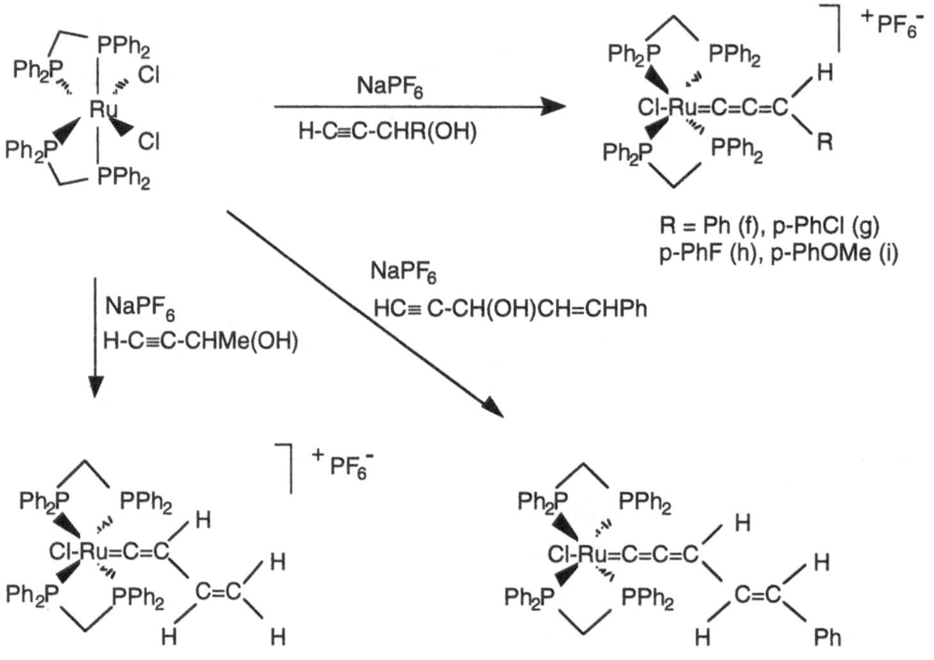
NaPF6
H-C≡C-CHR(OH)
R = Ph (f), p-PhCl (g)
p-PhF (h), p-PhOMe (i)
NaPF6
HC≡ C-CH(OH)CH=CHPh
NaPF6
H-C≡C-CHMe(OH)

Scheme 3

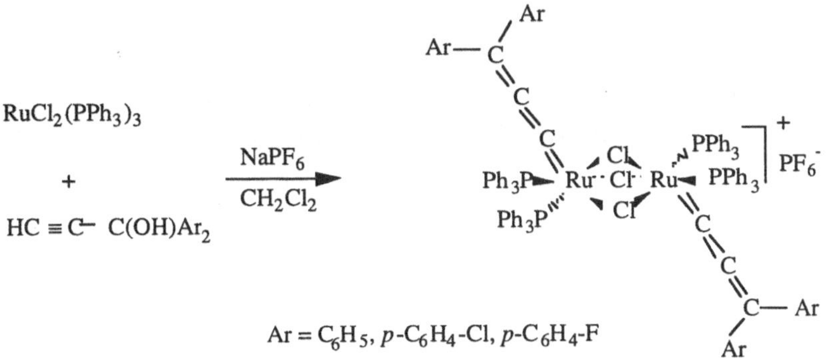

$$RuCl_2(PPh_3)_3$$

$$+$$

$$HC \equiv C-C(OH)Ar_2 \xrightarrow[\text{CH}_2\text{Cl}_2]{\text{NaPF}_6}$$

Ar = C$_6$H$_5$, p-C$_6$H$_4$-Cl, p-C$_6$H$_4$-F

Scheme 4

In order to reach higher metallacumulenes, the above results suggested that the activation of a terminal penta-1,3-diyne, having a leaving group at carbon C(5), should lead to a penta-1,2,3,4-tetraenylidene intermediate (Eq. 4).

Eq. 4

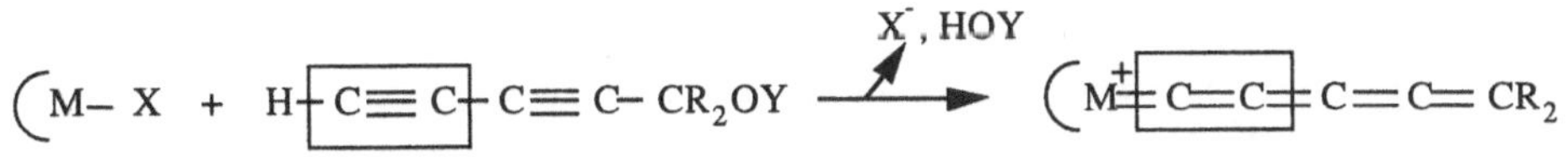

The activation of HC≡C-C≡C-CPh$_2$(OSiMe$_3$) by electrophilic RuCl$_2$(PMe$_3$)(arene) complexes[21-23], in the presence of methanol or a weak base, led to alkenyl allenylidene ruthenium complexes. The nature of these derivatives supported the formation of the corresponding cumulene Ru=(C=)$_4$CPh$_2$ (**I**) intermediate followed by addition to carbon C(3) (Scheme 5).

5

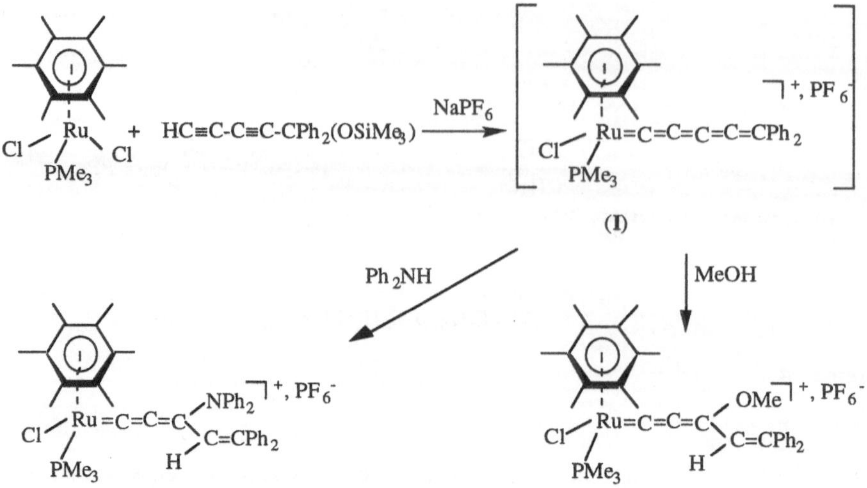

Scheme 5

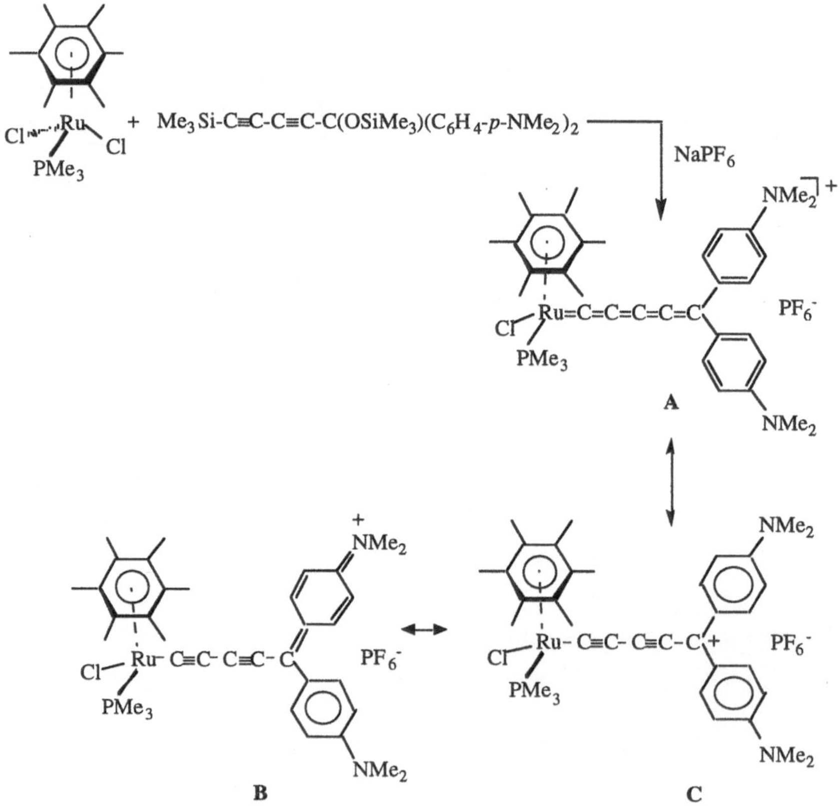

Scheme 6

The activation of $Me_3SiC_4C(OSiMe_3)(C_6H_4NMe_2)_2$[24], thus containing electron-releasing aryl groups at C(5) allowed to isolated a pseudo C(5)-cumulene, as its structure shows a diynyl character and thus a contribution of canonical forms **A**, **B** and **C** (Scheme 6).

To reach a real $Ru=C=C=C=C=CR_2$ assembly the electron-rich ruthenium precursor $RuCl_2(dppe)_2$ was used. The treatment of the corresponding $Ru(C_4)CPh_2OSiMe_3$ with $Ph_3C^+PF_6^-$ led to the isolation of the metallacumulene **II**, the X-ray structure of which showed the linear arrangement $Cl\text{-}Ru=(C=)_4CPh_2$[25] (Scheme 7).

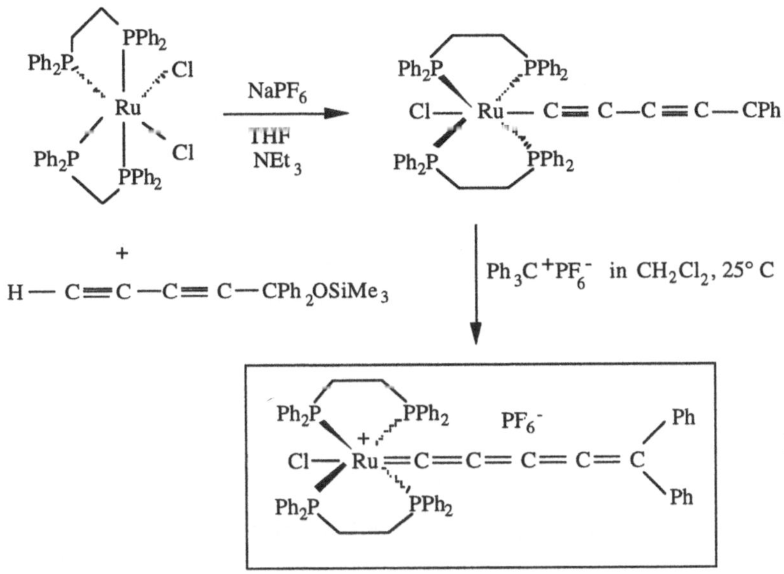

Scheme 7

The isolation of the first $M=(C=)_4CR_2$ complex allowed to show its reactivity and the electrophilicity of carbon C(3) which adds methanol or secondary amine, affording alkenyl allenylidenes, or undergoes electrophilic substitution on one phenyl group to give the bicyclic allenylidene ligand in **III**[25] (Scheme 8).

7

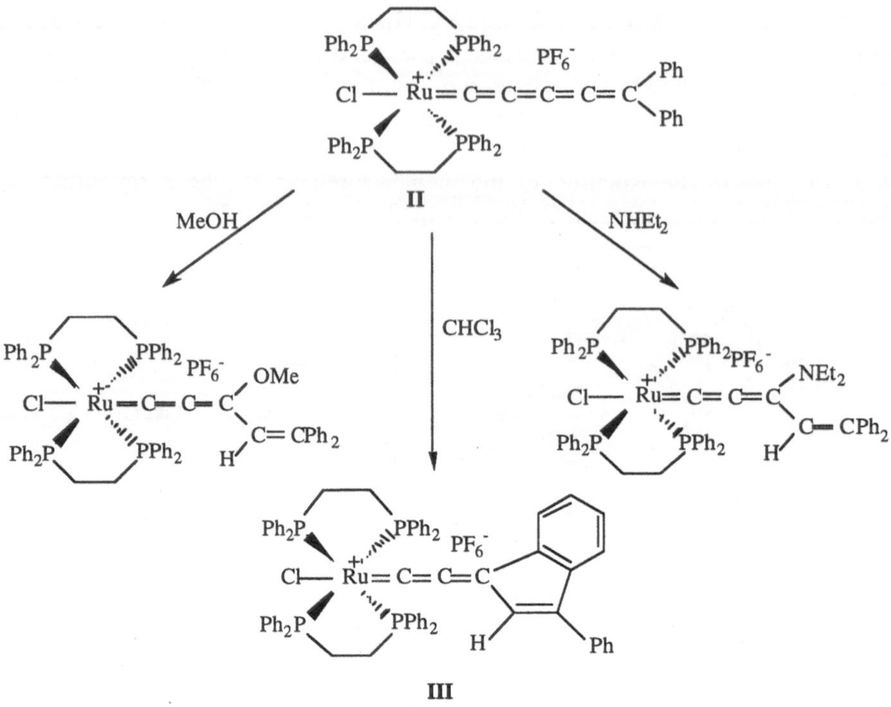

Ph$_2$P   PPh$_2$ PF$_6^-$

Cl — Ru = C = C = C = C = C — Ph / Ph

Ph$_2$P   PPh$_2$

II

MeOH   NHEt$_2$

CHCl$_3$

Ph$_2$P   PPh$_2$ PF$_6^-$ OMe

Cl — Ru = C = C = C
                              C = CPh$_2$
Ph$_2$P   PPh$_2$ H

Ph$_2$P   PPh$_2$ PF$_6^-$ NEt$_2$

Cl — Ru = C = C = C
                              C — CPh$_2$
Ph$_2$P   PPh$_2$ H

Ph$_2$P   PPh$_2$ PF$_6^-$

Cl — Ru = C = C = C

Ph$_2$P   PPh$_2$ H   Ph

III

Scheme 8

Since the report of the first $M=(C=)_4CPh_2$ complex **II**, three new penta-1,2,3,4-tetraenylidene metal complexes have just been isolated : *trans*-(P^{i}Pr$_3$)$_2$(Cl)Ir=(C=)$_4$CPh$_2$[26] and (OC)$_5$M=(C=)$_4$C(NMe$_2$)$_2$ (M = Cr, W)[27].

The search for higher metallacumulenes $LnM=(C=)_nCR_2$ (n ≥ 6) should involve electron-rich LnM unit and polyynyl-metal complexes $M(C)_nCR_2Y$ with a leaving group at the odd terminal carbon.

1.2 Diruthenium Complexes with Unsaturated Carbon-rich Bridges

The activation of terminal diynes by ruthenium complexes is relevant to the generation of bimetallic systems with unsaturated carbon-rich bridges linking two identical ruthenium moieties. The possibility offered by ruthenium complexes to give reversible redox processes has a potential to gather information on the communication capability of the bridge.

The double activation of the simple diyne HC≡C-C$_6$H$_4$-C≡CH by electron-rich ruthenium complex RuCl$_2$(dppe)$_2$ gives access to the diruthenium(II) complex with a vinylidene bridge. This very stable complex can however be easily deprotonated to afford the binuclear complex **V** with a bis alkynyl bridge (Scheme 9). The latter can be reversibly reduced by two one-electron processes (E$_{1/2}$ = +0.130 V$_{SCE}$ and +0.481 V$_{SCE}$) thus showing that the bridge favours reduction and establishes electronic communication between the two ruthenium moieties[28,29].

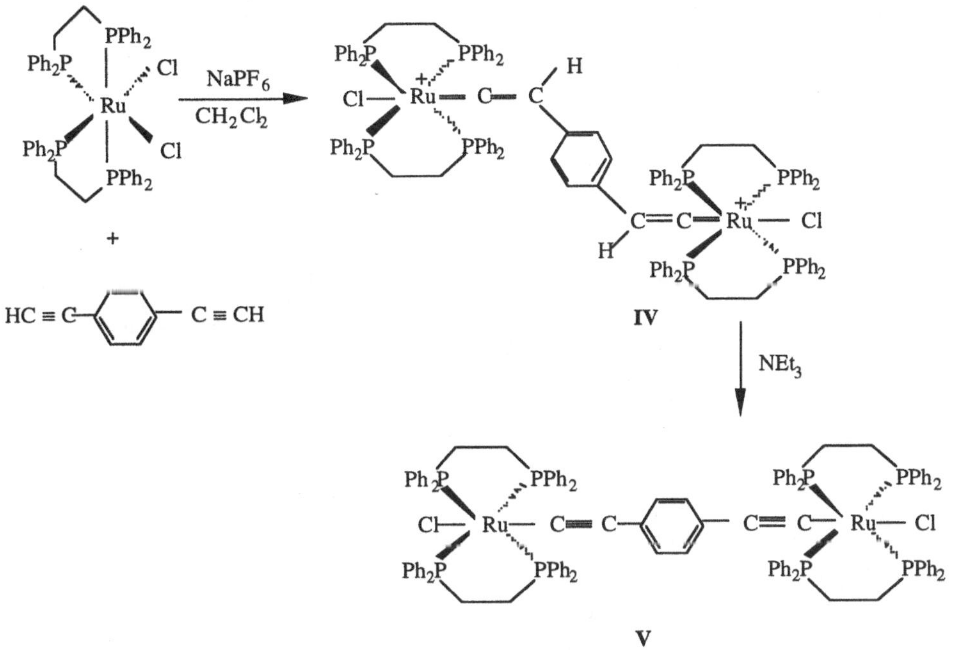

Scheme 9

The double activation of conjugated molecules containing two terminal prop-2-yn-1-ol functionalities by the 16-electron ruthenium species [RuCl(dppe)$_2$]PF$_6$ gives a selective access to either the mono ruthenium allenylidene derivatives of type **VI** or to the binuclear complexes with a bis η^1-allenylidene bridge **VII**[30] (Scheme 10).

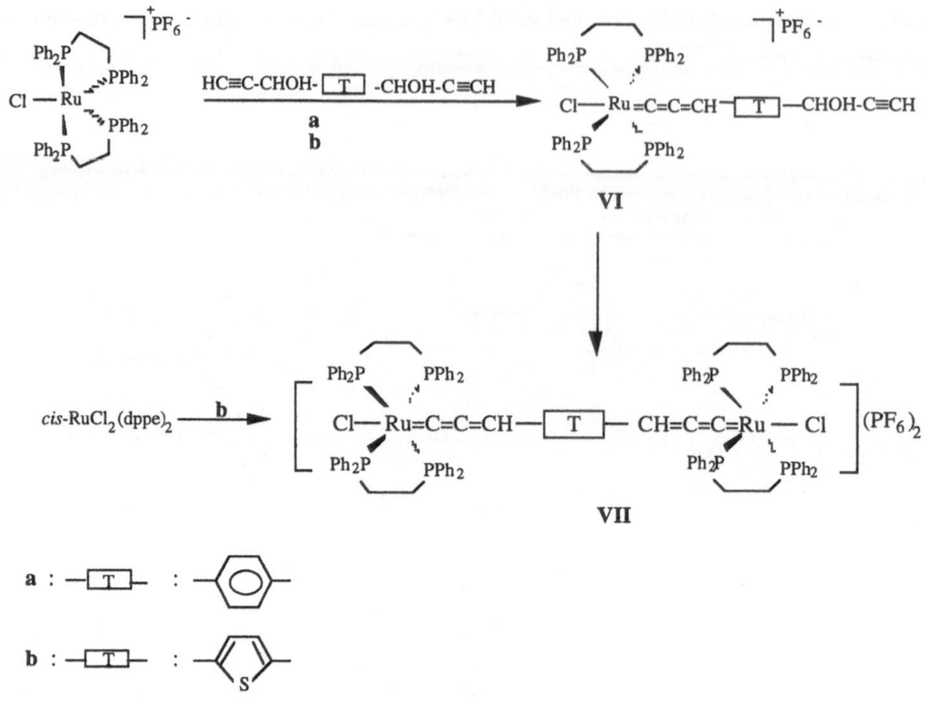

Scheme 10

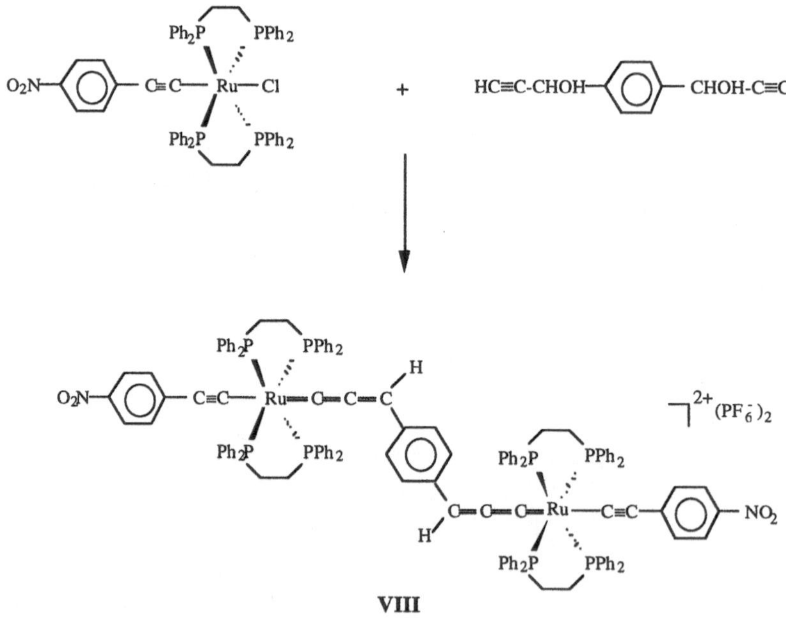

Scheme 11

The alkynyl ruthenium(II) derivatives are also able to activate molecules containing two bis propargylic alcohol groups. In that case diruthenium complexes containing a bis η^1-allenylidene bridge in the *trans* position of the functional alkynyl ligands **VIII** are selectively produced[30] (Scheme 11).

These bimetallic systems can be reversibly reduced and show two waves in cyclic voltammetry. **VIIa** : $E_{1/2}$ = -0.079 and -0.281 V_{SCE} ; **VIIb** : E= +0.095 and -0.176 V_{SCE}; **VIII** : $E_{1/2}$ = +0.082 and -0.074 V_{SCE}.

This study shows that (i) the bis η^1-allenylidene bridge establishes an efficient electronic communication between the two ruthenium moieties, (ii) the 2,5-C_4H_2S bridging group favours this communication with respect to the 2,4-C_6H_4 group and (iii) the electron withdrawing $O_2NC_6H_4C\equiv C$ group significantly favours the reduction of the ruthenium moieties in **VIII**.

The rigidity of the -C≡C-C_6H_4-C≡C- fragment and the selectivity of alkyne activation processes offered the access to organometallic monomers for the design of mixed metal oligomers. The tetrayne derivatives of ferrocene **IX** and ruthenium **X** have been prepared by cross-coupling reactions and *via* the vinylidene intermediates respectively[31,32] (Scheme 12). The organometallics containing two terminal alkynes can be used for the production of mixed iron-palladium or iron-nickel oligomers (Scheme 12). Analogously, the ruthenium tetrayne derivative **X** can be used for the preparation of unidimensional ruthenium-palladium oligomers (Scheme 13).

It is noteworthy that the monomer **X** oxidizes reversibly at $E_{1/2}$ = + 0.45 V_{SCE} and the oligomer **XI** at $E_{1/2}$ = + 0.36 V_{SCE}. In the latter, all ruthenium atoms oxidize at the same potential indicating that the $Pd(PBu_3)_2$ group disfavours the communication with respect to the -C≡C-C_6H_4-C≡C- bridge in a diruthenium complex[29].

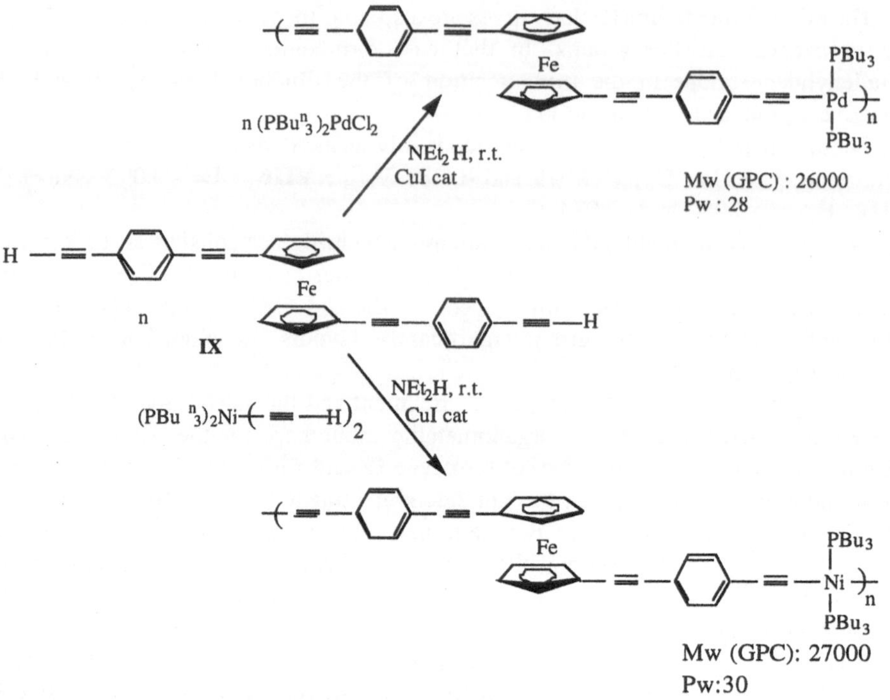

Fe

n (PBu$^n_3$)$_2$PdCl$_2$

NEt$_2$H, r.t.
CuI cat

PBu$_3$

Pd

PBu$_3$

Fe

Mw (GPC) : 26000
Pw : 28

H

n

IX

Fe

H

(PBu$^n_3$)$_2$Ni$(\equiv\!\!-\!\!H)_2$

NEt$_2$H, r.t.
CuI cat

Fe

PBu$_3$

Ni

PBu$_3$

Mw (GPC): 27000
Pw:30

Scheme 12

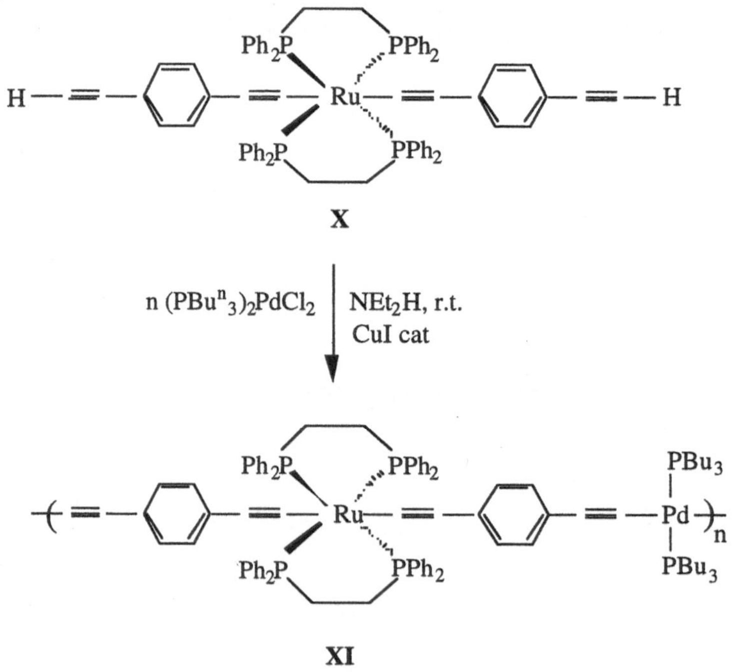

Ph$_2$P   PPh$_2$

Ru

Ph$_2$P   PPh$_2$

H   H

X

n (PBu$^n_3$)$_2$PdCl$_2$ | NEt$_2$H, r.t.
CuI cat

Ph$_2$P   PPh$_2$

Ru

Ph$_2$P   PPh$_2$

PBu$_3$

Pd

PBu$_3$

XI

Scheme 13

The most stable metallacumulenes, or diruthenium complexes with carbon-rich bridges, presented above have been obtained *via* activation of terminal alkynes or diynes with electron-rich ruthenium(II) derivatives. In these cases such ruthenium precursors provide only stoichiometric activation of alkynes. The activation of similar terminal alkynes, but with *electrophilic* ruthenium complexes, has offered selective, simple catalytic processes.

2 Ruthenium-catalysed Transformations of Alkynes

During the last decade ruthenium catalysts, which tolerate a variety of functional groups, have brought an important contribution to organic synthesis, especially *via* the activation of simple unsaturated hydrocarbon molecules, such as alkynes or alkenes[33-35].They have allowed the production of functional derivatives without the use toxic reagents and most processes undergo the combination of two substrates with atom economy[36].

The use of $RuCl_2(PR_3)$(arene) derivatives, which are reversibly oxidized at 0.75-0.95 V_{ECS}, for the stoichiometric activation of terminal alkynes showed that the activated species easily added weak nucleophiles such as methanol (Eq. 1). This evidence for an *electrophilic* activation of alkyne could be extended for simple catalytic additions to alkynes on the condition that excellent regioselectivity was reached.

2.1 Regioselective Additions of Carboxylic Acids to Alkynes

Enol esters, that are key intermediates as well for the formation of copolymers as for acylation and formylation reactions[37,38], can be produced by catalytic activation of terminal alkynes toward the addition of carboxylic acids in the presence of $RuCl_2(PR_3)$(arene) catalyst precursors[39-41]. The *regioselective Markovnikov* addition takes place and can be used for the synthesis of functional dienes from enynes in the presence of the $RuCl_2(PPh_3)$(*p*-cymene) as catalyst 42 (Eq. 5).

Eq. 5

The Markovnikov catalytic addition of a carboxylic acid containing a functional group to the C≡C bond is better accomplished by using the $[Ru(\mu\text{-}O_2CH)(CO)_2(PPh_3)]_2$ catalyst[43] as exemplified by the transformation of oxalic acid[44] and mandelic acid. The latter with alkynes gives a straightforward access to dioxolanones by successive addition to the C≡C and C=C bonds promoted by $[Ru(\mu\text{-}O_2CH)(CO)_2(PPh_3)]_2$[45] or $RuCl(C(OMe)=CHR)(PMe_3)$(arene)[46] precursors (Eq. 6).

Eq. 6

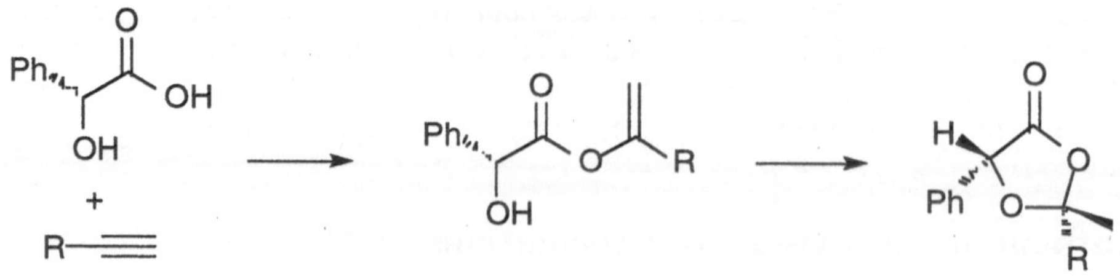

The ruthenium-catalyzed synthesis of enol esters to be profitable requires the recovery of the ruthenium catalyst. This has been achieved by the preparation of the ruthenium catalyst **XII** containing a polymeric phosphine[47]. The latter was simply obtained by polymerisation of ethene with BuLi, followed by treatment with Ph$_2$PCl. This catalyst **XII** containing Ph$_2$P(CH$_2$CH$_2$)$_{50}$CH$_2$CH$_3$ as ligand is soluble in toluene at the reaction temperature (100°C) and totally insoluble at room temperature. It is recovered by filtration and used six times without loss of activity[47].

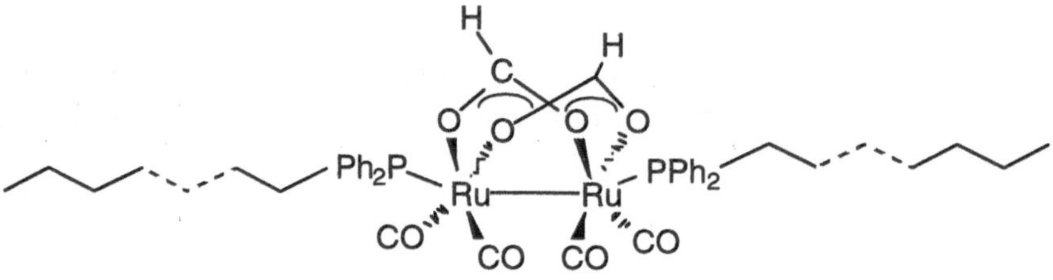

XII

The RuCl$_2$(Ph$_3$)(arene) catalyzed addition of carboxylic acids to pro-2-yn-1-ols apparently takes a different way as it directly gives β-oxopropyl esters[48,49]. Actually the carboxylate group also adds to the substituted C≡C carbon, before an internal transesterification takes place. This transformation applied to steroids having ethynyl and hydroxy groups at carbon C(17) is better promoted by [Ru(O$_2$CH)(CO)$_2$(PPh$_3$)]$_2$ and takes place with retention of configuration[50] (Eq. 7).

Eq. 7

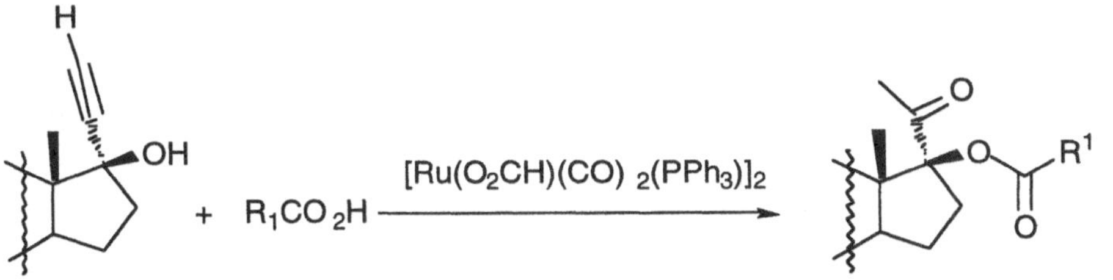

2.2 Stereoselektive anti-Markovnikov Addition of Carboxylic Acids to Alkynes

Previously, the search for phosgene derivative substitutes led to the direct catalytic addition of carbamic acid to the terminal carbon of alkynes to produce vinyl carbamates. The reaction was performed from ammonium carbamates (CO_2 + HNR_2) with $RuCl_2(PR_3)$(arene) catalysts and was specific of terminal alkynes[51] (Eq. 8).

Eq. 8 $R_2NH + CO_2 + HC{\equiv}CR \rightarrow R_2NCO_2\,CH{=}CHR$

Eq. 9 $Ru + HC{\equiv}CR \rightarrow \rightarrow (Ru{=}C{=}CHR$

The observed regioselectivity was attributed to the generation of a ruthenium vinylidene species $Ru{=}C{=}CHR$ expected to favour nucleophilic addition at the electrophilic carbon C(1) (Eq. 9).

In order to favour such a process, ruthenium(II) complexes containing both a bidentate diphosphine and easy to remove hydrocarbon (allyl) ligands have been prepared from a variety of $Ph_2P(CH_2)_nPPh_2$ ligands n = 1-6[52]. The length of the chain allowed to tune the properties of the complexes. Only the complex $(Ph_2P(CH_2)_4PPh_2)Ru(CH_2C(Me)CH_2)_2$ showed an efficient regioselectivity for the *anti*-Markovnikov, stereoselective *trans* addition of carboxylic acids to alkynes to produce a variety of (Z)-1-alkenyl esters in good yields[53] (Eq. 10-12).

Eq. 10

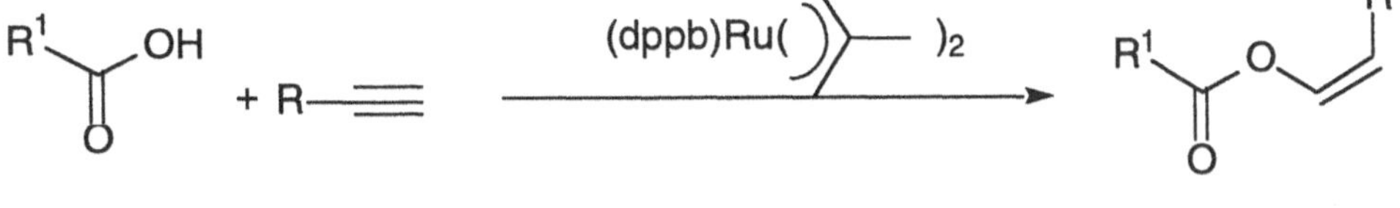

Eq. 11

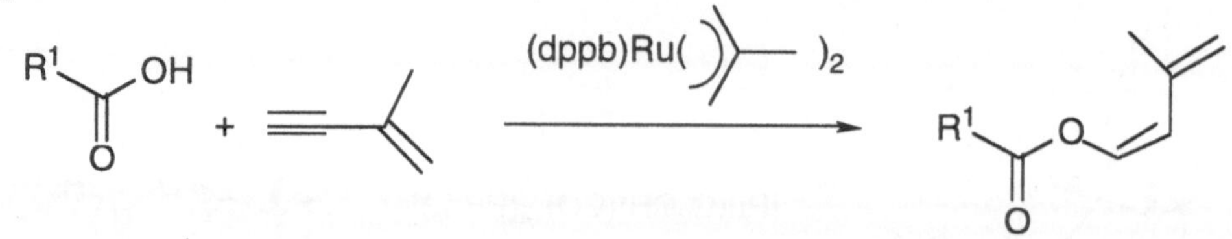

Eq. 12

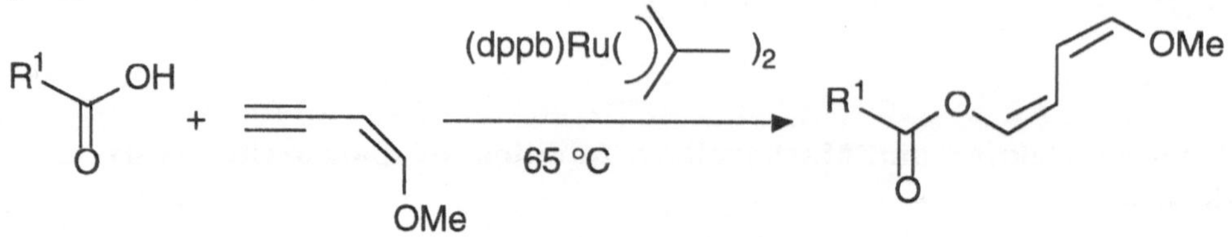

$$dppb = Ph_2P(CH_2)_4PPh_2$$

2.3 Catalytic Synthesis of Functional Furans

Although intermolecular addition of alcohols to alkynes could not be performed the electrophilic activation of 3-methylpent-2-en-4-yn-1-ol, but only in the Z configuration, led to functional furans[54] (Scheme 14). The $RuCl_2(PMe_3)(p$-cymene) catalyst operates at 80-100°C and tolerates functional groups.

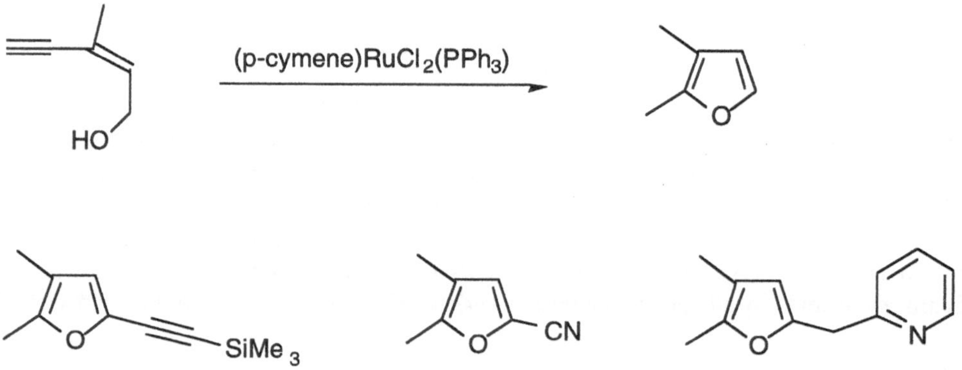

Scheme 14

2.4 Synthesis of α,δ-Unsaturated Aldehydes

As ruthenium(II) complexes allow the intramolecular addition of a OH group to the alkyne bond, the intermolecular addition of allyl alcohol to alkynes was attempted with the ruthenium(IV) complex $RuCl_2(allyl)(C_5Me_5)$. Actually it led to the unexpected carbon-carbon coupling reaction of both C≡C and C=C bonds to afford α,δ-unsaturated acetal and aldehydes[55] (Eq. 13, 14). It was assumed that the ruthenium(IV) complex was *in situ* reduced and shown that the ruthenium(II) complex $RuCl(COD)C_5Me_5$ was a more efficient catalyst. It operates in 10-15 minutes at room temperature in neat allyl alcohol to preferentially give the branch isomer (3: 1), which is the only derivative obtained from the bulky t-butylacetylene. The reaction is not limited to terminal alkynes, and symmetrically disubstituted alkynes afford the unsaturated aldehydes in good yields[56] (Eq. 14). These observations support, as the key activation step, the oxidative coupling of both C≡C and C=C bond at the ruthenium site.

Eq. 13

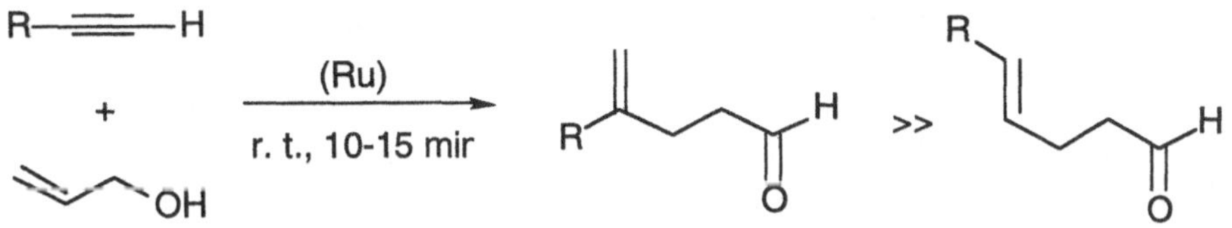

Eq. 14

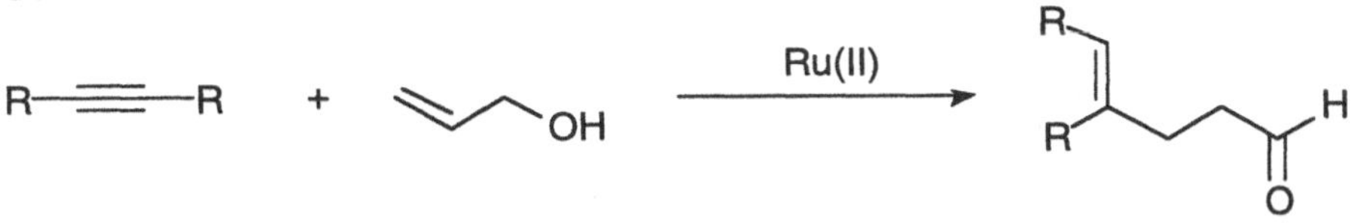

3 Conclusion

Electron-rich ruthenium(II) complexes provide at room temperature stoichiometric activation processes of unsaturated terminal alkynes and diynes to generate carbon-rich organometallics and bimetallic complexes. More *electrophilic* ruthenium(II) derivatives by contrast do not stabilize such unsaturated ligands but promote additions to the C≡C bond to produce functional alkenyl derivatives. The subtle, step-by-step modification of the ligands around the ruthenium(II) site of several families of catalysts allows to reach good regioselectivities.

References

1. F. Diederich, Y. Gubin, *Angew. Chem., Int. Ed. Engl.* **1992**, *31*, 1101; J. M. Tour, *Chem. Rev.* **1996**, *96*, 537; J. S. Miller, *Adv. Mater.* **1993**, *5*, 587 and 671.
2. U. Scherf, K. Müller, *Synthesis* **1992**, 23; T. Zyung, D.-H. Hwang, I.-N. Kang, H.-K. Shim, W.-Y. Hwang and J.-J. Kim, *Chem. Mater.* **1995**, *7*, 1499.
3. S. Takahashi, Y. Takai, H. Morimoto, K. Sonogashira, *J. Chem. Soc. Chem. Commun.* **1984**, 3; J. P. Rourke, D. W. Bruce, T. B. Marder, *J. Chem. Soc., Dalton Trans.* **1995**, 317; M. Altmann, V. Enkelmann, G. Leiser, U. H. F. Bunz, *Adv. Mater.* **1995**, *7*, 726.
4. C. Denault, I. Ledoux, D. W. Samuel, J. Zyss, M. Bourgault, H. Le Bozec, *Nature* *374*, 339 **1995**, ; S. R. Marder "Metal containing Materials for non linear optics" in *Inorganic Materials,* Eds. D. Bruce, D. O'Hare, J. Wiley, 115 ,**1992**.
5. U. H. F. Bunz, *Angew. Chem., Int. Ed. Engl.* **1996**, *35*, 969; J. E. C. Wiegelmann, U. H. F. Bunz, P. Schiel, *Organometallics* **1994**, *13*, 4649; T. Bartik, B. Bartik, M. Brady, R. Dembinski, J. A. Gladysz, *Angew. Chem., Int. Ed. Engl.* **1996**, *35*, 414 and references therein.
6. H. Werner, T. Rappert and J. Wolf, *Israel Chem.* **1990**, *30*, 377; N. Le Narvor, C. Lapinte, *Organometallics* **1995**, *14*, 634; M. S. Khan, A. K. Kakkar, S. L. Ingham, P. R. Raithby, J. Lewis, B. Spencer, F. Wittmann, R. H. Friend, *J. Organomet. Chem.* **1994**, 472 - 247; E. Viola, C. Lo Sterzo, R. Crescenzi, G. Frachey, *J. Organomet. Chem.* **1995**, *493*, C9; J. F. Capon, N. Le Berre-Cosquer, B. Leblanc, R. Kergoat, *J. Organomet. Chem.* **1996**, *508*, 31; G. Jia, H. P. Xia, W. F. Wu, W. Sang Ng, *Organometallics* **1996**, *15*, 3634.
7. O. Lavastre, M. Even, P. H. Dixneuf, A. Pacreau, J. P. Vairon, *Organometallics* **1996**, *15*, 1530; R. B. King, M. S. Saran, *Chem. Commun.* **1972**, 1052; *J. Am. Chem. Soc.* **1972**, *95*, 1817.
8. M. I. Bruce, A. G. Swincer, *Adv. Organomet. Chem.* **1983**, *22*, 59; M. I. Bruce, *Chem. Rev.* **1991**, *91*, 197.
9. H. Le Bozec, K. Ouzzine, P. H. Dixneuf, *Organometallics* **1991**, *10*, 2768.
10. M. I. Bruce, R. C. Wallis, *J. Organomet. Chem.* **1978**, *161*, C1; M. I. Bruce, R. C.Wallis, *Aust. J. Chem.* **1979**, *32*, 1471.
11. R. Le Lagadec, E. Roman, L. Toupet, U. Müller, P. H. Dixneuf, *Organometallics* **1994**, *13*, 5030.
12. D. Touchard, P. Haquette, N. Pirio, L. Toupet, P. H. Dixneuf, *Organometallics* **1993**, *12*, 3132.
13. D. Touchard, C. Morice, V. Cadierno, P. Haquette, L. Toupet, P. H. Dixneuf, *J. Chem. Soc., Chem. Commun.* **1994**, 859.
14. E. O Fischer, H.-J. Kalder, A. Franck, F. H. Kölher, G. Huttner, *Angew. Chem., Int. Ed. Engl.* **1976**, *15*, 623.
15. H. Berke, *Angew. Chem., Int. Ed. Engl.* **1976**, *15*, 624.
16. J. P. Selegue, *Organometallics* **1982**, *1*, 217.
17. D. Touchard, N. Pirio, P. H. Dixneuf, *Organometallics* **1995**, *14*, 4920.

18. D. Touchard, N. Pirio, L. Toupet, M. Fettouhi, L. Ouahab, P. H. Dixneuf, *Organometallics* **1995**, *14*, 5263.

19. D. Pilette, K. Ouzzine, H. Le Bozec, P. H. Dixneuf, C. E. F. Rickard, W. R. Roper, *Organometallics* **1992**, *11*, 809.

20. D. Touchard, S. Guesmi, M. Bouchaib, P. Haquette, A. Daridor, P. H. Dixneuf, *Organometallics* **1996**, *15*, 2579.

21. A. Romero, D. Peron, P.H. Dixneuf, *J. Chem. Soc., Chem. Commun.* **1990**, 1410.

22. D. Péron, A. Romero, P. H. Dixneuf *Organometallics* **1995**, *14*, 3319.

23. A. Wolinska, D. Touchard, P.H. Dixneuf, A. Romero, *J. Organomet. Chem.* **1991**, *420*, 217.

24. D. Peron, A. Romero, P. H. Dixneuf, *Gazz. Chim. Ital.* **1994**, *124*, 497 Volume L. Sacconi.

25. D. Touchard, P. Haquette, A. Daridor, L. Toupet, P. H. Dixneuf, *J. Am. Chem. Soc.* **1994**, *116*, 11157.

26. R. W. Lass, P. Steinert, J. Wolf, H. Werner, *Chem. Eur. J.* **1996**, *2*, 19.

27. G. Roth, H. Fischer, *Organometallics* **1996**, *15*, 1139.

28. S. Guesmi, D. Touchard, P. H. Dixneuf, Unpublished results.

29. O. Lavastre, P. H. Dixneuf, Unpublished results.

30. S. Guesmi, D. Touchard, P. H. Dixneuf, Submitted for publication.

31. O. Lavastre, M. Even, P. H. Dixneuf, A. Pacreau, J.-P. Vairon, *Organometallics* **1996**, *15*, 1530.

32. O. Lavastre, L. Ollivier, S. Sinbandhit, P. H. Dixneuf, *Tetrahedron* **1996**, *52*, 5495.

33. T. Kondo, A. Tanaka, S. Kotachi, Y. Watanabe, *J. Chem. Soc., Chem. Commun.* **1995**, 413; S. W. Zhang, T. Mitsudo, T. Kondo, Y. Watanabe, *J. Organomet. Chem.* **1995**, *485*, 55; T. Mitsudo, S. W. Zhang, M. Nagao, Y. Watanabe, *J. Chem. Soc., Chem. Commun.* **1991**, 598; T. Mitsudo, Y. Hori, Y. Yamakawa, Y. Watanabe, *J. Organomet. Chem.* **1987**, *52*, 2230.

34. N. Chatani, Y. Fukumoto, T. Ida, S. Murai, *J. Am. Chem. Soc.* **1993**, *115*, 11614; S. Murai, F. Kakiuchi, S. Sekine, Y. Tanaka, A. Kamatani, M. Sonoda, N. Chatani, *Nature* **1993**, *366*, 529; S. Murai, *J. Synth. Org. Chem. Jpn.* **1994**, *52*, 120; F. Kakiuchi, Y. Tanaka, T. Sato, N. Chatani, S. Murai, *Chem. Lett.* **1995**, 679; F. Kakiuchi, Y. Yamamoto, N. Chatani, S. Murai, *Chem. Lett.* **1995**, 681.

35. H. Nishiyama, Y. Itoh, Y. Sugawara, H. Matsumoto, K. Aoki, K. Itoh, *Bull. Chem. Soc. Jpn.* **1995**, *68*, 1247; A. Demonceau, E. Abreu Dias, C. A. Lemoine, A. W. Stumpf, A. F. Noels, *Tetrahedron Lett.* **1995**, *36*, 3519; S. H. Kim, W. J. Zuercher, N. B. Bowden, R. H. Grubbs, *J. Org. Chem.* **1996**, *61*, 1073.

36. B. M. Trost, *Angew. Chem., Int. Ed. Engl.* **1995**, *34*, 259.

37. M. Neveux, C. Bruneau, P. H. Dixneuf, *J. Chem. Soc., Perkin Trans.* **1991**, *1*, 1197.

38. Z. Kabouche, C. Bruneau, P. H. Dixneuf, *Tetrahedron Lett.* **1991**, *32*, 5359.

39. C. Ruppin, P. H. Dixneuf, *Tetrahedron Lett.* **1986**, *27*, 6323.

40. C. Bruneau, M. Neveux, Z. Kabouche, C. Ruppin, P. H. Dixneuf, *Synlett* **1991**.

41. C. Ruppin, P. H. Dixneuf, S. Lécolier, *Tetrahedron Lett.* **1988**, *29*, 5365.

42. K. Philippot, D. Devanne, P. H. Dixneuf, *J. Chem. Soc., Chem. Commun.* **1990**, 1199.

43. B. Seiller, D. Heins, C. Bruneau, P. H. Dixneuf, *Tetrahedron* **1995**, *51*, 10901.
44. M. Neveux, C. Bruneau, S. Lécolier, P. H. Dixneuf, *Tetrahedron* **1993**, *49*, 2629.
45. M. Neveux, B. Seiller, F. Hagedorn, C. Bruneau, P. H. Dixneuf, *J. Organomet. Chem.* **1993**, *451*, 133.
46. O. Lavastre, P. H. Dixneuf, *J. Organomet. Chem.* **1995**, *488*, C9.
47. O. Lavastre, P. Bebin, O. Marchaland, P. H. Dixneuf, *J. Mol. Catal. A* **1996**, *108*, 29.
48. C. Bruneau, Z. Kabouche, M. Neveux, B. Seiller, P. H. Dixneuf, *Inorg. Chim. Acta* **1994**, *222*, 155.
49. D. Devanne, C. Ruppin, P. H. Dixneuf, *J. Org. Chem.* **1988**, *53*, 925.
50. C. Darcel, C. Bruneau, P. H. Dixneuf, G. Neef, *J. Chem. Soc., Chem. Commun.* **1994**, 333.
51. R. Mahé, Y. Sasaki, C. Bruneau, P. H. Dixneuf, *J. Org. Chem.* **1989**, *54*, 1518.
52. H. Doucet, J. Höfer, C. Bruneau, P. H. Dixneuf, *J. Chem. Soc., Chem. Commun.* **1993**, 850.
53. H. Doucet, B. Martin-Vaca, C. Bruneau, P. H. Dixneuf, *J. Org. Chem.* **1995**, *60*, 7247.
54. B. Seiller, C. Bruneau, P. H. Dixneuf, *J. Chem. Soc., Chem. Commun.* **1994**, 493.
55. S. Dérien, P. H. Dixneuf, *J. Chem. Soc., Chem. Commun.* **1994**, 2551.
56. S. Dérien, D. Jan, P. H. Dixneuf, *Tetrahedron* **1996**, *52*, 5511.

Polymer Synthesis with Organometallic Intermediates

Uwe Bunz, Viola Francke, Markus Klapper, Frank Uckert, Klaus Müllen*

Max-Planck-Institut für Polymerforschung, Ackermannweg 10, D-55128 Mainz

1 Problems of Polymer Synthesis

This article is concerned with the role of organometallic intermediates in polymer synthesis. While chapter 2 deals with polycondensation reactions leading to conjugated polymers (aryl-aryl, aryl-vinyl, aryl-ethinyl coupling) chapter 3 describes anionic polymerization reactions using mono- and bifunctional initiators. There is sometimes the belief that, once a method of bond formation has been established, the repetitive use of this reaction in polymer formation is straightforward or even trivial. This is, however, not born out by experiment. Chapter 1 will therefore highlight a few typical problems which are encountered during polymer synthesis.

1. The polydisperse character and the molecular weight of macromolecules can render a full structural elucidation difficult, for example, end group analysis is essential for studying the possibility of further functionalization. In the last years, new experimental techniques have been developed. MALDI-TOF MS spectrometry in particular has dramatically widened the scope of polymer analytical studies[1] . Thus in many examples, given below, detailed information about the polymer structure could be obtained.
2. Even minor side reactions can produce structural defects in the molecule or the molecular weight of polycondensation products can be limited as a result of unbalanced stoichiometry (see chapter 2)[2].
3. Polymers and their oligomeric analogs often possess an extremely low solubility in organic solvents. This problem can sometimes be overcome by introducing alkyl substituents or structural irregularities and by making use of precursor polymers[3]. A typical example is the synthesis of poly(*para*-phenylene)s (PPP) which are known to be insoluble and intractable. Solubilizing alkyl groups produce a steric inhibition of π-conjugation which can be overcome by synthesizing step-ladder or ladder PPP's[4].
4. Oligomers formed initially during polycondensation or polyaddition processes often show reduced reactivity with increasing chain length. For higher degrees of oligomerization the increasing viscosity produces a limitation of molecular weights. The decreasing mobility of the oligomers is responsible for the reduced reactivity. The parameters mobility, starting concentration and reaction time play a key role in optimizing

molecular weights. Here it is important to find a compromise between sufficient mobility, good solubility of the monomers, high starting concentration and reaction time.

5. Polymer-analogous transformations often cannot proceed to completion, e.g., as a result of growing steric shielding. For example, although a polyetherketone reacts quantitatively with phenyllithium to yield a polyalcohol, when the same polymer is reacted with an excess of a macromolecular nucleophile such as living polystyrene anion, the resulting graft polymer exhibits a degree of conversion of only about 32 % (Chapter 3)[5].

2 Polycondensation Reactions Using Organometallic Intermediates

2.1 Aryl-Aryl Coupling

Polyarylenes are of considerable importance for a number of different reasons. Structurally, they allow the construction of polymers with varying topologies, displaying inherently different and often important morphologic, electronic, optic, and chiroptic properties[3]. To fully understand the properties of the desired polymers the step-wise synthesis of well defined and discrete oligomers is useful[6].

An illustrative example to this end is the synthesis of 2,7-oligo(4,9-dioctylpyrene)s. Starting from 2-bromo-4,9-dioctylpyrene (1), the dimer 2 was prepared by dehalogenative condensation with Ni(COD)$_2$ (Yamamoto method) in almost quantitative yield[7]. Having established that Ni(0)-coupling is an appropriate method for the construction of these oligomers, a mixture of the mono- 1 and the dibromopyrene 3 was coupled with Ni(COD)$_2$ in DMF/toluene to give the corresponding trimer 4 and tetramer after column separation[7]. Despite the octyl groups, the solubility of the tetramer is already so low that the synthesis of the corresponding 2,7-polypyrene is difficult.

The situation for the formation of a polymer is much better in the case of the less planar and more flexible 4,5,9,10-tetrahydropyrene **5** with solubilizing octyl groups in the 4- and 9-positions (THP). Here, the coupling of the corresponding 2,7-dibromo-*trans*-4,9-dioctyl-4,5,9,10-tetrahydropyrene gives (under similar coupling conditions employing Ni(COD)$_2$) polymer **6** with a respectable degree of polymerization (DP) of about **88**, showing the efficiency of this polymer forming reaction[8]. Polymer **6** is a completely soluble, new type of PPP with a step-ladder structure, in which each pair of neighboring aromatic rings is doubly bridged with ethano bridges. Since the solubilizing alkyl groups are attached at the periphery of the molecule, there is no inhibition of π-conjugation due to twisting of the main chain.

But even in the field of classic poly(*para*-phenylene)s (PPP) there are still unsolved problems such as the synthesis of optically active PPP's. Utilizing a Suzuki[9] coupling of **7** with **8** gives rise to the polymer **9**, containing optically active cyclophane units[10]. Polymer **9** is completely soluble and forms lyotropic nematic liquid crystalline (LC) mesophases. The obtained DP hereby is approximately **25**.

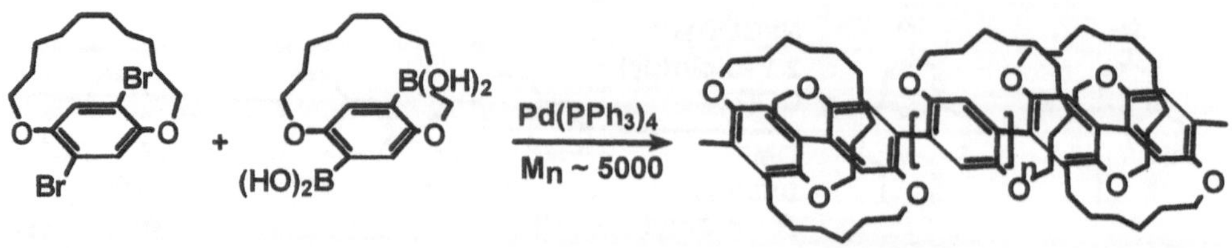

7   8   9

The Suzuki type couplings are exceedingly useful in another vein, giving rise to the formation of PPP with the structure **10**, which is converted into the polymeric ladder **11** in two steps. Depending upon the choice of substituents the yield of soluble material and its molecular weight vary dramatically (Table 1). The combination of hexyl and decyl side groups thereby seems to show the best improvement in solubility.

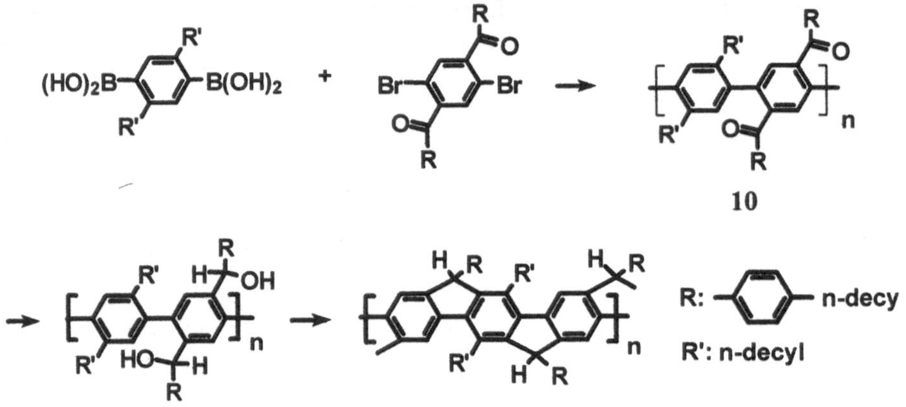

10

11

R: ──⟨benzene⟩── n-decy

R': n-decyl

Table 1. Preparation of ploy(*para*-phenylene)s by Suzuki coupling

R^1	R^2	M_n	Solubility			Yield (%)
			$CHCl_3$	CH_2Cl_2	toluene	(*soluble part)
H	$C(CH_3)_3$	-	-	-	-	-
H	$C(CH_3)_3$	-	(x)	-	-	6*
C_6H_{13}	$C(CH_3)_3$	6000	x	-	-	78
C_6H_{13}	$C_{10}H_{21}$	17000	x	x	x	80
C_6H_{13}	$C_{10}H_{21}$	25000	x	x	x	83

The Suzuki coupling fails completely though in the case of the sterically (by the peri-H atom!) more encumbered naphthalene derivatives while it is efficient in the construction of topologically different, star shaped objects, where a solubilized terphenyl boronic acid is coupled to a hexa(iodophenyl)benzene core[11]. The palladium catalyzed coupling in high yield gives the novel hexakis(quaterphenyl)benzene 12.

12

Along the same lines, our group is currently actively engaged in the chemistry of the rylenes, for which no useful homologization routes are described in the literature. Utilizing bromoperylene 13, palladium-catalyzed stannylation and subsequent Stille type coupling, accesses the symmetric dimer 14 in 76 % yield. Gratifyingly, this material cyclizes to the desired quaterrylenebisimide 15 in 80 % yield, using standard conditions[12]. A similar route can be applied for the synthesis of benzoylterryleneimide 17, again using a stannylated peryleneimide 16. The overall yield of the synthesis of 17 is, with 57 %, satisfying[13].

The development and use of different organometallic coupling reactions for the formation of aryl-aryl bonds allows the facile synthesis of large, oligomeric aromatic systems which otherwise can be prepared only with considerable difficulty. A good example of this is the successful synthesis of large graphite segments.

$$\text{Sn}_2\text{Bu}_6 \quad \text{Pd(0)} \quad 88\%$$

$$\text{R} = \text{i-propyl / i-propyl}$$

$$\text{Pd(II)} \quad \text{toluene} \quad 76\%$$

$$\text{EtOH / KOH / Ox.} \quad 80\%$$

13   **14**   **15**

$$\text{Pd(PPh}_3)_4 \quad \text{DMF, 90 °C} \quad \text{4 d, 73\%}$$

$$\text{KOH / EtOH ox., 70 °C} \quad \text{15 min} \quad 80\%$$

16   **17**

R = i-propyl / i-propyl , H   R' = —O—⟨⟩—+ , H

R = H: λ_{max} = 676 nm
ε = 61700 l mol^{-1} cm^{-1}

The successful photocycloaddition of polymeric distyrylbiphenyls **50** (vide infra) led to the idea to prepare polymers similar to **51** with six membered rings instead of cyclobutanes in the polymer backbone[14]. While of course a [2+2] cycloaddition route is useful only for the construction of four membered rings, the logical extension of the concept to Diels-Alder reactions, involving butadiene and styrene substituted biphenyls, may open new vistas for access to larger, fused, six membered ring systems which in due course could be dehydrogenated to large disc-shaped and ladder-type aromatics.

Starting from iodostilbene **18**, halogen metal exchange and addition of zinc bromide gives rise to the formation of the organozinc **19**, which cleanly couples to the aromatic iodide **20**, under the conditions of the Knochel-reaction[15], to give, in a 75 % yield, the desired precursor **21**.

27

Extension of this reaction utilizing the Suzuki-coupling leads to the formation of the terphenyl **22** in almost quantitative yield. Surprisingly, **22** is stable under the reaction conditions and does not perform the Diels-Alder reaction in the aqueous/organic two phase system encountered. However, the DA reaction proceeds cleanly and in quantitative yield, when **22** is heated in tetrachloroethane for 1 h at 110 °C.

22

The obtained product **23** is already composed of eleven cyclic units and furnishes the completely unsaturated graphitic rhombus **24** after aromatization with DDQ and six subsequent cyclizations under the influence of a mixture of AlCl$_3$ and CuCl$_2$ in CS$_2$[16]. The cyclization reaction deserves further comments in that similar reactions had been used by Kovacic[17] for the synthesis of linear poly(*para*-phenylene)s (PPP). In the intermolecular case, the reaction is not particularly selective and gives material which is crosslinked and insoluble and was never well characterizable.

22 → 23

1,1,2,2-tetra-chlorethane

110 °C / 1h

quantitative reaction

In contrast, the intramolecular version of the Kovacic reaction is highly selective and gives the desired cyclodehydrogenated product **24** in quantitative yield. This behavior can be regarded as an illustrative example of Eschenmosers principle, which states that reactions which are only moderately selective in an intermolecular case, can be highly selective and quantitative when conducted *intramolecularly*.

AlCl₃ / CuCl₂

CS₂

12 h at 20°C

99 % yield

24

Rhombus **24** is insoluble in common organic solvents, but can be sublimed at 550°C in ultrahigh vacuum. The obtained microcrystalline material is bright yellow and, despite its considerable size, still far away from the black and lustrous graphite. This again shows that there are a number of reasons to scrutinize these large sheet like aromatic particles besides pure structural interest, such as topological/spectroscopic relationships or the examination of π-π stacking interactions in these compounds. In a more advanced vein, these systems may at a future point in time be important building blocks for what is termed "molecular electronics" today, utilizing such materials as molecular wires or molecular conducting sheets[18].

Despite the fact that these reactions proceed cleanly and in high yields, it still would be of considerable interest to synthesize supernaphthalene decorated with *tert*-butyl groups in order to obtain a soluble and therefore conventionally characterizable derivative. Attempts to use the standard coupling conditions for the cyclization of **25** though led to the formation of a complex mixture of products, due to the fact that the *n*-alkyl groups do not survive these conditions. Fortunately, the use of the less Lewis-acidic, but still oxidizing iron trichloride at ambient temperature in dichloromethane solves the problem, giving an 85 % yield of the alkylated, well soluble supernaphthalene **26**.

The synthesis of the large sheet-like and disc-like aromatics demonstrates clearly that, for the synthesis of large organic molecular objects, the use of sophisticated organometallic strategies in combination with conventional aromatic chemistry can lead to spectacular successes.

2.2 Aryl-Vinyl Coupling

Poly(*para*-phenylenevinylene) (PPV) is an interesting material, due to the fact that it shows conductivity when doped and it can be used in light emitting diodes[19]. The quest for high molecular weight PPV is solved now by the advent of the Wessling-Zimmermann route; more recent developments allow the synthesis of this material even at low temperatures[20]. Extension of this route towards the corresponding *ortho*- or *meta*-phenylenevinylene polymers is not possible though. Additionally, the introduction of other arylene units into the conjugated polymer backbone is not feasible by this highly specialized route. Multiple potential applications of the poly(arylenevinylene)s (PAV) in electrooptics and molecular electronics fuel the interest in the development of efficient preparative accesses to oligomeric and polymeric arylenevinylene representatives in a defect-free state.

The Stille reaction is the coupling of iodo- or bromoarenes with unsaturated vinylic or aromatic tin reagents under the influence of catalytic palladium (typically $Pd(PPh_3)_4$) at enhanced temperatures in polar aprotic media[21]. While widely applicable, the need of elevated temperatures has a negative effect upon catalyst stability, so that Stille conditions are not optimal for the synthesis of polymers. The yields obtained in the Stille reaction are uniformly between 70 and 90 %; satisfactory for small molecule targets, but unacceptable for the synthesis of macromolecules, where the DP can be estimated to $DP < 1/[1-(X/X_0)]$, with X/X_0 defined as the conversion of all functional groups in the monomers relating to the chemical yield of the reaction. Recent efforts by Farina[22] led to the development of catalyst systems, such as $Pd(P(furyl)_3)_4$ or $Pd(AsPh_3)_4$, active even at ambient temperatures and generally giving quantitative yields, making the Stille reaction potentially useful for the synthesis of polymeric phenylenevinylenes.

In order to evaluate scope and limitation of the Stille coupling with respect to the formation of phenylenevinylenes, we conducted a model study. Reaction of iodobenzene **27** with the stannane **28**, after careful optimization of the reaction conditions with the use of the Farina catalyst system gives rise to the quantitative formation of the desired coupling product stilbene **29**. This experiment indicated that it might be possible to extend this coupling towards the synthesis of larger oligomers and polymers. When we tried the coupling of iodobenzene **27** with **30** under the same conditions, the yield of stilbene (**29**) was only 88 %. In addition, **31**, **32** and **33** were isolated as further unwanted products of the coupling, indicating that the close neighborhood of the two tin functionalities may lower the yield of **29**.

Using the styryl tin **28**, we were successful in constructing the corresponding trimers **35** and **37** by reaction with *para*- and *ortho*-diiodobenzene (**34**, **36**) in 92 % and 85 %, respectively. If **36** is then reacted with **30**, the corresponding mixture of oligomers **38** forms in 80 % with DP ≈ 5, indicating a similar situation, as in the coupling of **27** with **30**[14].

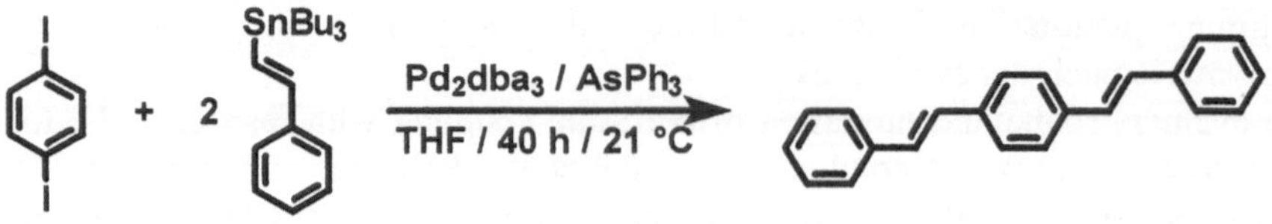

34 + 2 **28** → **35**

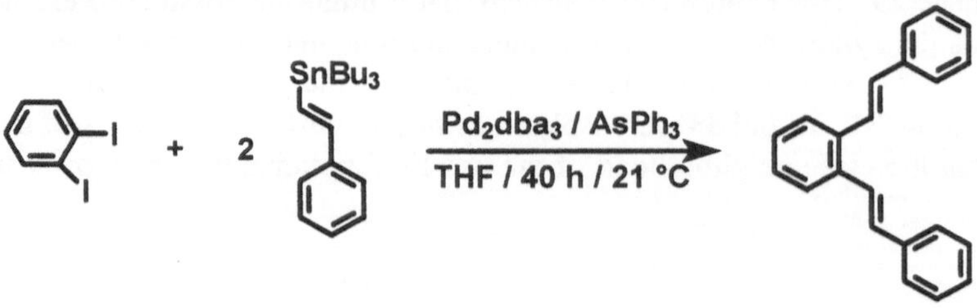

36 + 2 **28** → **37**

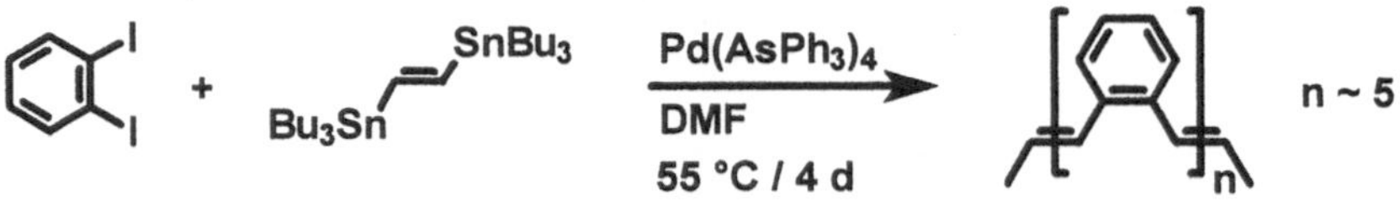

36 + **30** → **38** n ~ 5

A handy way out of this uncomfortable situation which would also provide access to higher molecular weight *ortho*-phenylenevinylenes (*ortho*-PV) would be the use of an enlarged monomer in the polycondensation reaction. Stille coupling of **39** with **30** (Table 2) gave, under optimized conditions, rise to polymer **40** with a DP ≈ 25 (regarding to the repeat unit), making the use of an enlarged monomer a more desirable strategy for the formation of higher oligomers and polymers in the case under consideration, the *ortho*-PV **40**. In a couple of typical polymerization results are shown using the two monomers **36** and **39**.

Table 2

Entry	Iodo compound (mmol)	Catalyst (mmol)	Solvent (ml)	Reaction time (temp.)	M_n	M_w
a.	**36** (1)[b]	Pd-As[d](0.04)	DMF (15)	96 h (55 °C)	450	750
b.[a]	**36** (1)	Pd-As (0.04)	DMF (15)	96 h (55 °C)	750	1350
c.	**39**[c](1.64)	Pd-P[e] (0.164)	DMF(10)	48 h (45 °C)	830	2400
d.	**39** (1.64)	Pd-P(0.164)	DMF (10)	168 h(45 °C)	890	2350
e.	**39** (1.64)	Pd-P (0.164)	THF (10)	168 h (45 °C)	960	2250
f.	**39** (1.64)	Pd-As (0.164)	DMF (10)	48 h (45 °C)	840	2200

[a] Addition of a second amount of monomers and catalyst after 48 h. [b] A = 1,2-Diiodobenzene. [c] B = Bis-(2'-Iodo)-1,2-distyrylbenzene. [d] Pd(AsPh₃)₂. [e] Pd(P(2-furyl)₃)₂.

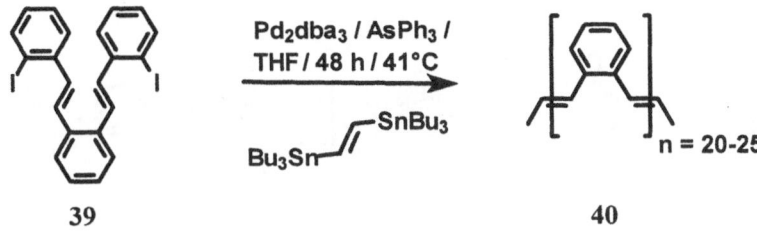

The question regarding the typical chain termination reactions in the Stille coupling is still open. To this end, a GC-MS study was conducted. Thereby we were able to discern the occurrence of six different compounds.

The main termination reactions from this study seem to be destannylation, dehalogenation (partial as well as complete), butyl transfer and the formation of large cycles. In conclusion, we can note that Stille methodology is well suited for all type of oligomer synthesis, but inappropriate for the formation of high molecular weight polymers. It is of course not necessary to stick slavishly to only one type of organometallic coupling in the construction of conjugated polymers. The Heck reaction[23], that is the reaction of aromatic halides with alkenes under the influence of palladium acetate in an amine as solvent, is a conceptually elegant synthetic tool, amenable for the synthesis of polymers and oligomers containing a distyrylbiphenyl unit. The main problems with the Heck reaction are accompanying dehalogenation reactions as well the regioselectivity of the 1,1- *versus* the 1,2-attack of the reactive palladium species leading to undesired exo-methylene groups in the polymer[24].

Yet reaction of divinylbiphenyl **41** with **42** yields polymer **43** with a DP ≈ 12. The use of a larger monomeric system, i.e. coupling of **44** with **41**, a strategy we have used to our advantage in the construction of *ortho*-PV, gives polymer **45** with a higher molecular weight[25].

If the functionalities in the monomers are permutated, i.e. dibromide **46** and divinyl-benzene **47** are subjected to the conditions of the Heck reaction, surprisingly, no polymeric material can be isolated. Instead, an inseparable mixture of the two deeply orange fluorene derivatives **48** and **49** is formed in a 66 % yield, indicating that merely the intramolecular 1,1-connection of the biphenyl units to form the five membered ring system is observed with **47**.

In polymer **50** - which of course is conjugated - irradiation leads to a complete bleaching of the band observed at 350 nm, indicating that the conjugation between the arene units is interrupted in the product **51** by the cyclobutane units. From crystalline monomeric distyrybiphenyls, it is known that irradiation leads to the clean formation of a cyclobutane in a topochemical fashion[14]. The ease of bleaching makes **50** a valuable candidate for material science applications[14].

2.3 Aryl-Ethynyl Coupling

The coupling of alkynes with aromatic bromides and iodides under the influence of a mixed palladium-copper catalyst system in the presence of an amine solvent gives the corresponding arylated alkynes. This coupling was discovered in the mid-seventies independently by Heck, Cassar, Sonogashira and Hagihara[26]. The typical catalyst systems used are either $PdCl_2(PPh_3)_2/CuI$, $Pd_2dba_3/PPh_3/CuI$ or $Pd(PPh_3)_4/CuI$. It was later shown (Alami, Linstrumelle) that CuI is not always necessary in these couplings, but its addition never leads to a decreased performance of the reaction[27]. With iodoarenes, the reaction normally is fast at ambient temperature and yields the corresponding coupling products in high to quantitative yields and almost independently of the structure of the iodoarene. Side reactions are rare and involve oxidative coupling of the alkyne group in cases where oxygen is not rigorously excluded[28]. Since 1993, improved coupling procedures were reported by Alami, Ferri and Linstrumelle, making the reaction even more reliable and clean. They particularly looked into the role of the amine in the coupling reaction and discovered that, in most cases, piperidine or pyrrolidine gave the most reliable results and the highest yields[27]. The reaction proceeded markedly better than when using the conventional triethylamine or butylamine recipes. The reason for this behavior is not explained yet and may lie in a subtle change of one or more elementary steps in the mechanism of the coupling.

Due to the quantitative yield, the arene-alkyne-coupling is ideal for utilization as a polymer forming reaction. Indeed, coupling the diiodide **52** with the diyne **53** in an AABB-type polycondensation gave rise to the isolation of a yellowish polymer of the structure **54**. The characterization of **54** was accomplished using all conventional spectroscopic techniques. The GPC indicated a DP of about 22, indicating that the yield of the reaction was >97 %. The M_w/M_n value, i.e. the polydispersity, was 2.43 in accordance with the Flory Schultz-theory, which predicts a polydispersity of 2 for a polycondensation[28].

These rigid rod polymers have, by virtue of their synthesis, two differing end groups which can both be used for further functionalization.

The motivation for the synthesis of the PPE's was twofold: a) PPE's could satisfy the need for molecular wires which bridge gaps in the nm-length between two conducting moieties[29]; b) PPE's show strong fluorescence and are thus useful for optical applications of all kinds, including their exploitation in light emitting diodes (LED)[30]. Of course, for the application in b) the end groups do not play a crucial role, because the large π-system will almost completely dominate the optical behavior. For the use in a), i.e. as bridging moiety in that what is termed molecular electronics, the end functionalization is crucial. A particularly useful end group would be a thiol, which directly and strongly interacts with gold surfaces, thus behaving similarly to an "alligator clip" in the case of a macroscopic wire[31]. In order to attach the alligator clips to the PPE backbone, a model reaction was developed for an oligomer.

To attach the endgroups, the free alkyne function of polymer **54** was reacted with the sulfur containing iodoarene **55**. In a second step, the free iodine end group of **56** was quenched in a Heck-Cassar-Sonogashira-Hagihara reaction with alkyne **57**. Surprisingly, neither here nor in the synthesis of the model compound does the thiocarbamate functionality disturb the reaction or poison the catalyst[32].

In the last step, the thiocarbamate **58** was hydrolyzed to the corresponding thiophenol **59** by stirring with potassium hydroxide in methanol and subsequent acidification with HCl. All characterizations show that the thiophenol capped polymer is structurally intact and well defined. Future experiments will show its utility as molecular wire. Thereby the formation of a polydisperse mixture is quite advantageous from the material science aspect; it means that the gaps which will be produced by some etching technique can be bridged, despite the fact that they will not all have the same width.

Until now the only reasonable way to make PPE polymers was the Heck-Cassar-Sonogashira-type approach, which is satisfying for many applications. Yet it may be valuable to have a second, independent process for forming these polymers which does not depend upon the use of catalytic palladium. One conceivable route to this type of polymer would involve the use of a metathetical reaction.

The Acyclic Diyne Metathesis reaction involves (in the simplest case) the reaction of two substituted alkynes by formal scission of the triple bond under statistical rearrangement. The reaction can be driven to the right, if one of the formed alkynes is much more volatile than all of the others and is removed *in vacuo*. A good choice is butyne with a bp of 27 °C. The only product formed in the reaction is thus the dimer **60**.

60

Application of this principle to a polymerization reaction utilizing dipropynylated benzenes should lead to PPE's. And indeed, reaction of 2,5-dihexyl-1,4-dipropynylbenzene **61** with $(tBuO)_3W\equiv C-tBu$ in dichlorobenzene at 80 °C *in vacuo* gave rise, after 18 h reaction time, to the corresponding PPE **62** with a DP of ≈ 35. The polydispersity was determined to be 2.1, a value expected for a polycondensation reaction. It was of importance that the polymer weight distribution was monomodal, with the implication, that no crosslinking occurs during the reaction (*vide infra*). The yield of **62** was 86 % with the formation of some high molecular weight and less soluble material. By ^{13}C NMR spectroscopy, it was possible to find out that the end groups in polymer **62** were still propynyl groups. Additionally, it was an important find, that no traces of cross linked products were detected. These findings imply that it should be possible to bring up DP to higher values if catalyst decomposition could be suppressed.

While it was not possible to eliminate catalyst decomposition, consecutive addition of further batches of catalyst after 24 and 48 h and a total reaction time of 78 h led to the isolation of a 68 % yield of polymer **62** and the formation of copious amounts of insoluble material. Investigation of the obtained polymer showed that the DP value increased remarkably to ≈ 150. Such a high DP is difficult to obtain when using a Pd-catalyzed coupling reaction and therefore the metathesis reaction is an extremely valuable supplementation of the known methodology of PPE-synthesis[33].

Preliminary results for other bispropynylated benzenes e.g. **63** and **65** show that the formation of PE-polymers by metathesis is a general process; the obtained unoptimized DP values are here in the order of 20-30. In the case of poly(terphenylene)ethynylene (**66**), the DP is due to the low solubilizing power of the phenyl groups.

The fact that the end groups are uniformly propynyl groups in **62**, makes their functionalization in the corresponding polymer possible. Addition of the thiocarbamate **67** and of fresh tungsten carbyne readily forms the desired, end capped polymer.

Here also, the thio functionality does not interfere with the catalytically active transition metal moiety. Again, further functional group manipulation gives rise to the formation of the desired thiophenol-capped PPE-polymer **58** (*vide supra*) as a valuable building block in material science and molecular wire-type applications.

3 Polymerization and Modification of Polymers Using Metallorganic Compounds

3.1 Metallorganic Intermediates as Mono- and Difunctional Initiators

Most commercial polymers are prepared by radical or condensation polymerization, the primary advantage of these techniques being the simplicity of preparation[34]. However, these polymers typically have broad molecular weight distributions and special polymer architectures, such as block copolymers are only in a few cases accessible. Anionic polymerisation permits better control of polymer structures, molecular weight and tacticity and allows the quantitative functionalization of polymer chains. In particular, living anionic polymerization techniques can be used for the synthesis of block copolymers[35]. Block copolymers are applied as thermoplastic elastomers (e.g. SBS-Polymers (= bl-polystyrene-bl-polybutadiene-bl-polystyrene)), polymeric surfactants in emulsion polymerization and compatibilizers in polymer blending.[36]

Typically, anionic propagation species are generated at low temperatures by the addition of a monomer M_1 to an organometallic compound such as butyllithium. After forming the A-block by complete conversion of the first monomer M_1, the B-block is generated by the addition of the second monomer M_2 to the still living anion.

Diblock copolymers are easily synthesized by initiation with a monofunctional initiator. To synthesize triblock copolymers, however, bifunctional initiators are essential. For several reasons, the situation for these initiators is more complicated. The stability of the initiator is much more important. In the case of the monofunctional initiators, deactivation of a part of the initiator only leads to higher molecular weights but not to a mixture of different types of polymers. However, if some of the anions in a difunctional initiator are protonated or deactivated by side reactions, the resulting material contains a mixture of AB and BAB-block copolymers negatively influencing the mechanical properties in comparison to the pure BAB-block copolymer[36].

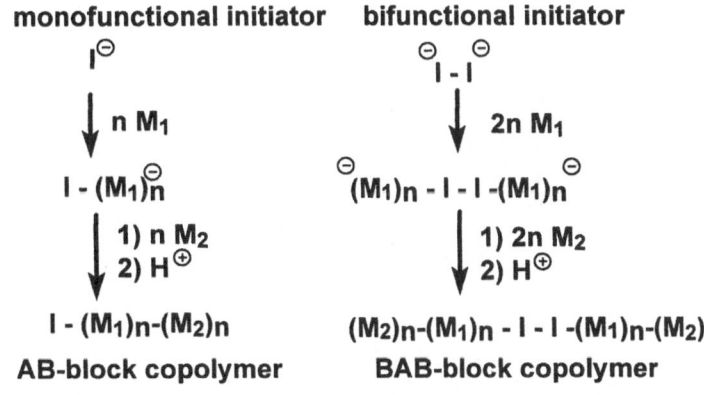

Furthermore block length and molecular weight distributions are strongly affected by the limited solubility of many bifunctional initiators and by different reactivity of the two anionic sites in a bifunctional initiator.

The synthesis of poly(bl-polystyrene-bl-polybutadiene-bl-polystyrene) (SBS-block-copolymers) may be used as an example to illustrate anionic initiator systems for triblock polymer formation. The advantages and disadvantages of various initiators will be demonstrated. In the commercial process, SBS-polymers were generated by the dianion, synthesized in a two-fold addition to butyllithium on p-diphenyldiethenylbenzene (68)[37]. However, the dianion 69 can undergo LiH elimination, to some extent, causing some monofunctionalities in the system.

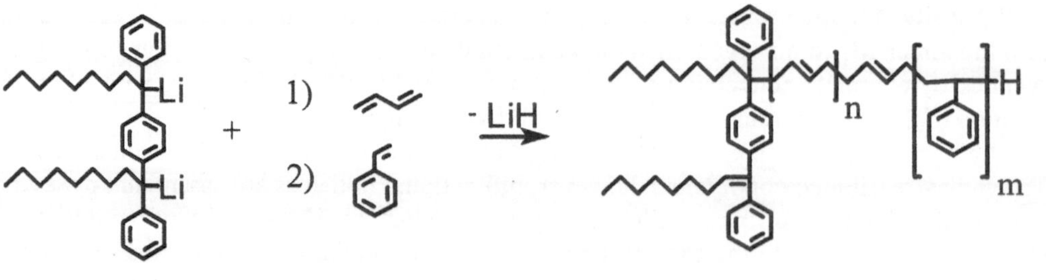

69

To supress this side reaction, a bifunctional initiator **70** was synthesized where the dianion **71** is generated by a reductive ether cleavage[38]. This new initiator is differentiated from **72** in that it lacks the possibility of the LiH elimination while increasing the initiation rate due to the reduced steric hindrance at the anionic sites.

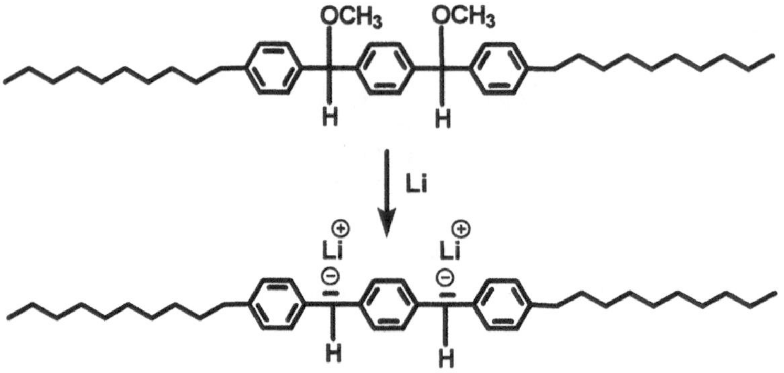

70

71

To overcome the problem of the limited solubility of the bifunctional initiator, the alkyl chains were placed in *para* positions to the outer phenyl rings. This allows anionic polymerization in nonpolar solvents, such as hexane, which is necessary to produce 1,4-polybutadiene and not the undesired 1,2-polybutadiene blocks[36].

Polymers with suitably attached chromophores have played an important role in the study of miscibility or phase separation of polymer blends by fluorescence techniques[39]. Conventional methods for introducing suitable chromophores require the polymer analogous modification of a precursor polymer or the copolymerization of labelled and unlabelled monomers[40]. With both methods, the content and the distribution of the luminescent labels is difficult to control. Deactivation of a living species by aldehydes or acid chlorides of chromophores is also problematic due to the limited efficiency of this reaction[41].

To obtain quantitatively monofunctionalized systems, dihydroanthracene mono anion (**73**) was used as initiator which was generated by deprotonation of the dihydroanthracene at -78 °C in THF. Addition of methyl methacrylate or styrene to the anion allows the synthesis of polystyrene or poly(methyl methacrylate) (PMMA) **74** .

73   74

Subsequent aromatization of the dihydroanthracene unit with o-chloranil results in polymer **75**, labelled quantitatively with anthracene[42] which enables us to study the chain conformation in solution by fluorescence measurements[43].

74   75

Furthermore, with the anthracene labelled PMMA **75** we were able to produce reversible switching of polymer properties. The sample viscosity or Tg can be changed by dimerization of two polymer chains in a [4+4]-cycloaddition reaction under irradiation. Heating the dimer of the anthracene labelled samples up to 100 °C or irradiation at shorter wavelengths results in the starting material again[44].

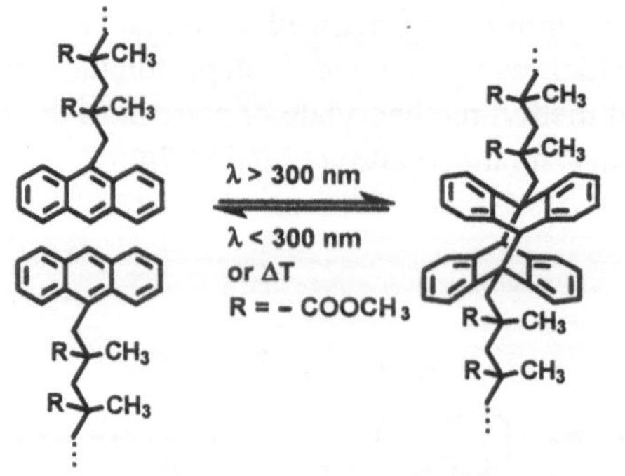

75

To extend the concept of anionic polymerization with anthracene derivatives as initiator, we designed a bifunctional initiator **76** consisting of two dihydroanthracene units, connected by an alkylene or oligoethyleneoxide spacer. Via two-fold deprotonation at -78 °C a dianion of **77** was quantitatively generated. Polymerization of tert-butyl methacrylate, followed by methyl methacrylate resulted in BAB-block copolymer **78**.

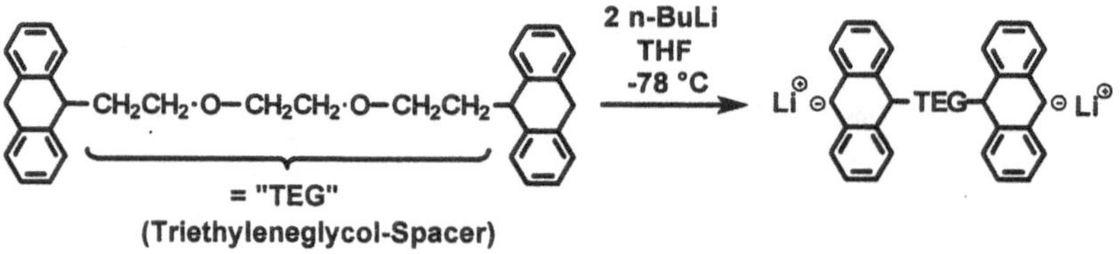

77

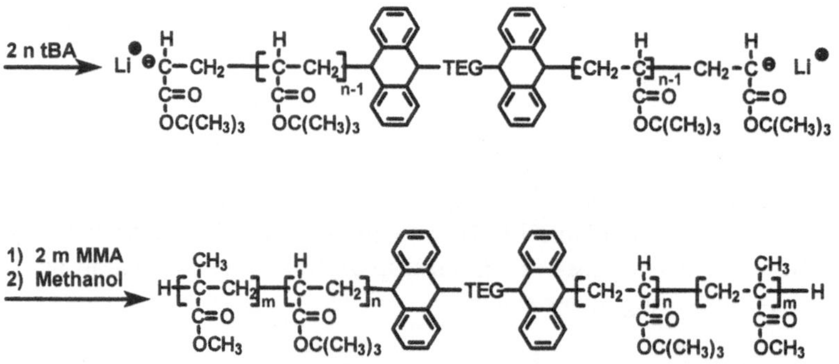

78

By variation of the amount of initiator and monomers one can control the overall polymer chain length as well as the block length. An amphiphilic block copolymer **79** with a

44

PMMA-bl-polyacrylic acid-bl-PMMA was isolated by selective hydrolysis of the tert-butyl esters.

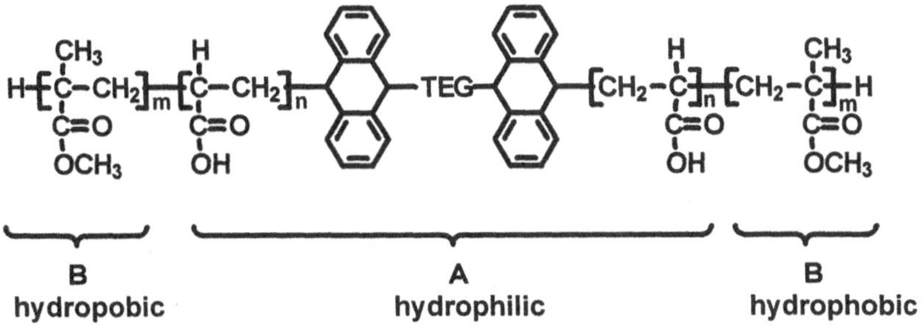

B
hydropobic

A
hydrophilic

B
hydrophobic

79

Polymer **79** forms micelles and can be used as polymeric surfactant in emulsion polymerization. Again, rearomatization of the dihydroanthracene moieties allows for the formation of a quantitative anthracene labelled polymer **80**. This polymer might be a helpful system for clarifying the mechanism of emulsion polymerization. Due to the fluorescence of the polymeric surfactant, several spectroscopic methods could be applied for studying the structure of latex particles.

3.2 Modification of Polymers via Organometallic Intermediates

Organometallic compounds were not only applied as initiator in anionic polymerization, but also be used as reagents in modifications of polymers to change the properties and architecture of polymers. In particular, carbonyl groups can be transformed by the nucleophilic addition of a carbanion. By adding phenyllithium to a solution of a poly(ether ketone), **81** could be quantitatively converted to the more polar polyalkohol **82**, which was then further modified to produce the red polycation **83**[45].

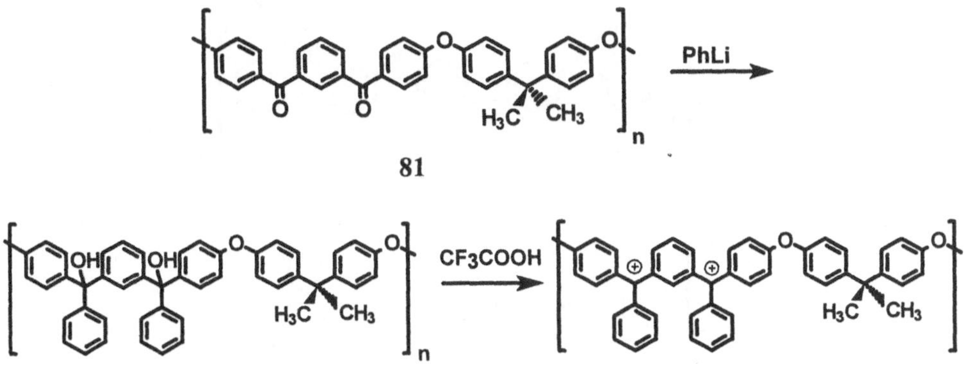

81

82   83

45

In a similar, quantitative transformation, a polyfulvene **84** could be achieved by the reaction of a tert-butyl substituted cyclopentadienyl monoanion with poly(ether ketone) **81**. These results encouraged us to use larger metallorganic systems for the modification of polycarbonyl systems.

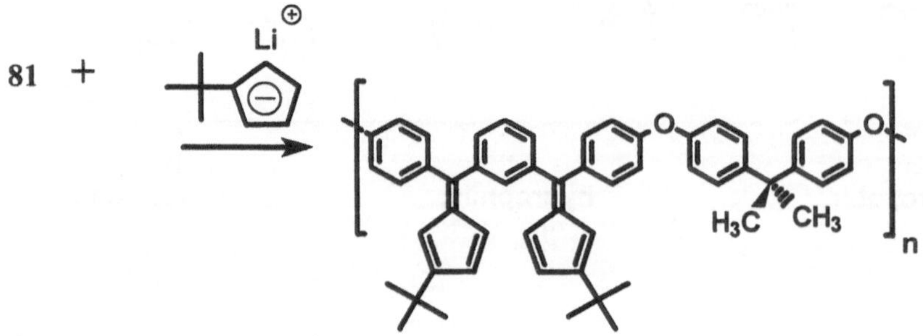

84

Using the living anions of polystyrene or polyisoprene, generated by anionic polymerization, we were able to graft side chains onto a poly(ether ketone) backbone. Through nucleophilc addition of the anions to the carbonyl groups of **81**, polystyrene or similarly polyisoprene chains could be attached[51].

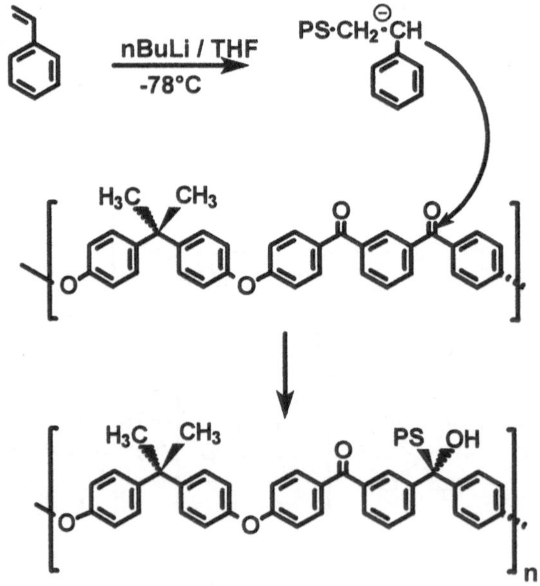

81

85

46

In this way, well defined comb-type polymer structures **85** were made accessible by a *grafting-onto* process wherein the properties of a technical polymer (polystyrene, polyisoprene) are modified by attaching to a so-called high-performance polymer (e.g. poly(ether ketone)). The great advantage of this method is that the degree of grafting can be controlled by the ratio of initiator to carbonyl groups in **81** and the length of the side chains by the ratio of initiator to monomer. This allows a better structure-property correlation in comparison to polymers prepared by *grafting-from* methods. Mechanical and thermal properties can be easily tuned by this process of anionic deactivation. For example, the Tg of a polyisoprene (M_n = 3000) increases from -48 °C up to -2 °C grafted onto **81** with a conversion of 23 % of the carbonyl groups. In comparison with the reactions with phenyllithium or cyclopentadienide monoanion, the conversion of carbonyl groups is limited to 32 %. This can be explained by a shielding effect from the already attached long polymer chains of polystyrene or polyisoprene on the unreacted carbonyl groups[5].

Acknowledgements

Financial support by the Bundesministerium für Bildung und Forschung, The Fonds der Chemie and the Volkswagenstiftung is gratefully acknowledged

References

1. a) P. O. Danis, D. E. Karr, F. Meyer, A. Holle, C. H. Watson, *Org. Mass Spectrom.* **1992**, *27*, 843; b) J. Spickermann, K. Martin, H. J. Räder, K. Müllen, R.-P. Krüger, H. Schlaad, A. H. E. Müller, *Eur. Mass Spectrom.* 2, **1996**, *161*; c) H. J. Räder, J. Spickermann, K. Müllen, *Macromol. Chem. Phys.* **1995**, *196*, 3967.
2. M. P. Stevens, *Polymer Chemistry: An Introduction*, Oxford University Press, 2nd ed., New York, 1990.
3. a) J. M. Tour, *Adv. Mater.* **1994**, *6*, 190; b) M. Rehahn, A. D. Schlüter, G. Wegner, W. J. Feast, *Polymer* **1989**, *30*, 1054; c) M. Rehahn, A. D. Schlüter, G. Wegner, W. J. Feast, *Polymer* **1989**, *30*, 1060; d) D. L. Gin, V. P. Conticello, *TRIP* 4 **1996**, 217.
4. a) U. Scherf, K. Müllen, *Adv. Polym. Sci.* **1995**, *123*, 1; b) M. Kreyenschmidt, F. Uckert, K. Müllen, *Macromolecules* **1995**, *28*, 4577; c) M Fukuda, K. Sawada, K. Yoshino, *J. Polym. Sci. A.* **1993**, *31*, 2465.
5. T. Wehrmeister, M. Klapper, K. Müllen, *Macromolecules* **1996**, *29*, 5805.
6. a) M. Baumgarten, U. Bunz, U. Scherf, K. Müllen, *Molecular Engineering for Advanced Materials*, J. Becher and K. Schaumburg (eds.), Kluwer Academic Press, **1995**, 159 b) J. Grimme, U. Scherf, *Makromol. Chem. Phys.* **1996**, *197*, 2297.
7. M. Kreyenschmidt, M. Baumgarten, N. Tyutyulkov, K. Müllen, *Angew. Chem.* **1994**, *106*, 2062; *Angew. Chem. Int. Ed. Engl.* **1994**, *33*, 1957.
8. a) M. Kreyenschmidt, F. Uckert, K. Müllen, *Macromolecules* **1995**, *28*, 4577; b) T. Yamamoto, A. Morita, Y. Miyazaki, T. Maruyama, H. Wakayama, Z. Zhou, Y. Nakamura, T. Kanbara, S. Sasaki, K. Kubota, *Macromolecules* **1992**, *25*, 1214.

9. N. Miyaura, A. Suzuki, *Chem. Rev.* **1995**, *95*, 2457.

10. J. Huber, U. Scherf, *Macromol. Rapid Commun.* **1994**, *15*, 897.

11. M. A. Keegstra, S. De Feyter, F. C. DeSchryver, K. Müllen, *Angew. Chem.* **1996**, *108*, 830; *Angew. Chem. Int. Ed. Engl.* **1996**, *35*, 774.

12. H. Quante, K. Müllen, *Angew. Chem.* **1995**, *107*, 1487; *Angew. Chem., Int. Ed. Engl.* **1995**, *34*, 1323.

13. a) F. Holtrup, Gerd Müller, H. Quante, S. DeFeyter, F. DeSchryver, K. Müllen, *Chem. Eur. J.*, accepted; b) H. Azizian, C. Eaborn, A. Pidcock, *J. Organometal. Chem.* **1981**, *215*, 49; c) J. K. Stille, *Angew. Chem.* **1996**, *98*, 504; *Angew. Chem. Int. Ed. Engl.* **196**, *25*, 508; d) I. P. Beletskaya, *J. Organometal. Chem.* **1983**, *250*, 551.

14. A. Böhm, M. Adam, H. Mauermann, S. Stein, K. Müllen, *Tetrahedron Lett.* **1992**, *33*, 2795.

15. a) C. Jubert, P. Knochel, *J. Org. Chem.* **1992**, *57*, 5425; b) A. Sidduri, M. J. Rozema, P. Knochel, *J. Org. Chem.* **1993**, *58*, 2694.

16. a) M. Müller, H. Mauermann-Düll, M. Wagner, V. Enkelmann, K. Müllen, *Angew. Chem.* **1995**, 107, 1751; *Angew. Chem. Int. Ed. Engl.* **1995**, *34*, 1583; b) M. Müller, J. Petersen, R. Strohmaier, C. Günther, N. Karl, K. Müllen, *Angew. Chem.* **1996**, *108*, 947; *Angew. Chem. Int. Ed. Engl.* 35 **1996**), 886.

17. a) P. Kovacic, *Chem. Rev.* **1987**, *87*, 357; b) P.Kovacic, A.Kyriakis, *J. Am. Chem. Soc.* **1963**, *85*, 454; c) P. Kovacic, A.Kyriakis, *Tetrahedron Lett.* **1962**, 467.

18. K. Müllen, S. Valiyaveettil, V. Francke, V. S. Iyer, *Atomic and Molecular Wires*, Kluwer Academic Press, in press

19. J. H. Burroughes, D. D. C. Bradley, A. R. Brown, R. N. Marks, K. Mackery, R. H. Friend, P. L. Burn, A. B. Holmes, *Nature* **1990**, *347*, 539.

20. a) R.A.Wessling *J. Polym. Sci., Polym. Symp.* **1986**, *72*, 55; b) R. O. Garay; U. Baier; C. Bubeck; K. Mullen, *Advan. Mater.* **1993**, *651*, 5.

21. J. K. Stille, *Angew. Chem.* **1986**, *98*, 504; *Angew. Chem. Int. Ed. Engl.* **1986**, 25, 508.

22. V. Farina, B. Krishnan, *J. Am. Chem. Soc.* **1991**, *113*, 9585.

23. A. de Meijere, F. E. Meyer, *Angew. Chem.* **1994**, *106*, 2473; *Angew. Chem. Int. Ed. Engl.* **1994**, *33*, 2379.

24. a) W. Heitz, W. Brügging, L. Freund, M. Gailberger, A. Greiner, H. Jung, U. Kampschulte, N. Nießer, F. Osan, H.-W. Schmidt, M. Wicker, *Makromol. Chem.* **1988**, *189*, 119; b) A. Greiner, W. Heitz, *Makromol. Chem., Rapid Commun.* **1988**, *9*, 581; c) M. Brenda, A. Greiner, W. Heitz, *Makromol. Chem.* **1990**, *191*, 1083; d) A. Greiner, W. Heitz, *Polym. Prepr.* **1991**, *32*, 333.

25. H. Mauermann-Düll, A. Böhm, G. Fiesser, K. Müllen, *Macromol. Chem. Phys.* **1996**, *197*, 413.

26. a) H. A. Dieck, R. F. Heck, *J. Organomet. Chem.* **1975**, *93*, 259; b) I. Cassar, *ibid.* **1975**, *93*, 253; c) K. Sonoshigara, Y. Thoda, N. Hagihara, *Tetrahedron Lett.* **1975**, 4467.

27. M. Alami, F. Ferri, G. Linstrumelle, *Tetrahedron Lett.* **1993**, 6403.

28. Z. Xu, J. S. Moore, *Angew. Chem.* **1993**, *105*, 261; *Angew. Chem., Int. Ed. Engl.* **1993**, *32*, 246.

29. J. S. Schumm, D. L. Pearson, J. M. Tour, *Angew. Chem.* 106 **1994**), 1445; *Angew. Chem. Int. Ed. Engl.* **1994**, *33*, 1360.

30. T. Mangel, A. Eberhardt, U. Scherf, U. H. F. Bunz, K. Müllen, *Macromol. Rapid Commun.* **1995**, *16*, 571.

31. A.-A. Dhirani, R. W. Zehner, R. P. Hsung, P. Guyot-Sionnest, L. R. Sita, *J. Am. Chem. Soc.* **1996**, *118*, 3319.

32. V. Francke, T. Mangel, A. Eberhardt, K. Müllen, to be published.

33. K. Weiss, A. Michel, E.-M. Auth, U. H. F. Bunz, T. Mangel, K. Müllen, *Angew. Chem.*, in press.

34. G. Moad, H. Solomon, *The Chemistry of Free Radical Polymerization*, Pergamon Press, New York **1995**

35. M. Morton, *Principles of Anionic Polymerisation*, Academic Press New York, **1983**, *103*. A. Noshay, J.E. McGrath, *Block Copolymers, Overview and Critical Survey*, Academic Press, New York **1977**.

36. M. Morton, L.J. Fetters in W.M. Saltman, *The Stereo Rubbers*, John Wiley & Sons, New York **1977**

37. a) L. H. Tung, G. Y. S. Lo, D. E. Beyer, *Macromolecules* **1978**, *11*, 616; b) P. Guyot, J. C. Favier, H. Utterhoeven, M. Fontanille, P. Sigwalt, *Polymer* **1981**, *22*, 1724.

38. T. Schäfer, PhD thesis, to be published.

39. a) F. C. De Schrijver, *Makromol. Chem. Suppl.* **1979**, *3*, 85.

    b) F. Mikes, H. Moravetz, *Macromolecules* **1984**, *17*, 60;

    c) Y. Zhao, R. E. Prud'homme, *Polym. Bull.* **1991**, *26*, 101.

40. T. Miyashita, M. Ohsawa, M. Matsuda *Macromolecules* **1986**, *19*, 585.

41. L. Leemans, R. Fayt, P. Teyssie, *Macromolecules* **1990**, *23*, 1554.

42. T. Bartz, M. Klapper, K. Müllen, *Makromol. Chem.* **1994**, *195*, 1097.

43. K. Haimer, N. Khelfallah, M. Klapper, K. Müllen, to be published.

44. T. Bartz, M. Klapper, K. Müllen, *Acta Polymer.* **1994**, *45*, 248.

45. S. Mullins, M. Klapper, K. Müllen, to be published.

Organoaluminium and Organogallium Complexes as Efficient and Selective Alkylating Agents[1]

Waël Baidossi[a], Jochanan Blum*[a], Michael Frick[b], Dmitri Gelman[a], Bernd Heymer[b], Herbert Schumann[b], Stefan Schutte[b] and Eduard Shakh[a]

a) Department of Organic Chemistry, Hebrew University, Jerusalem 91904, Israel
b) Institute für Anorganische und Analytische Chemie, Technische Universität Berlin, Strasse des 17. Juni 135, D-10623 Berlin, Germany

Although trialkylaluminum compounds are known since 1865[2] and widely used in the industry as catalysts for Ziegler-Natta, Phillips and ring opening metathesis polymerization processes[3], they are seldom employed in lab-scale alkylations owing to their pyrophoric nature. The observation that substitution of one alkyl group in trialkylaluminum by a chelating ligand reduces the sensitivity to moisture and oxygen[4], arose our interest in studying the application of the stabilized *di*alkylaluminum complexes as inexpensive substituents for Grignard and organolithium compounds.

The following six stabilized dimethylaluminum complexes

$$Me_2Al(CH_2)_3NMe_2 \; (1)^{[5]},$$
$$[Me_2Al(\mu\text{-}O(CH_2)_2NMe_2)]_2 \; (2)^{[5]},$$
$$[Me_2Al(\mu\text{-}O(CH_2)_2OMe)]_2 \; (3)^{[6,7]},$$
$$[Me_2Al(\mu\text{-}O(CH_2)_2CMe_2OMe)]_2 \; (4)^{[6]},$$
$$[Me_2Al(\mu\text{-}O\text{-}2\text{-}C_6H_4OMe)]_2 \; (5)^{[6]},$$
$$[Me_2Al(\mu\text{-}OCH_2\text{-}2\text{-}C_6H_4OMe)]_2 \; (6)^{[6]}$$

have recently been prepared in our laboratory, and, as expected, proved to methylate various ketones in very high yield. [3-(Dimethylamino)propyl]-dimethylaluminum, **1**, was found to be the most efficient among these reagents and therefore, it was chosen for our standard experiments. In a typical methylation experiment 1.3 meq of **1** and 1 mmol of benzophenone in 1.5 ml of toluene was heated under N_2 for 2 h at 80°C. After quenching with dilute hydrochloric acid, 93% of 1,1-diphenylethanol was obtained as the only isoluble product.

Likewise, fluorenone, cyclopropyl phenyl ketone, and even the bulky 2-adamantone could be methylated under these conditions. Thus, we believe that the process outlined in eq 1 is of general applicability, although some ketones react only at substantially higher temperature than 80 °C. (Cyclohexyl phenyl ketone, e.g., reacted only at 155°C).

$$\underset{R \quad R'}{\overset{O}{\big|\big|}} \xrightarrow[91-96\%]{1,\ 80\ °C,\ 1\text{-}3\ h} \underset{R \underset{Me}{\overset{OH}{\big|}} R'}{} \tag{1}$$

Diketones, such as 1,2-acenaphthenedione and 1,9-phenanthrenequinone were found to undergo bis-methylation at 80°C and to give the corresponding dimethyl diols in quantitative yield within 1-2 h. While the phenanthrenequinone forms solely the *trans* dimethyl diol, the acenaphthenedione gives a 4:1 mixture of *cis-* and *trans*-isomers. When, however, the methylation reagent **1** is substituted by **5**, *cis*-1,2-dihydro-1,2-dimethylacenaphthenediol becomes the only alkylation product.

Simple α,β-unsaturated ketones (e.g., benzalacetone and benzalacetophenone) are methylated by **1**, according to eq 2, i.e., at the C-C double bond rather than at the carbonyl function.

$$\underset{R \quad R'}{\overset{O}{\big|\big|}} \xrightarrow[76-96\%]{1,\ 80\ °C,\ 1\text{-}2\ h} \underset{R \quad R'}{\overset{O \quad Me}{\big|\big|}} \tag{2}$$

α,β-Alkynones (e.g., 1,3-diphenylprop-1-yn-3-one, PhC≡CCOPh) as well as sterically hindered ketones (tetraphenylcyclopentadieneone), undergo only 1,2-methylation. Alicylic aldehydes (e.g., cyclohexanecarboxaldehyde) are alkylated by **1** in good yield even at room temperature. Aromatic aldehydes form however in benzene or toluene, chalcones as the result of initial methylation of the aldehyde (eq 3) followed by hydrogen transfer from the methyl aryl carbinol to the unreacted starting aldehyde (eq 4). The ketone, so formed, undergoes then a base-mediated condensation with the aldehyde according to eq 5.

$$\text{Ar-CHO} \quad + \quad \mathbf{1} \quad \longrightarrow \quad \underset{\text{Me}}{\overset{\text{OH}}{\text{Ar-CH}}}\text{-H} \tag{3}$$

$$\text{Ar-CHO} \quad + \quad \underset{\text{Me}}{\overset{\text{OH}}{\text{Ar-CH}}}\text{-H} \quad \rightleftharpoons \quad \text{Ar-CO-Me} \quad + \quad \text{Ar-CH}_2\text{OH} \tag{4}$$

$$\text{Ar-CHO} \quad + \quad \text{Ar-CO-Me} \quad \longrightarrow \quad \text{Ar-CH=CH-CO-Ar} \quad + \quad \text{H}_2\text{O} \tag{5}$$

These undesired transformation of aromatic aldehydes could be eliminated by addition of ca. one equivalent of $AlCl_3$ to the aldehyde prior to its interaction with **1**. In the presence of the trichloride, the aldehydes are smoothly converted into the expected methyl aryl carbinols (eq 1, R = Ar, R' = H). Thus, treatment of 2 mmol of benzaldehyde in 2 ml of toluene for 30 min at 30°C with 2.1 meq of $AlCl_3$, followed by stirring for 2 h at 60°C, yielded 88% of 1-phenylethanol. By the same procedure 4-tolualdehyde, 1-, and 2-naphthaldehyde and 9-anthracenecarboxaldehyde gave the corresponding carbinols in 71-95% yield.

$$\mathbf{1} + \text{R}-\!\!\equiv\!\! \quad \xrightarrow[\text{15-30 min}]{50\ ^\circ\text{C}} \quad \left[\begin{array}{c} \text{R} \\ \\ \text{R} \end{array} \text{Al} \underset{\text{Me}^\diagup \diagdown \text{Me}}{\overset{\text{N}}{}} \right]^- \quad \begin{array}{l} \xrightarrow[\text{70-80}\ ^\circ\text{C, 3 h}]{\text{R}^1\text{R}^2\text{CO}} \quad \underset{\text{R}^2}{\overset{\text{OH}}{\text{R}^1\text{C}}}\!\!-\!\!\equiv\!\!-\text{R} \\[2em] \xrightarrow[\text{80}\ ^\circ\text{C, 2-3 h}]{\text{R}^3\text{COCH}=\text{CHR}^4} \quad \text{R}^3\text{CO-}\underset{\equiv\!-\text{R}}{\overset{\text{R}^4}{\text{CH}}} \end{array} \tag{6}$$

Heating of **1** for 15-30 min with terminal acetylenes proved to convert the methylation agents into dialkynylaluminum complexes that efficiently alkynylate saturated carbonyl compounds (1,2-alkynylation) and α,β-unsaturated ketones (1,4-addition) (eq 6). E.g., when a toluene solution of $PhC\equiv CH$ and **1** is heated for 15 min at 50°C, and the resulting $(PhC\equiv C)_2Al(CH_2)_3NMe_2$ reacted with benzophenone for 3 h at 70°C, 93% of $PhC\equiv CC(Ph_2)OH$ is obtained. Benzalacetophenone gives under similar conditions 78% of $PhC\equiv CCH(Ph)CH_2COPh$.

Arylbromides and -iodides do not react by themselves with the dimethylaluminum complexes **1-6**. However, when the halides are treated with a Pd(0) or Pd(II) catalyst prior to the addition of the aluminum compounds, the halogen atoms are substituted by methyl groups. In a typical experiment a solution of 1 mmol of 1-bromonaphthalene and 0.02 mmol of either $Pd(PPh_3)_4$ or $PdCl_2(PPh_3)_2$ in 4 ml of benzene was initially heated at 50°C for 30 min. A quantity of 0.54 meq **1** was added and the mixture heated for 3 h at 80°C. Quenching with dilute hydrochloric acid afforded 1-methylnaphthalene in quantitative yield. In a similar manner bromobenzene, iodobenzene and 9-bromoanthracene were methylated in 91-97% yield (eq 7).

$$RX \quad + \quad Pd(PPh_3)_4 \quad \xrightarrow[80\,°C]{30\ min} \quad \xrightarrow[80\,°C,\ 2\text{-}3\ h]{\textbf{1}} \quad RMe \qquad\qquad (7)$$

$$X = Br, I$$

Dihaloarenes (1,8-diiodonaphthalene, 2,2'-diiodo-1,1'-biphenyl, bis(2-iodophenyl)-methane, 3,4-dibromothiophene) are methylated stepwise, forming initially the mono-methylated, and were shown to undergo then the dimethylated arene in exceedingly high yield. Vinyl- and benzyl bromides were shown to undergo similar methylation by 1 in the presence of the palladium catalyst. E-β-Bromostyrene, Z-bromostilbene and 2-bromoindene, e.g., gave under the conditions of eq 7, 97, 96 and 88% of the corresponding methylated products. Benzyl bromide, 9-bromo-9-phenylfluorene and 2-(α-bromobenzyl)naphthalene yielded ethylbenzene, 9-methyl-9-phenylfluorene and 2-[1-(phenyl-1-ethyl)]naphthalene, respectively, in 99, 89 and 65% yield. Since the dimethylaluminum complexes **1-6** alkylate both carbonyl- and bromofunctions, brominated benzophenones undergo bis-methylation. 4-Bromobenzophenone, e.g., afforded under the conditions of eq 7 a mixture of 19% of $4\text{-MeC}_6\text{H}_4\text{COPh}$ and 81% of $4\text{-MeC}_6\text{H}_4\text{C(Me)(Ph)OH}$.

In analogy to the above mentioned organoaluminum complexes, the corresponding stabilized dimethylgallium $Me_2Ga(CH_2)_3NMe_2$ (**7**)[8], $Me_2GaO(CH_2)_2NMe_2$ (**8**)[9], and $[Me_2Ga(\mu\text{-}O(CH_2)_2OMe]_2$ (**9**)[6] have been synthesized. In contrast to the aluminum compounds **7-9** do not alkylate ketones and aldehydes, but methylate, in the presence of a palladium catalyst, aryl bromides and iodides, as well as the corresponding vinyl halides. The yields were shown to be uniformly high although the reaction rates proved to be usually lower than those observed with the analogous aluminum complexes. The relative initial rates of methylation of 1-bromonaphthalene by **1-9** at 88°C after treatment with $PdCl_2(PPh_3)_2$ for 30°C] were as follows: **3** - 1.00; **4** - 0.87; **6** - 0.56; **2** - 0.47; **1** - 0.35; **7** - 0.20; **8** - 0.15; **9** - 0.04 and **5** - 0.03. It is notable that while the oxygen-free aluminum complex **1** is the most efficient carbonyl-methylation complex the oxygen-containing compounds **2**, **3**, **4** and **6** are superior bromine-substitution agents. Among the gallium complexes [3-(dimethylamino)-propyl]dimethylgallium (**7**) is the most efficient one.

The methylation rate of 1-bromonaphthalene was found to depend strongly also on the palladium catalyst. When e.g., $1\text{-BrC}_{10}H_7$ was methylated by 3 at 88°C (ratio of [substrate] : [3] : [Pd] = 50 : 27.5 : 1) the relative activities for [Pd] = $Pd(PPh_3)_4$, $PdCl_2(PPh_3)_2$, $[NH_4][PdCl_4]$, $PdCl_2$, $Pd(CN)_2$, $Pd(OAc)_2$ were 1.00, 0.93, 0.40, 0.15, 0.03 and 0.02, respectively.

Since our stabilized gallium complexes **7-9** react neither with carbonyl compounds nor with some other vulnerable functions, including benzylic halides, it is feasible to alkylate many functionalized aryl bromides and iodides in a highly selective manner. Thus, e.g., 4-bromobenzaldehyde, 4-bromobenzophenone, 4-bromoacetophenone, methyl 4-bromobenzoate, 4-bromobenzyl bromide, 4-bromochlorobenzene, 4-bromo-benzonitrile, 4-bromoacetanilide, 4-bromonitrobenzene and 4-bromo-chalcone could be converted according to eq 8 exclusively into 4-MeC_6H_4CHO, 4-MeC_6H_4COPh, $4\text{-MeC}_6H_4COCH_3$, $4\text{-MeC}_6H_4CO_2Me$, 4-MeC_6H_4CHBr, 4-MeC_6H_4Cl, 4-MeC_6H_4CN, 4-MeC_6H_4NHAc, $4\text{-MeC}_6H_4NO_2$ and $4\text{-MeC}_6H_4CH\text{=}CHCOPh$, respectively, in nearly quantitative yields (93-100%).

$$\text{Br}\!-\!\!\langle\text{C}_6H_4\rangle\!-\!X \quad \xrightarrow[\text{0.5 h, 88 °C}]{PdCl_2(PPh_3)_2} \quad \xrightarrow[\text{4-18 h, 88 °C, 93-100\%}]{Me_2Ga(CH_2)_3NMe_2} \quad \text{Me}\!-\!\!\langle\text{C}_6H_4\rangle\!-\!X \qquad (8)$$

All the stabilized dimethylaluminum and -gallium complexes prove to utilize both their methyl groups in the alkylation processes though the first Me was found to react faster than the second one.

Preliminary mechanistic studies revealed that the alkylation of aryl bromides follows the mechanism outlined in Scheme 1.

In conclusion, our experiments demonstrate that the stabilized dimethylaluminum and -gallium complexes are a new class of efficient alkylation reagents that can be employed under standard laboratory conditions, and that the methylation by the gallium compounds proceeds in greater selectivity than by any other current alkylating agent.

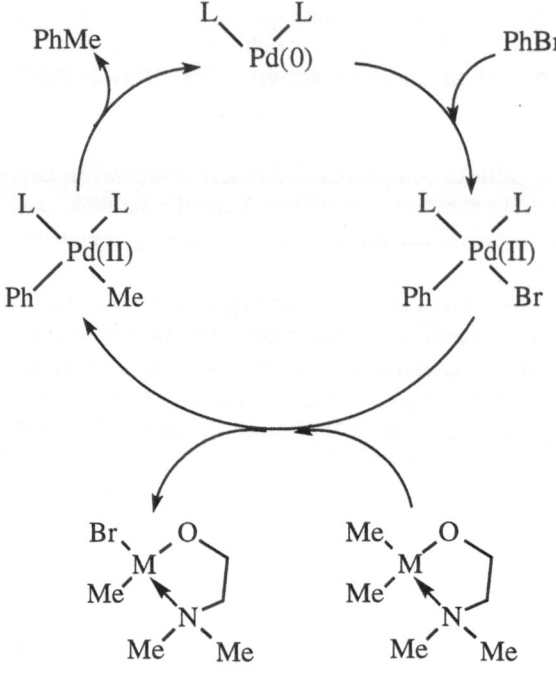

M = Al, Ga

Scheme 1. Possible mechanism of methylation of aryl bromides by complexes **1-9**

Acknowledgment

We acknowledge the support of this study by a travel grant of the Technische Universität Berlin within the partnership agreement of the TU and the Hebrew University of Jerusalem.

References

1. Part of these results have been published in: Baidossi, W.; Rosenfeld, A.; Wassermann, B.C.; Schutte, S.; Schumann, H.; Blum, J. *Synthesis* **1996**, 1127.
2. Buckton, G.B.; Odling, W. *Proc. Roy. Soc.* **1865**, *14*, 19.
3. See e.g., Zietz, Jr. J.R.; Robinson, G.H.; Lindsay, K.L.: In Comprehensive Organometallic Chemistry; Wilkinson, G.; Stone, F.G.A.; Abel, E.W., Eds.; Pergamon Press: Oxford, **1982**, Vol. 7, pp. 365-464.
4. See e.g., Fisch, J.J.: In Comprehensive Organometallic Chemistry; Wilkinson, G.; Stone, F.G.A.; Abel, E.W., Eds.; Pergamon Press: Oxford, **1982**, Vol. 1, pp. 555-682.
5. Schumann, H.; Schutte, S.; Seuss, T.D.; Wassermann, B.C.; Weinmann, R. *Organometallics*, in press.

6. Schumann, H.; Frick, M.; Heymer, B.; Girgsdies, F. *J. Organomet. Chem.* **1996**, *512*, 117.
7. Benn, R.; Rufinska, A.; Lehmkuhl, H.; Jannsen, E.; Krüger, C. *Angew. Chem.* **1983**, *95*, 808; *Angew. Chem., Int. Ed. Engl.* **1983**, *22*, 779.
8. Schumann, H.; Hartmann, U.; Dietrich, A.; Pickardt, J. *Angew. Chem.* **1988**, *100*, 1119; *Angew. Chem., Int. Ed. Engl.* **1988**, *27*, 1077.
9. Rettig, S.J.; Storr, A.; Trotter, J. *Can. J. Chem.* **1975**, *53*, 58.

Nickel Mediated Cyclobutabenzenes Syntheses

Amnon Stanger*, Alona Schachter, Nissan Ashkenazi, Roland Boese* and Peter Stellberg

Department of Chemistry, Technion - Israel Institute of Technology, Haifa 32000, Israel, and Institut für Anorganische Chemie der Universität-GH Essen, Universitätsstrasse 5-7, D-45117 Essen, Germany.

1 Introduction

Our interest in the Mills-Nixon effect[1, 2] and the need for functionalysed tricyclo-butabenzenes for its study,[3] led us to look for a better method[4] for their preparation. As model studies, we decided to prepare cyclobutabenzenes,[5] compounds that turn to be interesting by themselves. Here we report the nickel mediated synthesis of *trans*-1,2-dibromocyclobutabenzenes (**1**), *syn* and *anti*-*trans*-tetrabromo-[a,b],[d,e]-bicyclo-butabenzene (**2a** and **2b**) and *anti*-all-*trans*-hexabromotricyclobutabenzene (**3**).

1   **2 a**: *syn*-all *trans*. b:*anti*-all-*trans*   **3**

2 Cyclobutabenzenes

We decided to use the nickel mediated coupling of benzyl bromides in an intramolecular fashion to create the four membered ring. The initial attempts were disappointing; Thus, when α,α',-dibromo-o-xylene was treated with a Ni(0) complex ((Bu$_3$P)$_2$NiL, where L=anthracene or COD) only dimeric products, mainly dibenzocyclooctane were obtained (equation 1).

None of the intramolecular cyclization product was observed, even when the substrate was introduced by a syringe pump in order to maintain high dilution conditions. In order to get intramolecular ring closure instead of the intermolecular coupling it was thus necessary to attenuate the bimolecular coupling. Two possibilities were considered. One was to bind a large substituent near the reaction center in order to prevent easy approach of two molecules.

The second was to stabilize an intermediate that precedes the intermolecular coupling, and let entropy (that favors an intramolecular reaction) to play a more important role in the rate constant. As this intermediate is probably an insertion product of a Ni to a C-Br bond, an electronegative substituent should stabilize it. A substituent that fulfills these two demands is Br, and therefor α,α,α',α'-tetrabromo-o-xylene was subjected to one equivalent of (Bu$_3$P)$_2$NiL in DMF at 65-70 °C (equation 2). The products were **1** and **4** in 98.5:1.5, and after one recrystallization **1** was obtained pure an a yield better than 80%.

The use of this reaction to prepare **2** and **3** is discussed in sections 3 and 4, respectively. Here we discuss a synthetic methodology based on the use of **1** and its aromatic ring substituted derivatives as key compounds.

When **1** is heated, it electrocyclic ring opens to give the *E,E*-dibromo-*o*-xylylene (**5**) that can be trapped by Diels-Alder reaction. Model studies showed that this reaction is regio- and stereospecific (equation 3).[6]

$$(3)$$

In order to study this aspect we have prepared β-substituted α,α,α',α'-tetrabromo-*o*-xylenes and subjected them to nickel mediated ring closure conditions. The results are given in table 1. As can be observed, the reaction is fairly general, and all the substituted system that were experimented gave the desired cyclobutabenzenes.[7] Recently, we have also prepared α-substituted systems, where the substituents are methoxy, methyl-ester and sultam.

Table 1. Formation of β-substituted cyclobutabenzenes

X	method	yield (%)[a]
C(O)OCH$_3$	(Bu$_3$P)$_2$Ni-C$_{14}$H$_{10}$	80
OCH$_3$		
OC(O)CH$_3$	Ni	30
OC(O)CH(CH$_3$)OAc	(Bu$_3$P)$_2$Ni(COD)	10[b]
C(O)OH	Ni	20
C(O)-sultam	Ni	60

(a) Unoptimized yields.
(b) Only one diastereoisomer observed.

Some of these systems were subjected to a Diels-Alder reaction with several dienophiles. Table 2 summarizes some of these results. As can be observed, some of the reactions yield the addition products, whereas others yield a double HBr eliminated products.

Table 2. Diels-Alder reaction of cyclobutabenzene with dienophiles

X	dienophile	product
H	MAH	
H	TCNE	
H	styrene	
H	$\equiv\!\!-\!\!\text{Ph}$	
H		
H	1/2	
H		
C(O)CH$_3$	TCNE	
C(O)CH$_3$	MeO$_2$C$\!-\!\!\equiv\!\!-\!$CO$_2$Me	
C(O)CH$_3$	Ph	
C(O)CH$_3$	C$_6$F$_5$	
C(O)CH$_3$		

When DDQ is used as a dienophine elimination of BrCN takes place. In cases that asymmetric dienophile was used the structure of the product was determined by X-ray crystallography that established the stereo- and regiochemistry. Note that when substituted, the *trans*-dibromocyclobutabenzenes are chiral, and preliminary results suggest that when the substituent is chiral (e.g., sultam) a diasteriomeric excess is observed. Thus, these chiral systems could be used for substitution and ring expansion chemistry.

3 Bicyclobutabenzenes

These systems have been prepared by the cyclization of octabromodurene (equation 4).

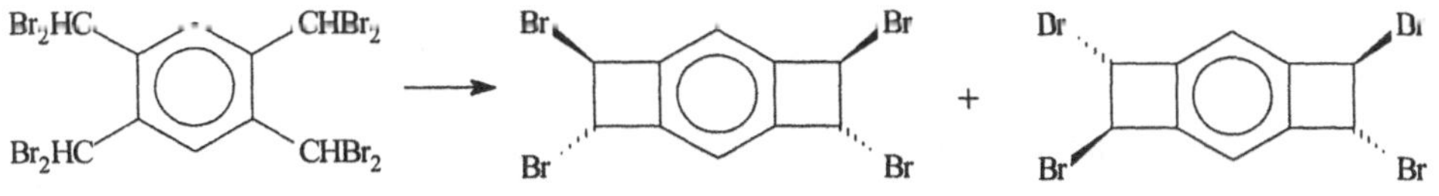

2a   **2b**

The fact that **2a** and **2b** are obtained in equal amounts suggests that the cyclization of the two rings is independent from each other. Counter to intuition, but in accordance to the reactivity of other similar systems,[8] **2** is less reactive than **1**. However, with a strong dienophile or under harsher conditions **2** reacts as two independent dienes, as demonstrated by the reaction with TCNE (equation 5).

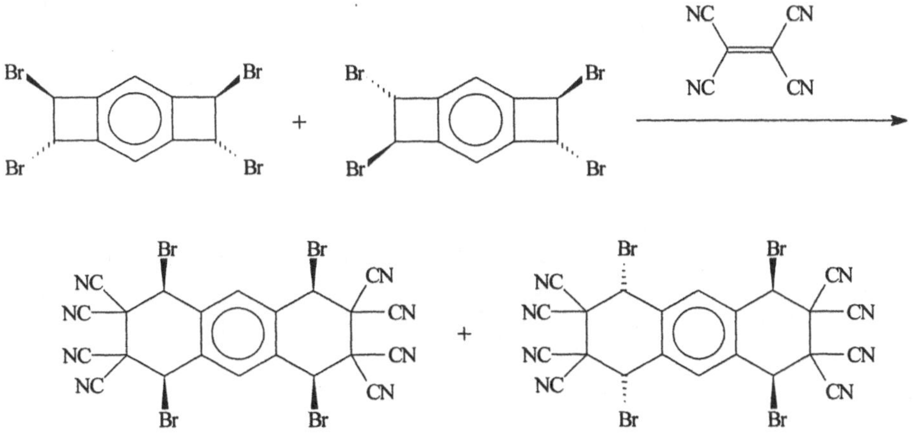

(5)

4 Tricyclobutabenzenes

Subjecting hexakis(dibromomethyl)benzene to nickel cyclization conditions (equation 6) results in the formation of **3** and **5** in 24 and 16% isolated yields, respectively[9]

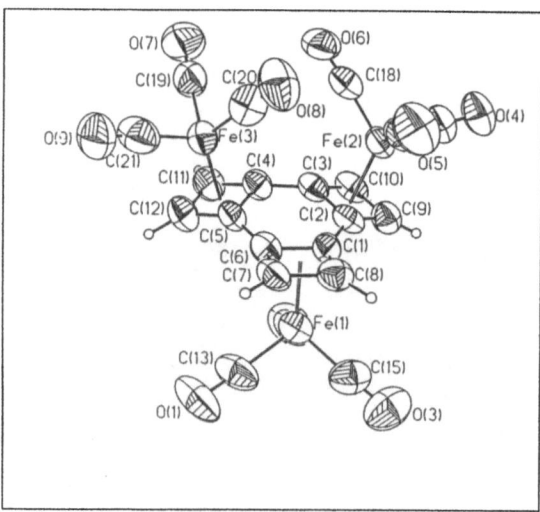

3   **5**   (6)

There are two interesting points here: (a) It seems that **5** is more stable than **3**, in contrast to the single-ring case (i.e., **1**) where the *o*-xylylene is less stable than the isomeric cyclobutabenzene. (b) Out of nine possible isomers of **3**, from which two have all the bromine atoms in *trans* arrangement, only one is obtained. Based on the results for a single ring closure, it is clear why any of the seven isomers that have at least one *cis* arrangement between two bromine atoms is not obtained, but the fact that only the *syn*-all-*trans* isomer is detected, and non of the symmetric *anti*-all-*trans* isomer is observed needs explanation. This is probably due to the fact that when two of the rings have already been formed, the approach to the bromine on the CHBr$_2$ that is *syn* to the bromine on the adjacent four membered ring is sterically hindered. Thus, the nickel can insert only to the *anti* C-Br bond, which leads to the observed *syn* arrangement of the bromine atoms.

Figure 1. X-ray structure of tris(tricarbonyliron)benzotricyclobutadiene

Preliminary results suggest that **3** is stable enough to allow chemical transformation without affecting the tricyclobutabenzene skeleton. Thus, when reacted with AgOAc or "super hydride" (LiBEt$_3$H) the hexaacetate and the parent hydrocarbon are obtained. Reacting **3** with Fe$_2$(CO)$_9$ results in the formation of benzocyclobutadiene-tris(irontricarbonyl) complex (figure 1).

5 Conclusions

Nickel mediated cyclobutabenzene synthesis have been proven to be useful in the stereoselective syntheses of *trans*-dibromocyclobutabenzenes as well as for the preparation of bi- and tricyclobutabenzene. The systems obtained are functionalyzed in the four-membered ring(s), thus allowing to carry out further chemical transformations on these systems. The methodology allows new entries to organic synthesis of polycyclic compounds, to polymer chemistry and to the preparation of compounds relevant to the study of the Mills-Nixon effect.

References

1. See R. Boese, D. Bläser, W. E. Billups, M. M. Haley, A. H. Maulitz, D. L. Mohler, K. P. C. Vollhardt, *Angew. Chem.* **1994**, *106*, 321; *Angew. Chem. Int. Ed. Engl.* **1994**, *33*, 313; and references therein.
2. (a) A. Stanger, *J. Am. Chem. Soc.* **1991**, *113*, 8277; (b) A. Stanger, K. P. C. Vollhardt, *J. Org. Chem.* **1988**, *53*, 4889.
3. It was predicted that organometallic complexes of cyclobutabenzene should show an essential bond fixation, see reference 2a.
4. There are two published procedures for the preparation of tricyclobuta-benzene. Both, however, give only the parent, non-functionalysed hydrocarbon in yields lower than 1% from commercial available starting materials. See (a) W. Nutakul, R. R. Thummel, A. D. Taggart, *J. Am. Chem. Soc.* **1979**, *101*, 770; (b) E. Heilbronner, B. Kovac, W. Nutakul, A. D. Taggart, R. P. Thummel, *J. Org. Chem.* **1981**, *46*, 5279; (c) C. W. Doecke, P. J. Garratt, H. Shahariazi-Zavareh, R. Zahler, *J. Org. Chem.* **1984**, *49*, 1412.
5. The most frequently way used for making cyclobutabenzene is NaI mediated cyclizatioon of α,α,α',α'-tetrabromo-*o*-xylene. See (a) Cava, M. P; Napier, D. R. *J. Am. Chem. Soc.* **1956**, *78*, 500. (b) *ibid* **1958**, *80*, 2255. This method, however, yields a mixture of *cis* and *trans* dibromo and doiodo cyclobutabenzenes in medium to low yields, and is completely unuseful for making **2** and **3** from 1,2,4,5-tetrakis(dibromomethyl)benzene and hexakis(dibromomethyl)benzene, respectively. See (c) M. K. Shepherd, *J. Chem. Soc. Perkin Trans. 1*, **1988**, 961.
6. Stanger, A.; Ashkenazi, N.; Shachter, A.; Bläser, D.; Stellberg, P.; Boese, R *J. Org. Chem.*, **1996**, *61*, 2549.
7. The yields reported in table 1 are unoptimized. However, when the reaction is carried out in an NMR tube the conversion of the starting material to the respective cyclobutabenzene is quantitative. Thus, the low yields observed in some cases are the result of the workup. We

are currently trying to optimized the workup conditions, and preliminary results show that the isolated yields can be substantially increased.

8. Bläser, D.; Boese, R.; Brett, W. A.; Rademacher, P.; Schwager, H.; Stanger, A.; Vollhardt, K. P. C. *Angew. Chem. Int. Ed. Enl.* **1989**, *28*, 206.
9. Stanger, A; Ashkenazi, N.; Boese, R.; Bläser, D.; Stellberg P. *Chem. Eur. J.*, accepted for publication.

Organometallic-Aided "SO", "S$_2$" and "S" Transfer in the Synthesis of Small-Membered Heterocycles

Uri Zoller[1]

Department of Science Education-Chemistry, Haifa, University-Oranim, Kiryat Tivon 36006, Israel

Introduction

The few known examples of four-membered 1,2-oxathietanes (**1a**),[1] 1,2-dithietanes (**1b**),[2] 1,2-dithietes (**2a**)[3] and the hitherto unknown 1,2,3-thiadioxetanes (**3**)[4] and oxathiete (**2b**) are of chemical[5] and practical significance, fundamental theoretical interest, and biological as well as medicinal importance.

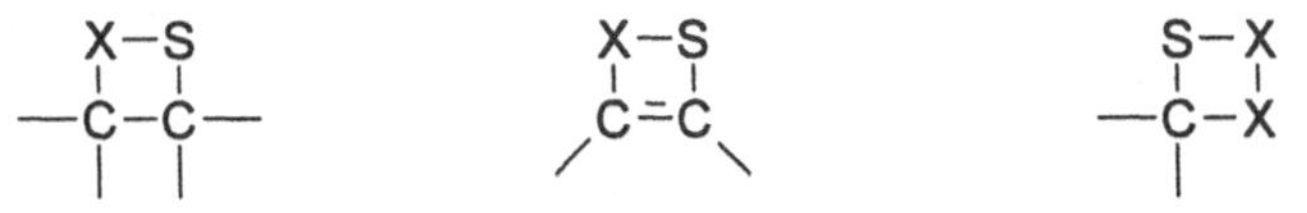

1	**2**	**3**
a X = O	**a** X = S	**a** X = O
b X = S	**b** X = O	**b** X = S

The practical importance is associated with both the analogy of these systems to that of the extensively studied 1,2-dioxetanes which are of importance in bioluminescence and chemiluminescence reactions[6] and their being potential reactive intermediates in chemical reactions,[7-9] photochemical in particular.[8,9] From a theoretical point of view, the interest in these heterocycles stems from their being potential 6π aromatic systems (e.g. **2**)[10] and the effects of the sulfur atom on the direction, rates, energetics, and energy distribution in their anticipated undergoing pericyclic reactions of the formal [σ2s + σ2a] type[11] (applicable in **1, 2** and **3**). The biological interest is related to the anticancer activity of some of the 1,2-

---

[1] Dedicated to Prof. Waldemar Adam on the occasion of his 60th birthday

oxathietanes[1] and their ability (i.e. **1a**) to induce pyrimidine dimers in DNA[12] as well as the biological activity associated with cyclic disulfides[13] (i.e. **1b** and, possibly, **3b**).

Saturated and unsaturated three-membered ring heterocycles continue to be a subject of considerable interest among chemists. Systems of type **4** and **5** are no exception[14,15] so that the preparation and isolation of the thus far elusive thiaziridine oxide (**4a**), thiadiaziridine oxide (**4b**), thiirene oxide (**5a**) and thiazirine oxide (**5b**) constitute an intriguing synthetic challenge.[16]

$$\overset{\overset{X}{\underset{\|}{}}}{\underset{\underset{Y\,{-}\,N}{\diagdown}}{S}}\;R$$

4

a X = O; Y = CR$_2$

b X = O; Y = NR

$$\overset{\overset{X}{\underset{\|}{}}}{\underset{Y\,{=}\,Z}{S}}$$

5

a X = O; Y = Z = C-alkyl

b X = O; Y = CR; Z = N

Such (and similar) rings are the foremost chemically reactive members among the small-ring compounds. The presence of one or more heteroatoms in these strained systems potentially alter their reactivity and stability, increases their content of strain energy[17] and substantially decreases the availability of methods for their synthesis.[13]

Both theoretical and experimental studies on the structure, energies, strain energies, bonding, and charge distribution in small-ring propellanes have been important in the development of modern organic chemistry.[17] The smallest conceivable member of this series of hydrocarbons is [1.1.1]propellane which was synthesized some years ago and was found to be remarkably stable.[18] Since the "normal" bond angles at sulfur are considerably smaller than those at the corresponding (tetrahedral) carbon, one might expect that small sulfur-containing rings might be less strained than their hydrocarbon analog (e.g. S.E.$_4$<S.E.$_3$).[14] Accordingly, the sulfur analog (**6**) of the propellane should be less strained than the latter and, therefore, an attractive synthetic target. Indeed, trithia[1.1.1]propellane **6** is predicted by MP2/6-31G* level calculations[19] to be a tightly-bound molecule which should be experimentally accessible.

Finally, the chemical and biological consequences of the replacement of the oxygen atom in the polycyclic aromatic hydrocarbon (PAH) epoxides[20] and the ultimate carcinogen PAH 7,8-*trans*dihydroxy- 9,10-epoxide (DE-2)[21] **7a** by the "bulkier", more polarizable and more nucleophilic sulfur atom, would provide a unique opportunity for study and in-depth understanding of the nature of bonding, thermostability and stereospecific structure-chemical/biochemical reactivity relationships in the diol episulfides (DeSs) (e.g. **7b**) once accessible.

Clearly, the synthesis of the four- and three-membered systems **1-7** followed by the study of their structure, chemistry, medicinal and carcino-toxicological potential[22,23] are of considerable interest and significance. All of the above heterocycles contain either the "S=X" moiety regiospecifically added to multiple bonds (e.g. alkenes, acetylenes or heteroalkenes) or a transferred "S" unit to bridge between two adjacent carbon atoms.

Such a synthetic methodology should provide, if successful, a straightforward direct entry into these systems. The realization of both, the efficient and convenient generation of "S=X" and the successful synthesis of the above unique four- and three-membered rings, by means of appropriate methodology(s), will not only open the door for the study of the chemistry of these unique systems - several of which are not known thus far - but also would facilitate the synthesis of many interesting S-X, S=X and S-containing larger-size heterocycles as well as other sulfur-containing organic compounds *via* their initial trapping followed by a multitude of possible transformations.

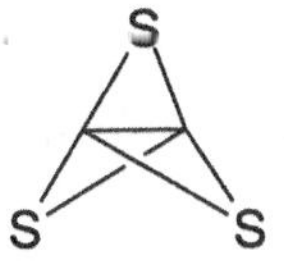

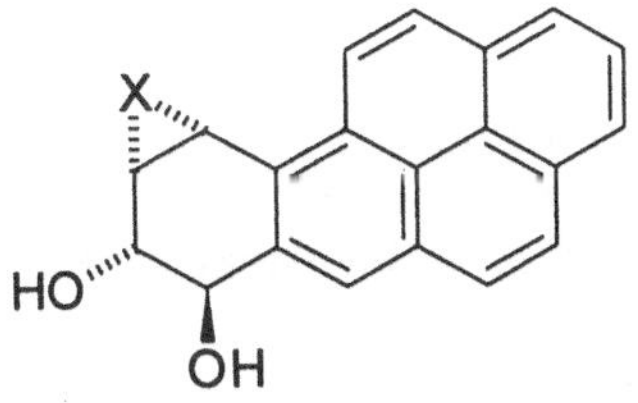

6

7   **a** X = O,
          b X = S

1 Synthesis Strategies

Based on the reported trapping of diatomic sulfur "S_2"[24] and "SO"[25,26] by dienes in little to moderate success, the trapping of these species by alkenes and alkynes was attempted, *via* their thermal or ligand exchange generation from transition metal complexes.

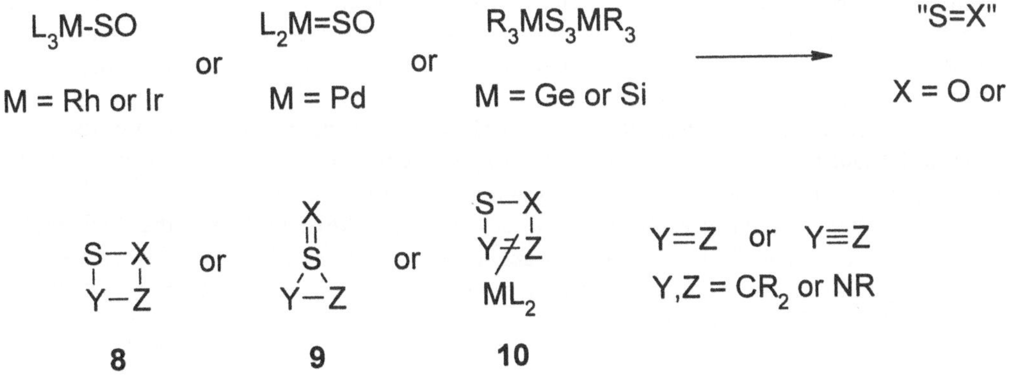

L_3M-SO L_2M=SO $R_3MS_3MR_3$ "S=X"
        or   or
M = Rh or Ir   M = Pd   M = Ge or Si   X = O or

8

9

10

Y=Z   or   Y≡Z

Y,Z = CR_2 or NR

Scheme 1

69

In the case of the alkynes, the expected labile unsaturated products were hoped to be isolable as stabilized organometallic complexes such as 10. Since metal carbonyl moieties are known to stabilize antiaromatic systems[27], the use of $M_n(CO)_m$ [e.g. $Fe_2(CO)_9$] for both the transfer and complexation-stabilization of thermodynamically unstable products was attempted too.

Although the acid-catalyzed synthesis of the bay-region tetrahydrobenz[a]pyrene (THBaP) was recently reported,[28] the well-established extreme susceptibility of the bay-region DEs to protic acids,[29] prompted us to investigate the transition metal complex Ti(i-OPr)$_4$ as a Lewis acid substitute for the synthesis of **7b** from its corresponding epoxide using nucleophilic S-transfer agents. Selected results of our recent synthetic work are presented and briefly discussed below.

2 Results and Discussion

Our initial working scheme followed the procedure depicted in eq. 1,[30] expecting **8** or **9** to be formed.

$$\underset{\textbf{11a, b}}{\underset{L^2}{\overset{Cl}{\diagdown}}\overset{L^1}{\underset{SO}{M{\diagup}}}} + R^1_2C=ZR^1_{2(1)} \xrightarrow{CO} \underset{L^2}{\overset{Cl}{\diagdown}}\overset{L^1}{\underset{CO}{M{\diagup}}} + \underset{R^1_2C-ZR^1_{2\,(1)}}{\overset{S-O}{|\quad|}} \ or\ \underset{R^1_2C-ZR^1_{2\,(1)}}{\overset{O}{\overset{\|}{S}}} \quad (1)$$

M = Rh, Ir   Z = C or N

11a, b   **8**   **9**

In our hands, in the reaction of either the Rh or the Ir complex **11a, b** with the nucleophilic ethyl- or trimethylsilylether, PhC=NPh, norbornene, diphenylacetylene, the dienes tetraphenylcyclopentadienone, 2,5-dimethyl-tetrahydrofurane, 2,3-dimethylbutadiene, and 3,5-ditertbutylorthoquinone, no trace of any of the expected cycloadducts was detected. No isolable products other than iPr$_3$PO and/or iPr$_3$PS were present in the filtrate after the removal of the organometallics by filtration.

Based on the experimental results reported above, our conclusion is, that the use of the Rhodium-SO and the Iridium-SO complexes for the generation of "SO" − to be trapped by alkenes, alkynes and/or conjugated dienic systems to form the corresponding [2+1]-,[2+2]-, or [2+4]-cycloadducts − is very problematic, to say the least. Most probably, the cycloaddition does not occur to any noticable extent, or does not take place at all, at least under the conditions employed by us. In fact, even the very modest "positive" results reported by Schenk et al.,[25] for the case of selected conjugated dienes[30-32] could not be reproduced by us.

A paper,[25] which was published while our project was well underway, reported the preparation of the palladium complex **13** and its SO-ligand undergoing a cycloaddition reaction with dienes; e.g. dimethylbutadiene, to give the thiophen **15**:

$$(2)$$

We have extensively explored this route in an attempt to extend its scope, by reacting the palladium complex **13** with alkenes, alkynes and heteroalkenes as well as with various conjugated dienes and heterodienes, including those multibond-containing compounds which were used with the rhodium and iridium complexes **11a, b**. However, except for recovering the unreacted starting materials, some $(C_6H_5)_3PO$ and $(C_6H_5)_3PS$ in the case of the norbornene, and the corresponding hydroquinone in the case of the 3,5-di-*tert*butylorthobenzoquinone, no reaction was detected between the complex **13** (the potential source for *in-situ* SO-generation) and either of the following: 3,5-di*tert*butyl-ortho-benzoquinone, acetylene, ethyl-vinyl ether, benzal aniline, tetraphenyl butadienone and norbornene. These negative results are remarkably similar to those observed for the Rh- and Ir-SO complexes. The palladium complex **13** did react with 3,4-dihydro-2*H*-pyran and 2,5-dimethylfurane. The tentative structures of the products **16** and **17** (both symmetric) are based on spectroscopic data (MS, ^{1}H- and ^{13}C-nmr) of the purified adducts (column chromatography and crystallization).

In both cases, the cheletropic mode of cycloaddidion of the SO moiety across the multibond system (1,2 or 1,4) is clearly preferred. This is in contrast with the exclusive [2+2]- or [2+4]-mode of addition in the case of the analogous diatomic sulfur "S_2".[24] Whereas the mechanism for the formation of the α-disulfoxide ring system appears to be straightforward, the elucidation of that for the formation of the oxidative dimerization requires further study and "hard" experimental evidence. Compound **17** is of particular interest, since very few α-disulfoxides are known to be formed, nor to be stable enough to survive isolation.[33]

The rather low yields of both **16** and **17** *via* the procedure depicted in eq. 2,[26] requires optimization of the reaction conditions in order to make this route synthetically practical.

$$\text{18} \quad \xrightarrow[\text{(Ph)}_3\text{PBr}_2/\text{ CH}_2\text{Cl}_2]{\text{Olefin/ Diene } ?} \quad \text{Olefin} \overset{S}{\underset{S}{<}} \quad (3)$$

18   **19**

In applying organometallics-involved synthetic methodologies within our persuit of stable strained three- and four-membered sulfur-containing heterocycles, we have extensively used the diatomic singlet "S_2"-generating agents (e.g. **18**) of the type developed by Steliou *et al.*[34] (eq. 3), and the monoatomic sulfur agents [(CH$_3$)$_3$Si]$_2$S and (CH$_3$)$_2$NCHS for four- (eq. 3) and three-membered rings[35,36] respectively.

Following our calculations,[19] predicting the trithia[1.1.1]propellane **6** be experimentally accessible (comparable in its strain energy to its surprizingly stable homologous hydrocarbon ([1.1.1]propellane),[18] we treated the readily accessible tetrabromo-1,3-dithietane[37] with hexamethyldisilathiane (or its Sn anolog), expecting to obtain the bridgehead dibromotrithia[1.1.1]propellane. Rather, we obtained the 1,3-dithietane thiaketone **21** (meaning double displacement of bromine, by sulfur, on the same carbon atom) *via* which we are currently trying to get to the target compound **6**, using organometallics (e.g. BuLi) to induce this transformation (eq. 4).

$$\text{20} \quad \xrightarrow[\text{CH}_3\text{CN/ CsF}]{\text{Me}_3\text{Sn-S-SnMe}_3} \quad \text{21} \quad \xrightarrow[?]{\text{BuLi}} \quad \text{6} \quad (4)$$

20   **21**   **6**

Finally, the treatment of ditrimethylsilyl-DE-2[38] with excess of N,N-dimethylthiaformamide in the presence of titanium tetraisopropoxide, afforded the target DeS compound **23** in 81% yield as colorless crystals[39] (eq. 5).

In conclusion, "SO" appears not to be reactive towards multiple single bonds and, therefore, not very useful, synthetically, in this respect. If cycloadds, the cheletropic mode appears to be preferred (exclusively?). At any rate, the chemistry of the thermally generated "SO" is different from that of the analogous "S_2" species. Organometallic complexes are very useful in syntheses whenever monoatomic "S"-transfer agents are used for providing the sulfur atom to the constructed small-ring, sulfur-containing heterocycles.

$$(5)$$

Acknowledgement

The financial support of the Volkswagen Stiftung for a substantial portion of the research work here presented is highly appreciated.

References

1. Lown, J.W.; Koganty, R.R. *J. Am. Chem. Soc.* **1986**, *108*, 3811.
2. Nicolaou, K.C.; De Frees, S.A.; Hwang, C.-K.; Stylianides, N.; Carroll, P.J. *J. Am. Chem. Soc.* **1990**, *112*, 3029.
3. Schulz, R.; Schweig, A.; Hartke, K.; Köster, J. *J. Am. Chem. Soc.* **1983**, *105*, 4519.
4. Coyle, J.D. *Tetrahedron* **1985**, *23*, 5393.
5. Zoller, U. Four-membered Rings with Two Sulfur Atoms: In Padwa, A. (Ed.), *Comprehensive Heterocyclic Chemistry*, Elsevier: England, **1996** (Forthcoming).
6. Adam, W: In Adam, W.; Cilento, G. (Eds.), *Chemical and Biological Generation of Electronically Excited States*, Academic Press: New York, **1982**, Chap. 4.
7. Wong, G.S.K. *J. Org. Chem.* **1979**, *44*, 1977.
8. Kusters, W.; De Mayo, P. *J. Am. Chem. Soc.* **1974**, *96*, 3502.
9. Jacobsen, N.; De Mayo, P.; Weedon, A.C. *Nouv. J. Chim.* **1978**, *2*, 331.
10. Calzaferri, G.; Gleiter, R. *J. Chem. Soc. Perkin Trans. 2*, **1975**, 559.
11. Woodward, R.B.; Hoffmann, R. *The Conservation of Orbital Symmetry*, Verlag Chemie; Weinheim, **1970**, p. 72.
12. Lamola, A.A. *Biochem. Biophys. Res. Comm.* **1971**, 43, 893.

13. Merck Index, 10th ed., Merck Co., Inc.; Rahway, NJ, 1983; entries 4301 and 9166, respectively, and references cited therein.
14. Zoller, U: In Hassner A., (Ed.), *Small Ring Heterocycles*, John Wiley Sons: New York, **1983**, pp. 333 - 449.
15. Zoller, U: In Patai, S.; Rappoport, Z.; Stirling, C. (Eds.), *The Chemistry of Sulphones and Sulphoxides*, John Wiley & Sons: Chichester, **1988**, pp. 379-481.
16. Zoller, U. Thiaziridines and Thiazirines. In Padwa, A. (Ed.), *Comprehensive Heterocyclic Chemistry*, Elsevier, Science: England, **1996** (Forthcoming).
17. Wiberg, K.B., *Chem. Rev.* **1989**, *89*, 975.
18. Wiberg, K.B.; Waker, F.H. *J. Am. Chem. Soc.* **1982**, *104*, 5239.
19. Riggs, N.V.; Zoller, U.; Nguyen, M.T.; Radom, L. *J. Am. Chem. Soc.* **1992**, *114*, 4354.
20. Zoller, U.; Shakkour; E. Pastersky, I. *Phosphorus, Sulfur and Silicon* **1994**, *95-96*, 453.
21. Yagi, H.; Thakker, D.R.; Hernandez, O.; Koreeda, M.; Jerina, D.M. *J. Am. Chem. Soc.* **1977**, *99*, 1604.
22. Adam, W.; Hadjiarapoglou, L.; Mosandl, T.; Saha-Möller, C.R.; Wild, D. *J. Am. Chem. Soc.* **1991**, *113*, 8005.
23. Harvey, R.G. *Polycyclic Aromatic Hydrocarbons: Chemistry and Biochemistry,* Cambridge University Press: Cambridge, **1991**.
24. Tardif, S.L.; Williams, C.R.; Harpp, D.N. *J. Am. Chem. Soc.* **1995**, *117*, 9067.
25. Leissner, J.; Schenk, W.A. *Z. Naturforsch.* **1987**, *B42*, 799.
26. Heyke, O.; Neher, A.; Lorenz, I.-P. *Z. Anorg. Allg. Chem.* **1992**, *608*, 23.
27. Emerson, G.F.; Watts, L.; Pettit, R. *J. Am. Chem. Soc.* **1965**, *87*, 131.
28. Shalom, Y.; Kogan, W.; Badria, H.; Harvey, G.; Blum, *J. Heterocycl. Chem.* **1996**, *33*, 53.
29. Walen, D.L.; Ross, A.M.; Montemarano, T.A.; Taker, D.R.; Yagi, H.; Jerina, D.M. *J. Am. Chem. Soc.* **1979**, *101*, 5086 .
30. Schenk, W.A., *Angew. Chem. Int. Ed. Engl.* **1987**, *26(2)*, 98.
31. Schenk, W.A; Leissner, J. *Z. Naturforsch.* **1987**, *B42*, 967.
32. Schenk, W.A.; Karl, U. *Z. Naturforsch.* **1989**, *44b*, 988.
33. Folkins, P.L.; Harpp, D.N. *J. Am. Chem. Soc.* **1991**, *113*, 8998.
34. Steliou, K.; Gareau, Y.; Harpp, D.N. *J. Am. Chem. Soc.* **1984**, *106*, 799.
35. Cappozzi, F.; Cappozzi, G.; Menichetti, S. *Tetr. Lett.* **1988**, *26*, 4177.
36. Takido, T.; Kobayashi, Y.; Itabashi, K. *Synthesis* **1986**, 779.
37. Reude, U.; Sundermeier, W. *Chem. Ber.* **1985**, *118*, 2208.
38. Jheingan, A.K.; Meehan, T. *J. Chem. Res.* (s), **1991**, 122.
39. Zoller, U.; Yagi, H.; Jerina, D.M. (Forthcoming publication).

Stereoselective Reactions of Transition Metal Carbonyl Complexes

Stephen A. Benyunes, Susan E. Gibson (née Thomas)*, Gary R. Jefferson,
P. Caroline V. Potter, Mark A. Peplow, Ellian Rahimian, Mark H. Smith and
Mark F. Ward

Department of Chemistry, Imperial College of Science, Technology and Medicine, South
Kensington, London SW7 2AY, UK

There is currently widespread interest in defining and controlling the stereochemical outcome of chemical reactions. Much attention has been lavished to date on organic reactions and, as a result, our control and understanding of these processes is now very sophisticated. In contrast there are still many areas of organometallic chemistry where the stereochemical consequences of basic transformations remain uncharted. In this paper, two recent studies on stereochemical aspects of two classes of transition metal carbonyl complex are described and discussed.

1 Stereochemical Studies of Vinylketene Iron Tricarbonyl Complexes

Some time ago, we discovered that iron tricarbonyl complexes of vinylketones are transformed into iron tricarbonyl complexes of vinylketenes on addition of an alkyllithium reagent under an atmosphere of carbon monoxide (Figure 1).[1, 2] This gave us rapid access to these stable crystalline complexes and enabled us to explore the fundamental reactivity of vinylketenes bound to a transition metal.

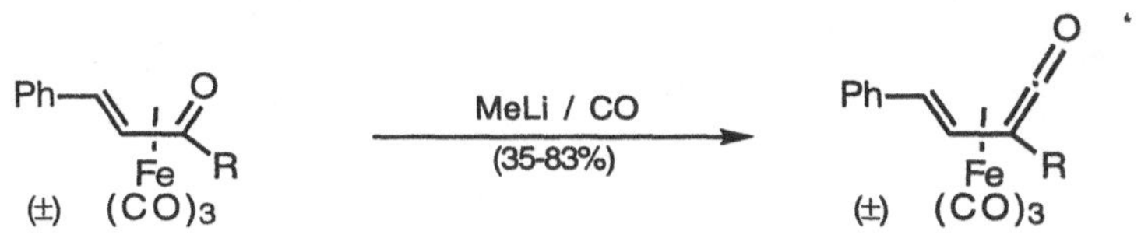

Figure 1

In the course of our studies, we found that isonitriles,[3] phosphonoacetate anions,[4] alkynes,[5] alkenes[6] and nucleophiles[7] all react with the vinylketene complexes to produce between them a rich diversity of both first generation organometallic complexes and second generation organic products by a variety of routes all involving fundamental organometallic processes (Figure 2).

Recently, we reasoned that access to enantiomerically enriched samples of iron tricarbonyl complexes of vinylketenes would enable us to probe whether or not their reactions proceed with retention, inversion or racemisation of configuration and thus should ultimately yield a wealth of stereochemical and mechanistic information about a range of organometallic processes.

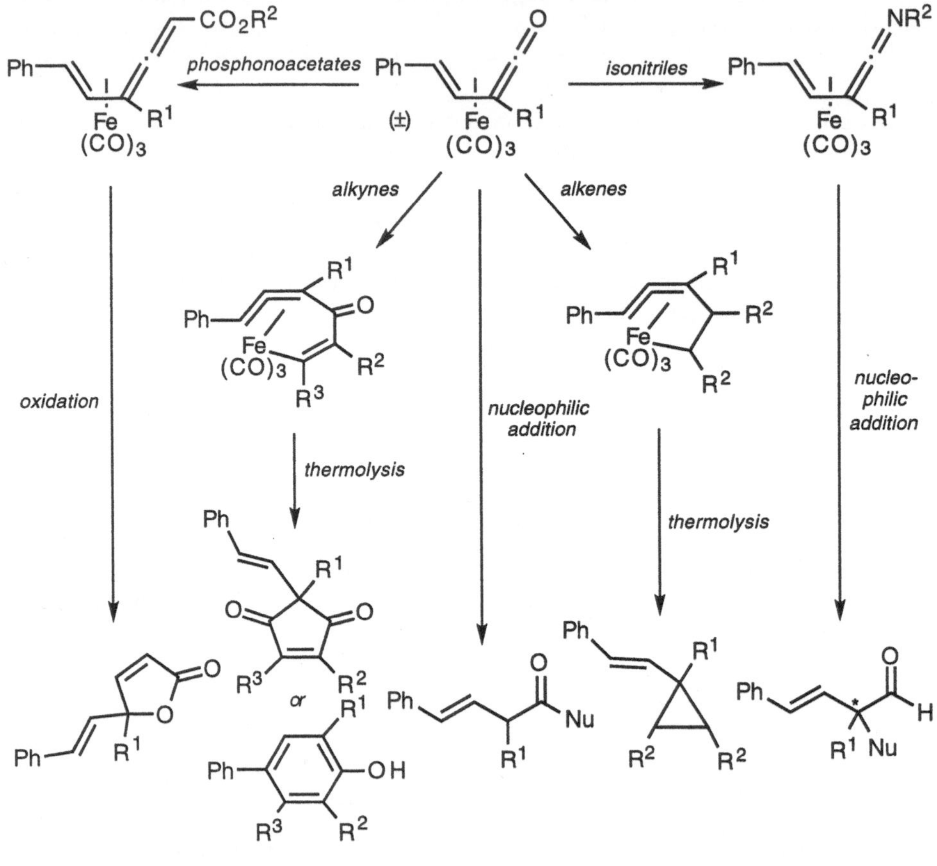

Figure 2

For example, iron vinylketene complexes react with electron-poor alkenes to give decarbonylated adducts which may then be thermolysed or oxidised to tetrasubstituted cyclopropanes[6] as exemplified in Figure 3.

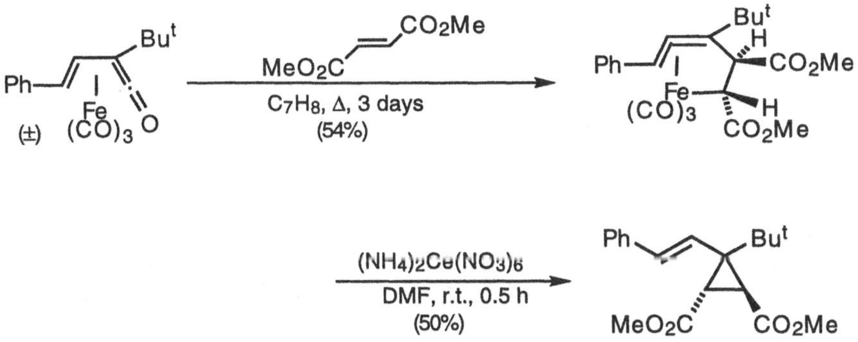

Figure 3

Two plausible mechanistic pathways for the decarbonylation/insertion reaction are presented in Figure 4, one of which would lead to the conservation of stereochemical information and the other of which would lead to racemisation at the iron centre. Thus reaction of an enantiomerically enriched vinylketene complex with an alkene and stereochemical analysis of the product should allow us to differentiate between the two pathways.

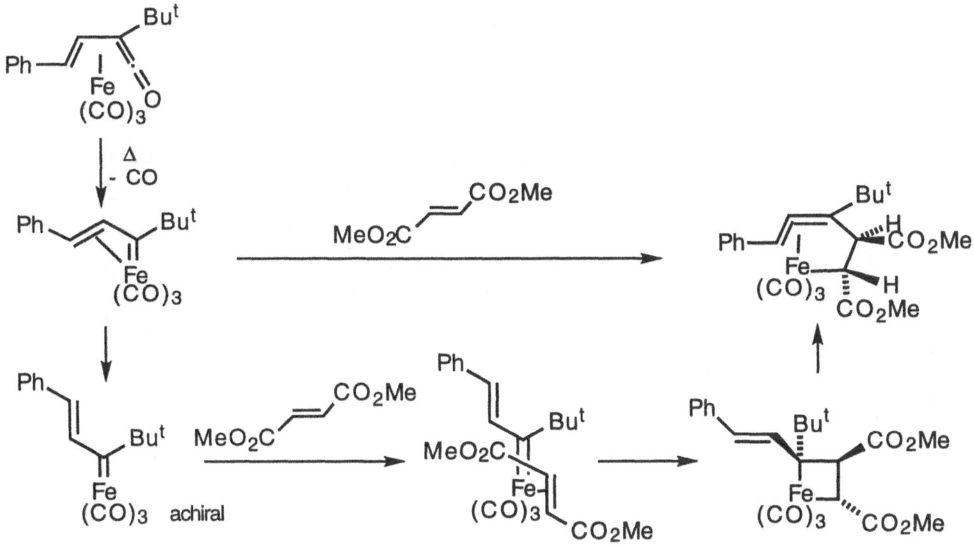

Figure 4

As the route to vinylketene complexes from vinylketone complexes depicted in Figure 1 had been found to be straightforward and reliable, this approach was the first to be considered for development into a route to the required enantiomerically enriched systems. The stereochemical outcome of this reaction, however, was of course unknown and could conceivably involve retention, racemisation or inversion of planar configuration. Although our proposed mechanism for this process (Figure 5) suggests that the process should proceed with retention of configuration, several of the steps in this mechanism as yet do not have a firm experimental basis. We thus decided to proceed with caution!

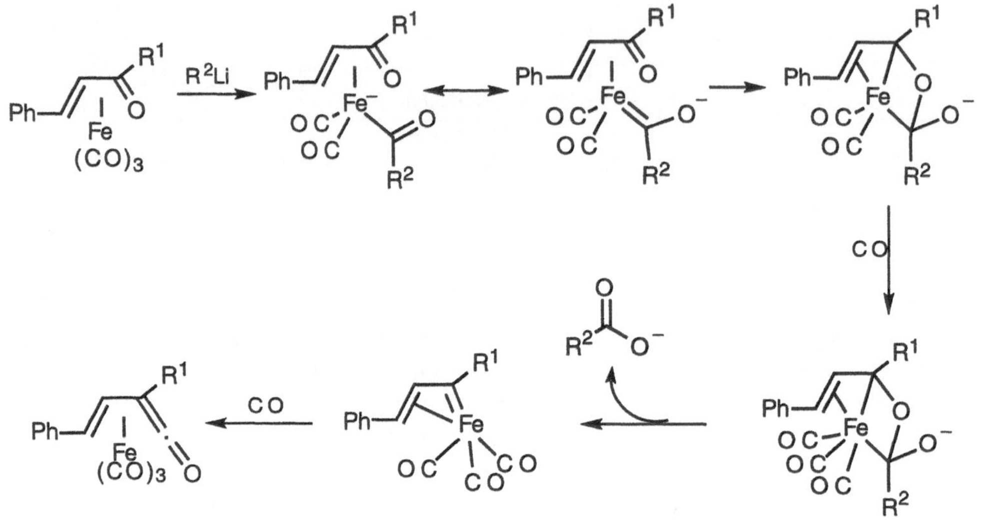

Figure 5

An enantiomerically enriched sample of a vinylketone substrate of known absolute configuration was obtained using a route based on a reported synthesis of enantiomerically enriched (benzylideneacetone)tricarbonyliron(0) (Figure 6).[8]

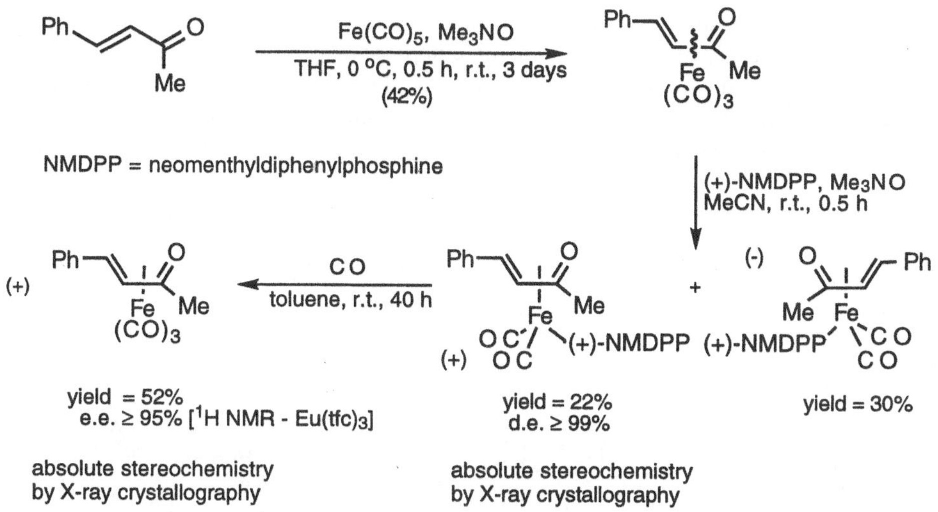

Figure 6

The enantiomerically enriched vinylketone complex thus obtained was then reacted with methyllithium under an atmosphere of carbon monoxide to give the corresponding vinylketene complex (Figure 7). Analysis of the product by HPLC revealed that its ee was 95% *i.e.* essentially identical to that of the starting material. Thus the conversion had occurred without any loss of stereochemical information, although at this point it could be argued that the process may involve either clean retention or clean inversion of planar configuration. Clearly this information needed to be firmly established before any stereochemical studies based on this complex would be worthwhile.

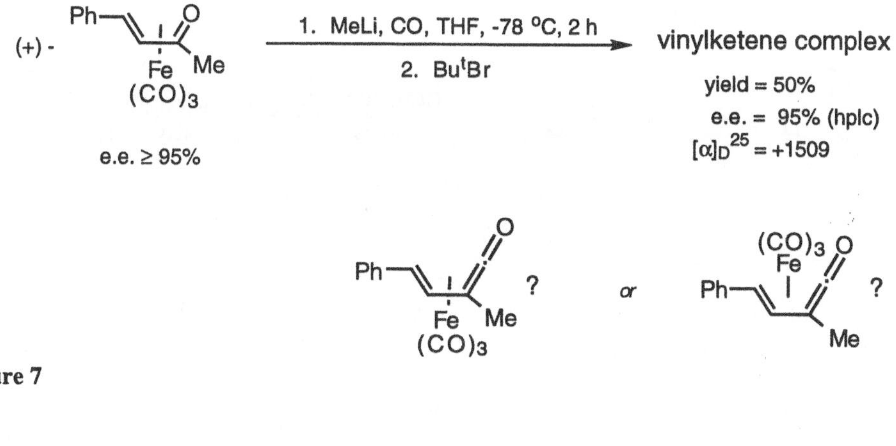

Figure 7

Unfortunately, determination of the absolute stereochemistry of the vinylketene product proved to be exceedingly difficult. Innumerable approaches were investigated and met with various problems, the most frustrating of which was the difficulty encountered in obtaining well-formed crystals from enantiomerically pure samples, even though the corresponding racemic systems gave rise to high quality crystals. Eventually, however, the study described below not only revealed the absolute stereochemistry of the vinylketene complexes, but also provided us with a more efficient route to enantiomerically enriched vinylketene complexes.

We recently demonstrated that vinylketene complexes react with the anions of diethyl N-alkyl- or N-arylphosphoramidates to give vinylketenimine complexes in moderate to excellent yield (Figure 8).[9]

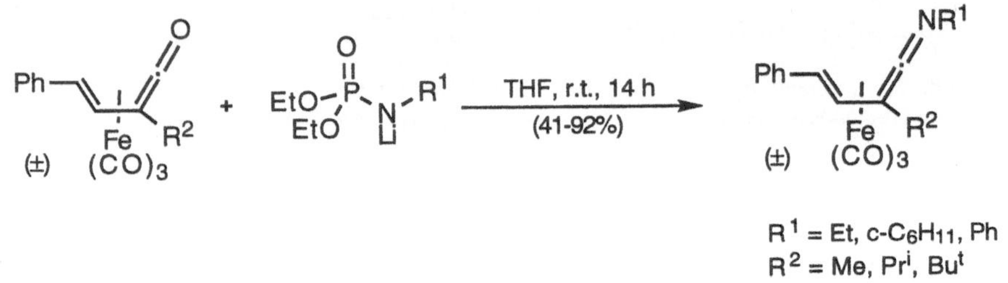

R^1 = Et, c-C$_6$H$_{11}$, Ph
R^2 = Me, Pr^i, Bu^t

Figure 8

We thus postulated that a non-racemic chiral N-alkylphosphoramidate may react preferentially with one of the enantiomers of vinylketene complexes to generate enantiomerically enriched vinylketene complexes by a kinetic resolution procedure. In order to test this hypothesis, a phosphoramidate was prepared from diethyl phosphite and (S)-α-methylbenzylamine using a literature procedure (Figure 9).[10]

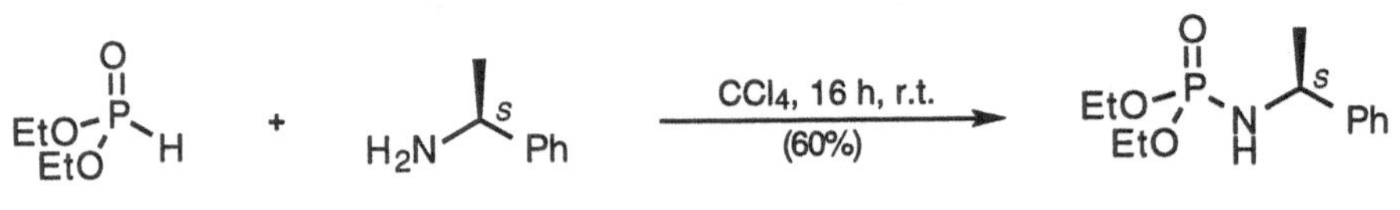

Figure 9

The phosphoramidate was initially reacted with an isopropyl-substituted vinylketene complex. Treatment of the complex with half an equivalent of the phosphoramidate anion gave, after work-up, a sample of unreacted vinylketene complex in 37% yield and 80% ee and a sample of the two diastereomers of the vinylketenimine complex in 47% yield. Crystallisation of racemate from the enantiomerically enriched vinylketene complex gave material of 99% ee in 23% yield (Figure 10).

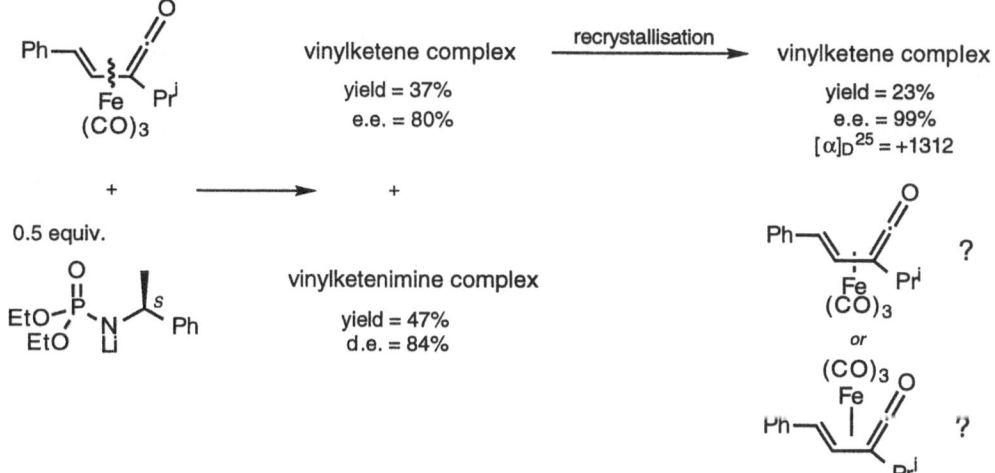

Figure 10

Although this kinetic resolution approach to enantiomerically enriched vinylketene complexes appeared synthetically promising, the absolute stereochemistry of the complexes remained undefined. This matter was finally settled when the phenyl-substituted vinylketene complex was resolved (Figure 11). Treatment of this complex with half an equivalent of phosphoramidate not only gave a sample of enantiomerically enriched vinylketene complex that could be crystallised to high ee, but it also gave a sample of vinylketenimine complex which yielded crystals of the major diastereomer of suitable quality for X-ray crystallography. This defined their absolute configuration as 5pR as shown in Figure 11. Thus, assuming that the ketene to ketenimine reaction occurs with retention of configuration with respect to the planar chirality of the system, the absolute configuration of the optically enriched dextrarotatory vinylketene complex is defined as 5pS.

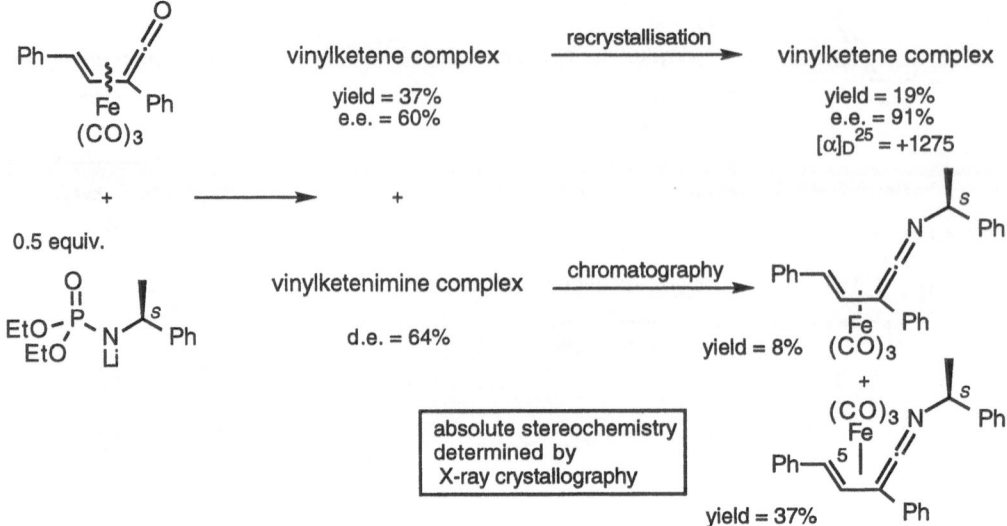

Figure 11

It is of note at this point that the kinetic resolution procedure now provides a relatively efficient route to vinylketene complexes of high optical purity and of known configuration.[11] The preparation of several complexes and their stereochemical data are summarised in Figure 12.

From this data, it is apparent that all the enantiomerically enriched complexes have the same absolute configuration as the phenyl-substituted complex and that this is the same as the configuration of the material derived from the methyllithium promoted preparation (Figure 13). Thus, it can be stated with confidence that the conversion of vinylketone complexes to vinylketene complexes occurs with overall retention of planar configuration.[12]

With good supplies of enantiomerically enriched vinylketene complexes in hand of known absolute configuration we were ready to start to examine the stereochemical consequences of the reactions of vinylketene complexes and their derivatives depicted in Figure 2. To date we have examined the reactions of the vinylketene complexes with alkenes and with isontriles and these two areas are discussed in the remainder of this first part of this paper.

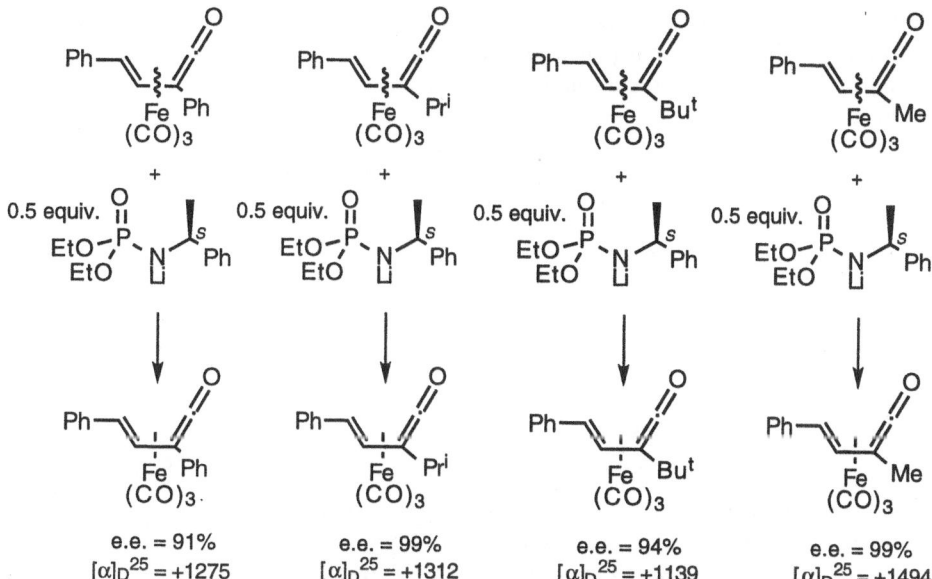

Figure 12

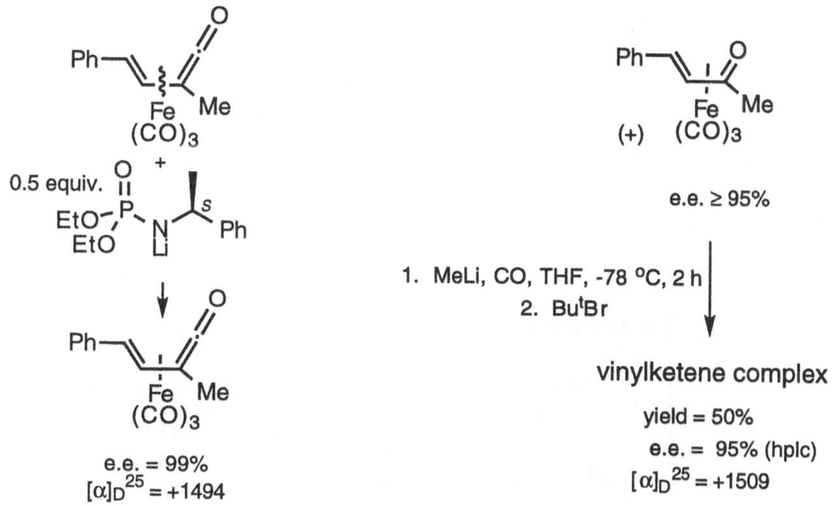

Figure 13

Our examination of the reaction between vinylketene complexes and alkenes proved to be brief. Heating a butyl-substituted vinylketene complex of 94% ee with dimethyl fumarate gave the expected decarbonylated adduct depicted in Figure 14. Examination of the product by [^1]H NMR spectroscopy in the presence of a chiral shift reagent revealed that it was racemic and thus the upper pathway outlined in Figure 4 may now be discounted as the reaction pathway clearly passes through a symmetrical intermediate.

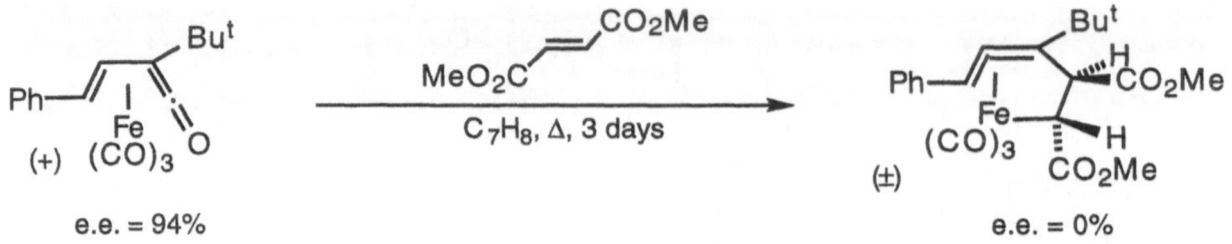

Figure 14

Our attention then turned to the reaction between vinylketene complexes and isonitriles. Some time ago, we discovered that heating tricarbonyl(vinylketene)iron(0) complexes in the presence of isonitriles led to the formation of tricarbonyl(vinylketenimine)iron(0) complexes (Figure 15).[3] Based on the isolation of dicarbonylisonitrile(vinylketene)iron(0) complexes from three of the reaction mixtures and the conversion of each of these compounds into vinylketenimine complexes under the reaction conditions, it was established that dicarbonylisonitrile(vinylketene)iron(0) complexes were intermediates in the conversion.

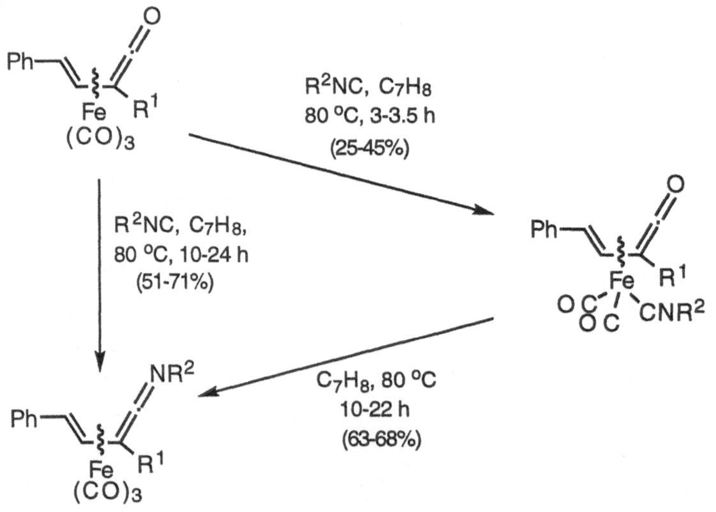

Figure 15

84

Furthermore, a careful examination of the product mixture generated when two different dicarbonylisonitrile(vinylketene)iron (0) complexes were heated in the same pot did not detect any evidence for the production of crossover products (Figure 16), thus suggesting that the carbonyl/isonitrile exchange process that occurs in the transformation proceeds without dissociation of the isonitrile ligand from the iron centre.

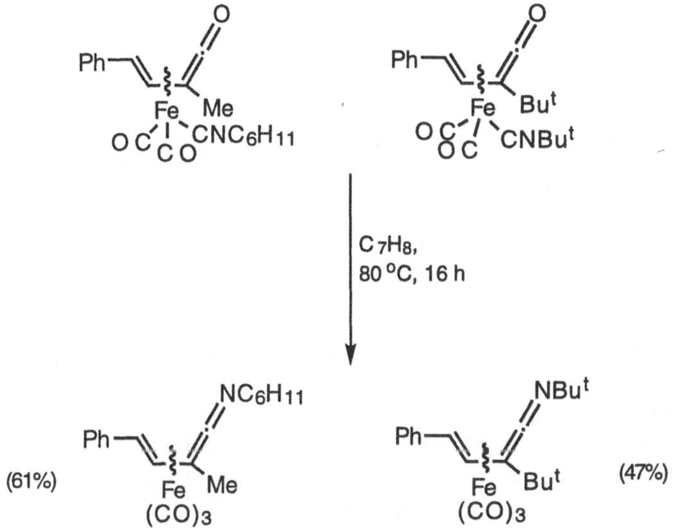

Figure 16

Based on (i) this latter result, (ii) an earlier observation by others that heating a tricarbonyl(vinylketene)iron(0) complex at 60 °C in toluene produces a tricarbonyl(η^3-vinylcarbene)iron(0) complex (a process which is reversed under one atmosphere of carbon monoxide at 20 °C),[13] and (iii) the knowledge that the coupling of metal-carbenes and isonitriles is a well-established method of generating ketenimines,[14] we proposed the mechanism shown in Figure 17 for the transformation of vinylketene complexes to vinylketenimine complexes. This pathway passes through a symmetrical intermediate and thus racemisation is expected during the course of the reaction.

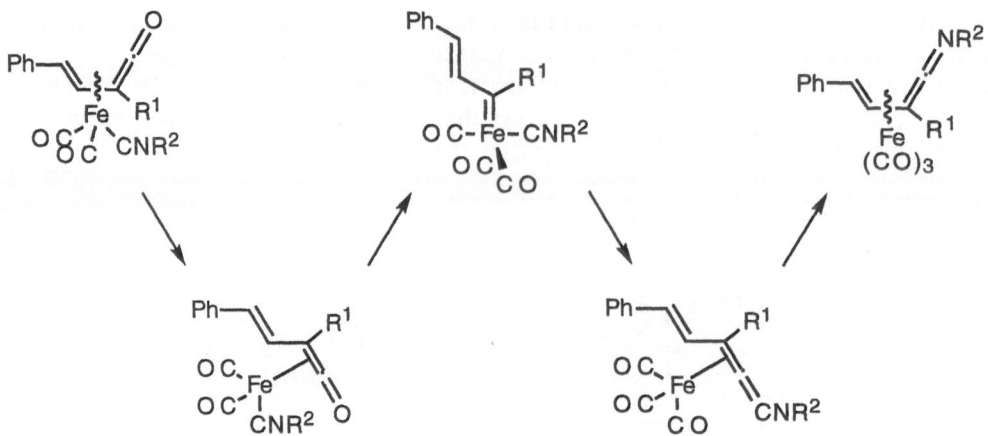

Figure 17

To test this hypothesis, an isopropyl-substituted vinylketene complex of 99% ee was transformed into a vinylketenimine complex using *tert*-butylisonitrile (Figure 18). The ee of the product was measured by ¹H NMR spectroscopy in the presence of a chiral solvating reagent and, to our surprise, found to be 96%, *i.e.* the reaction had occurred without loss of stereochemical information and clearly could not proceed *via* a symmetrical intermediate. Comparison of the CD spectrum of the product with the CD spectra of closely related vinylketenimine complexes of known absolute configuration,[15] revealed that the absolute configuration of the product was 5p*S* and that the conversion had occurred with overall retention of planar chirality.

In order to determine whether the formation of the dicarbonyl isonitrile intermediate and its subsequent isomerisation to the vinylketenimine complex proceeded with double retention or double inversion of configuration, the stereochemical consequences of these two steps were examined and both were found to proceed with retention of configuration.[16]

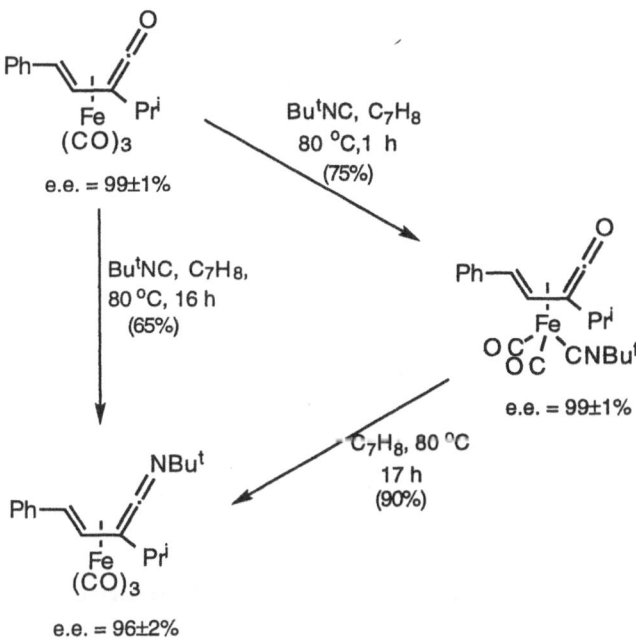

Figure 18

Clearly the results described above are inconsistent with the proposed intermediacy of a symmetrical intermediate and thus our mechanistic interpretation of the conversion of vinylketene complexes to vinylketenimine complexes needed to be refined (Figure 19). Simultaneous co-ordination of three carbon monoxide and one isonitrile ligands to the iron centre can only be achieved by dissociation of a two electron donor derived from C(3-5) of the organic ligand. Whilst dissociation of C-4 and C-5 and deinsertion of the ketene carbonyl led to the plausible but subsequently discredited symmetrical carbene intermediate illustrated in Figure 17, dissociation of C-3 and C-4 and deinsertion of carbon monoxide leads to a new intermediate in which the stereochemical information inherent in the starting material is preserved at C-5. Irreversible insertion of the isonitrile ligand into the iron C-(3) bond then gives a further chiral intermediate, and finally re-coordination of the iron to the same face of C-3 and C-4 that it dissociated from generates the observed product (coordination of the iron to the opposite face of C-3 and C-4 would generate a relatively high energy diasteromer and is therefore disfavoured).

To conclude this section, the availability of vinylketene complexes of high enantiomeric purity and known absolute configuration has enabled us to begin to probe the stereochemical consequences of some of their reactions. Our current work in this area is focussed on elucidating the stereochemical outcome of the reaction between vinylketene complexes and alkynes.

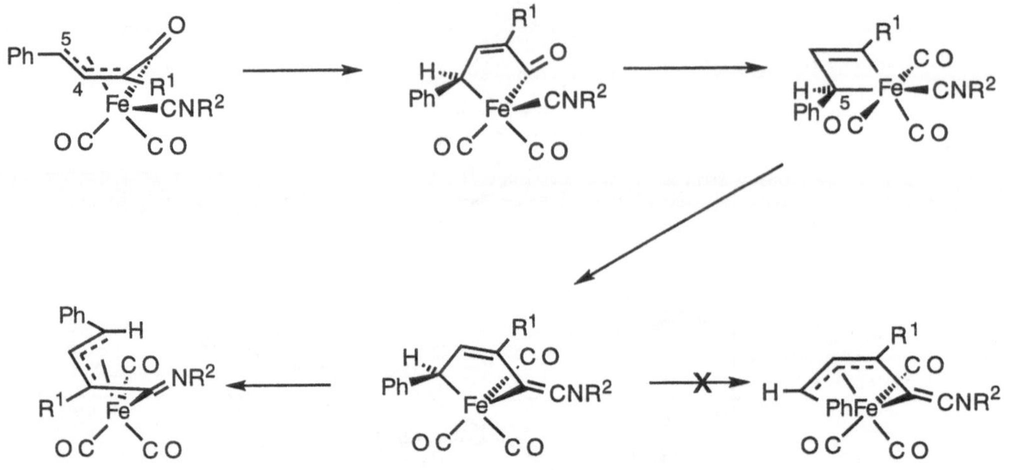

Figure 19

2 Applications of Chiral Bases in Arene Chromium Tricarbonyl Chemistry

Organic chemists have utilised enantiomerically pure chiral amines and amides to considerable effect in recent years and some of their successes are illustrated by the examples depicted in Figure 20.[17-19]

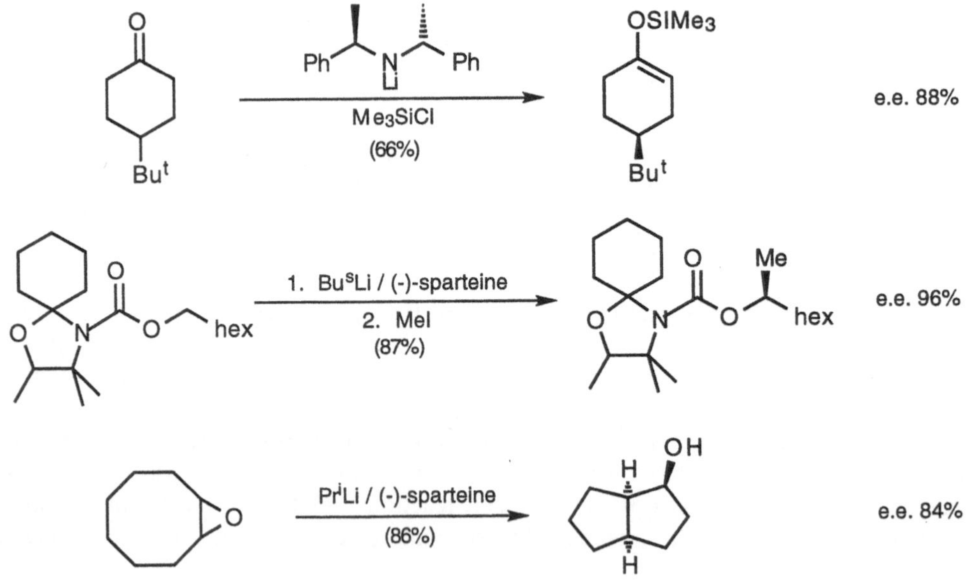

Figure 20

Some time ago we initiated a programme designed to exploit chiral amines and amides in organometallic chemistry. We reasoned that use of these reagents in conjunction with organometallic complexes may lead to i) new synthetic routes to enantiomerically enriched organometallic reagents and catalysts, and ii) the discovery of novel chiral base / substrate interactions. Our studies have involved several classes of organometallic complex to date, as indicated in Figure 21, but as our most successful and recent results have been obtained with tricarbonylchromium(0) complexes of alkyl benzyl ethers, these complexes alone will be the subject of the following discussions.

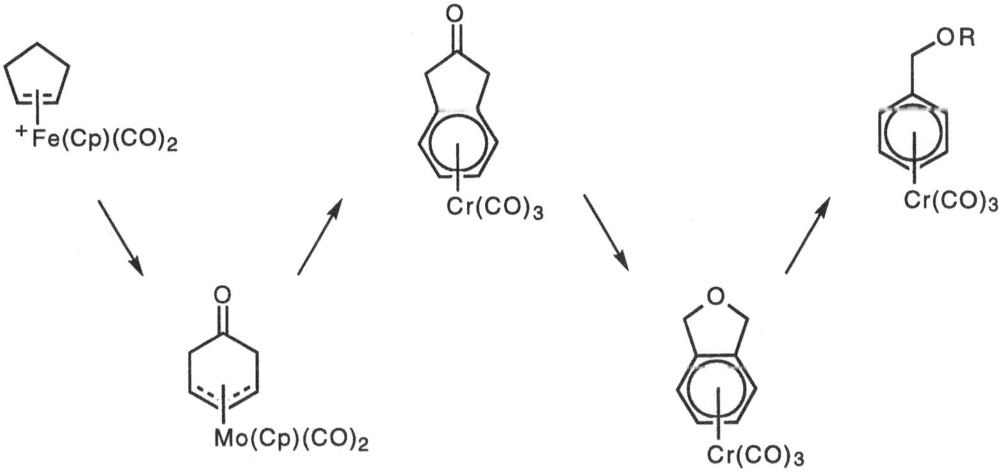

Figure 21

(Benzyl methyl ether)tricarbonylchromium(0) is synthesised by converting benzyl alcohol into (benzyl alcohol)tricarbonylchromium(0) followed by treatment with acidic methanol (Figure 22).[20]

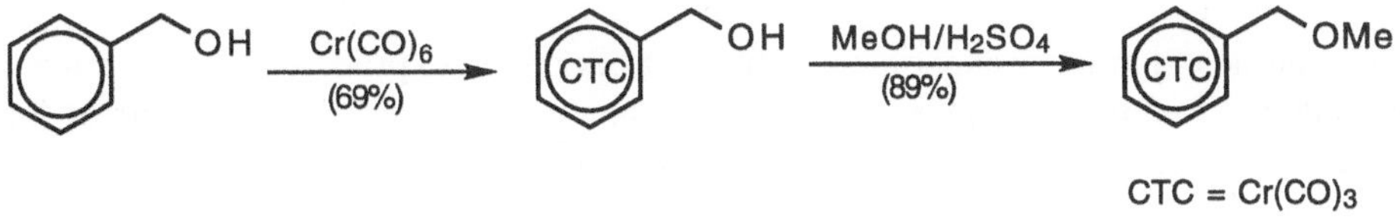

Figure 22

89

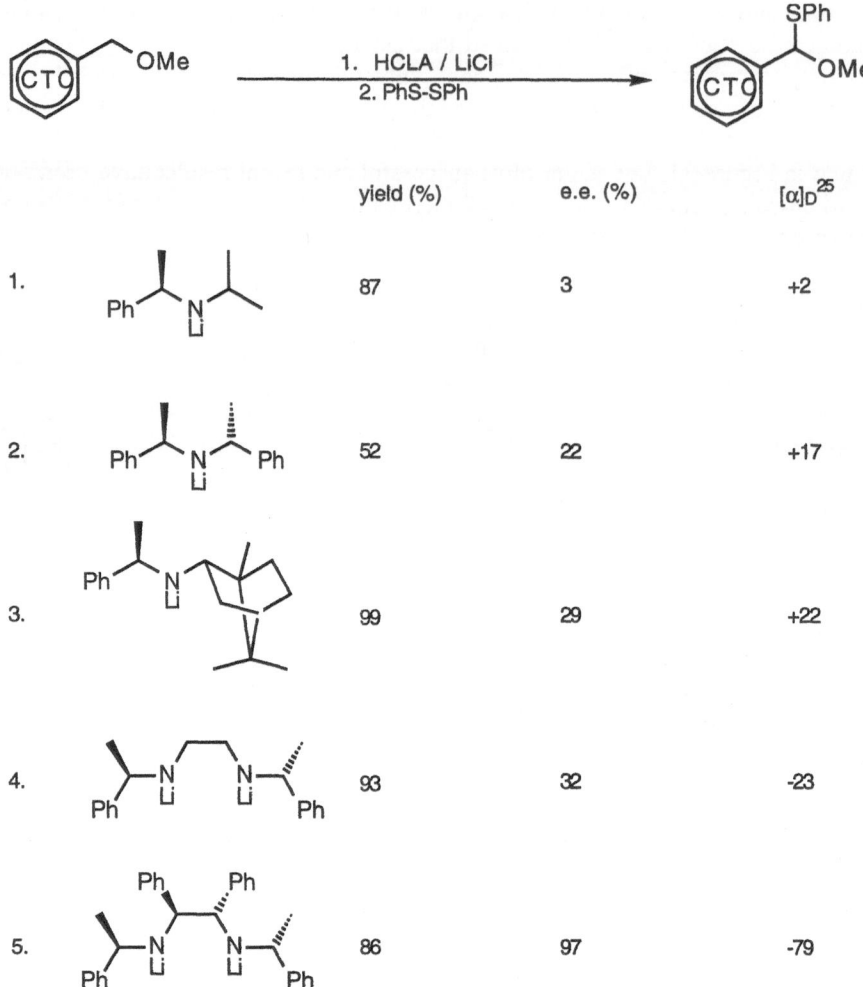

Figure 23

In an initial exploratory experiment, which was designed to probe whether or not a chiral base could be used to differentiate between the enantiotopic benzylic hydrogens of (benzyl methyl ether)tricarbonylchromium(0), a well-established chiral base was used to deprotonate the complex and diphenyl disulfide was used to quench the reaction. Workup gave a novel α-(phenylsulfenyl)benzyl methyl ether complex in 52% yield which was readily analysed by chiral HPLC and was found to have a moderate ee of 22% (Figure 23, Line 2). The next base to be employed was a recently introduced C_2-symmetric vicinal diamide.[21] In this case the reaction proceeded in 86% yield and to our delight, the ee of the product was found to be 97% (Figure 23, Line 5).[22]

Experiments using different mono- and diamide bases are now underway in an effort to understand the nature of the interaction between the chiral amides and (benzyl methyl ether)tricarbonylchromium(0). The most striking result of our experiments to date is that monoamide bases derived from (R)-α-methylbenzylamine lead to product of one configuration (Figure 23, Lines 1-3) whilst related diamide bases lead to product of the opposite configuration (Figure 23, Lines 4 and 5). Thus, on the basis of these initial results, it is postulated that two quite different mechanistic pathways are operating.

In order to determine the absolute stereochemistry of the products of the deprotonation / electrophilic quench reactions, the reaction between the most stereochemically discriminating diamide base and (benzyl methyl ether)tricarbonylchromium(0) was quenched with iodomethane (Figure 24). This gave an α-methylbenzyl methyl ether complex in 96% yield and 97% ee Comparision of the $[\alpha]_D$ of this material with that obtained from the same complex obtained by an independent route[23] revealed that its absolute configuration was R.

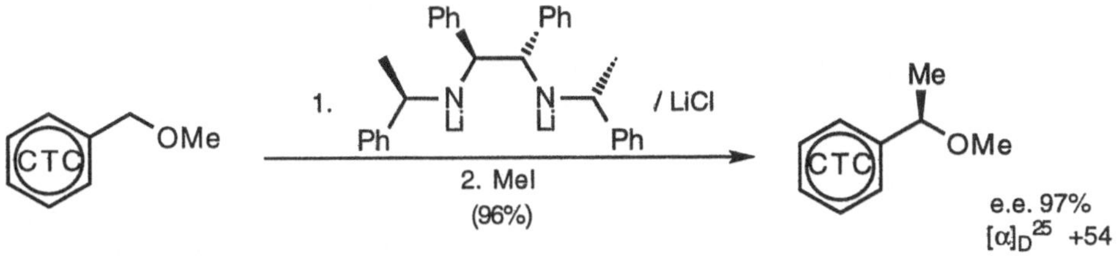

Figure 24

Although definition of the precise nature of this asymmetric process requires much more experimentation, the stereochemical outcome of the reactions with the mono- and diamide bases may be rationalised as indicated in Figure 25.

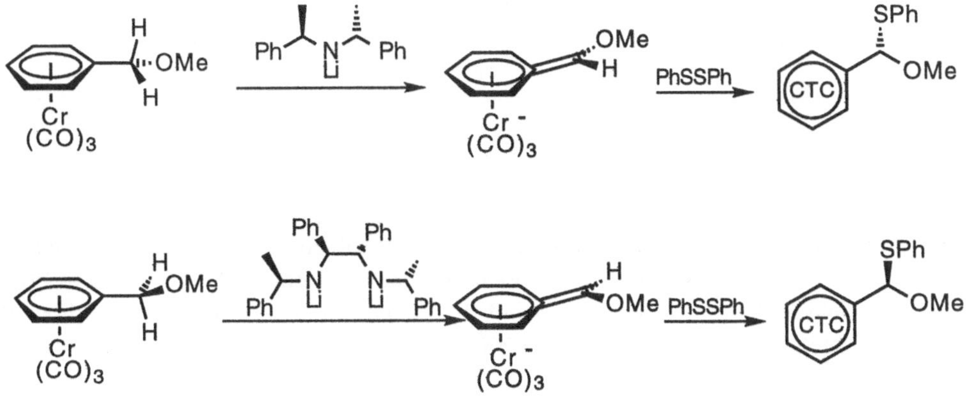

Figure 25

The monoamide bases remove the pro-*S* hydrogen from the *exo* face of the complex, with poor selectivity, to give an anion that is selectively quenched by diphenyl disulfide on the *exo* face to give product of *S* configuration. In contrast, the diamide bases remove the pro-*S* hydrogen from the *endo* face of the complex, a process facilitated by prior coordination between lithium and the chromium carbonyl moiety, to give an anion which is subsequently selectively quenched from the *exo* face to give, in this case, product of *R* configuration.

We have recently examined the possibility of applying the reactivity described above to chemistry designed to lead to the synthesis of enantiomerically pure tertiary benzyl alcohols. In contrast to enantiomerically pure secondary benzyl alcohols, which are now available *via* a number of excellent routes, enantiomerically pure tertiary benzyl alcohols are still difficult to synthesise. Perhaps the best route available to date, published earlier this year,[24] is illustrated in Figure 26.

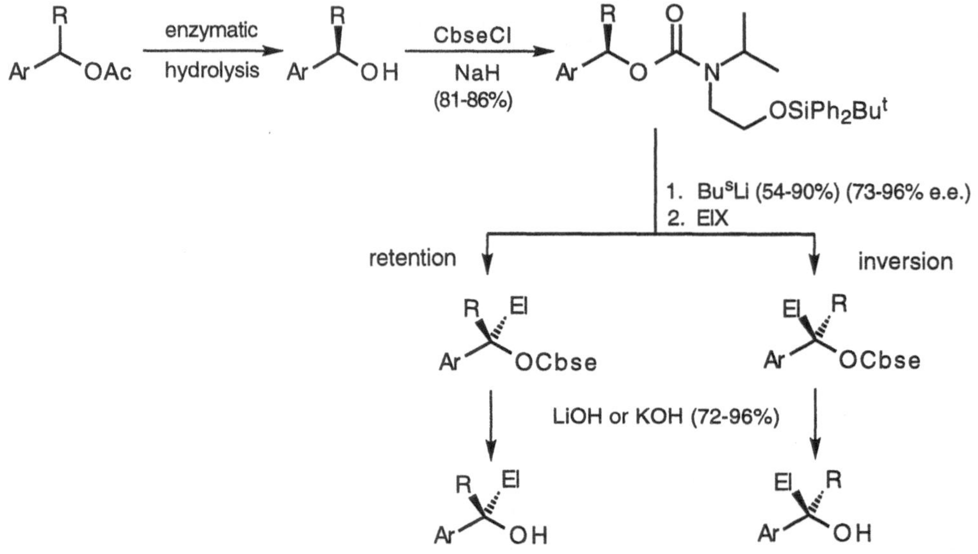

Figure 26

In order to apply our chemistry to the synthesis of enantiomerically pure tertiary benzylic alcohols, it was necessary first to examine the transformation highlighted in Figure 27. We wished to determine whether or not this transformation was chemically feasible, and if so, whether or not it occurred with good stereochemical control. If this was the case, then it was clearly important to determine whether it proceeded with retention or inversion of configuration.

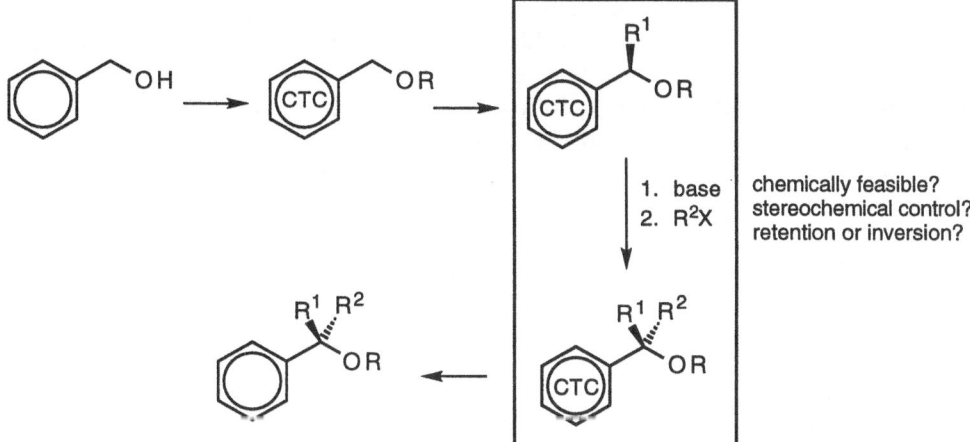

Figure 27

In a preliminary study, a secondary benzyl ether complex of 98.5% ee and *R* configuration was deprotonated with *tert*-butyllithium at -78 °C and subsequently quenched with DOCD$_3$ (Figure 28). Work-up gave a 90% yield of a deuterated complex which HPLC analysis revealed had an ee of 98.5% and was also of *R* configuration. Thus the deprotonation/quench sequence had clearly proceeded with overall retention of configuration presumably *via* a configurationally stable anion. Early studies on the *sec*-butyllithium promoted deprotonation of *N,N*-diisopropyl carbamate derivatives of secondary alcohols revealed that reprotonation of these systems with methanol led to the return of the starting material with retention of configuraion whilst reprotonation with acetic acid led to material of inverted configuration.[25] Quenching the reaction depicted in Figure 28 with DO$_2$CCD$_3$, however, gave product in which the stereochemical configuration had been retained suggesting that in the case of the chromium-stabilised anion, the two different quench processes follow essentially the same mechanistic pathway.

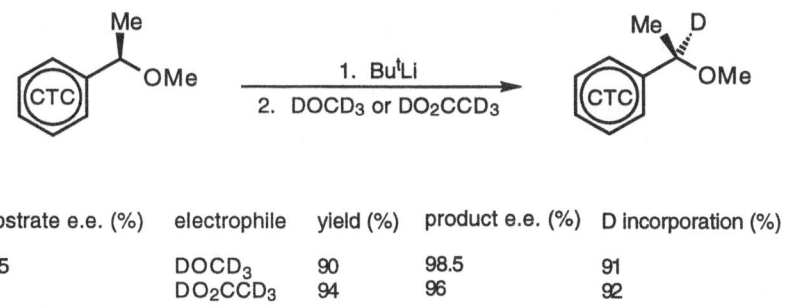

substrate e.e. (%)	electrophile	yield (%)	product e.e. (%)	D incorporation (%)
98.5	DOCD$_3$	90	98.5	91
96	DO$_2$CCD$_3$	94	96	92

Figure 28

Having ascertained that the lithium anion of tricarbonyl(methyl α-methylbenzyl ether)chromium(0) is configurationally stable at -78 °C, our attention turned to synthetically more interesting electrophiles. Whilst deprotonation of racemic tricarbonyl(methyl α-methylbenzyl ether)chromium(0) followed by quenching with iodomethane and chlorotrimethylsilane has been shown to proceed in good yield,[26] the stereochemical consequences and the synthetic scope of this type of process needed to be determined. Thus tricarbonyl(methyl α-methylbenzyl ether)chromium(0) (ee 96%) was deprotonated and quenched with benzyl bromide (Figure 29). Work-up led to the isolation of product that was essentially of the same enantiomeric purity as the starting material. Furthermore its absolute configuration was determined by an X-ray crystallographic analysis to be S, indicating that the deprotonation/alkylation sequence had proceeded with overall retention of configuration.[27]

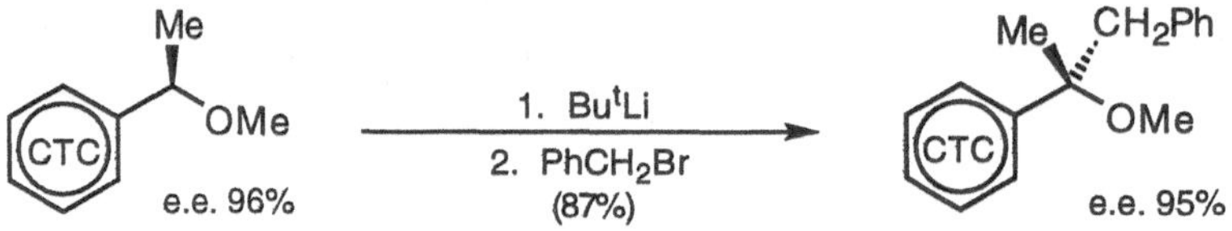

Figure 29

The overall retention of configuration observed in the deprotonation/quench sequences performed on tricarbonyl(methyl α-methylbenzyl ether)chromium(0) is rationalised by the process depicted in Figure 30. Removal of the benzylic proton of tricarbonyl(methyl α-methylbenzyl ether)chromium(0) from a conformation that places it antiperiplanar to the tricarbonylchromium(0) unit gives a chromium-centred and configurationally-locked anion. The incoming electrophile then approaches the sterically less-hindered *exo* face of the complex to give the observed products.

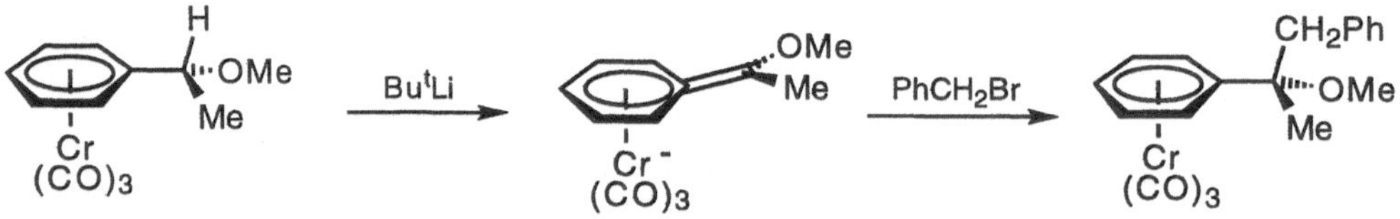

Figure 30

Finally, in recent weeks, we have synthesised the complexes depicted in Figure 31 using the methodology described above. In the near future we intend to examine a range of different oxygen substituents and decomplexation methods in order to determine the optimum system for the production of enantiomerically pure tertiary benzyl alcohols by this approach.

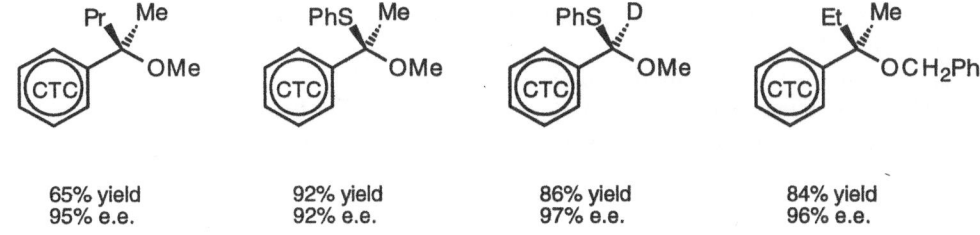

| 65% yield
95% e.e. | 92% yield
92% e.e. | 86% yield
97% e.e. | 84% yield
96% e.e. |

Figure 31

References

1. C.J. Richards and S.E. Thomas, *J. Chem. Soc., Chem. Commun.* **1989**, 21.
2. N.W. Alcock, C.J. Richards and S.E. Thomas, *Organometallics* **1991**, *10*, 231.
3. C.J. Richards and S.E. Thomas, *J. Chem. Soc., Chem. Commun.* **1990**, 307.
4. L. Hill, S.P. Saberi, A.M.Z. Slawin, S.E. Thomas and D.J. Williams, *J. Chem. Soc., Chem. Commun.* **1991**, 1290; S.P. Saberi and S.E. Thomas, *J. Chem. Soc., Perkin Trans. 1* **1992**, 259.
5. K.G. Morris, S.P. Saberi, A.M.Z. Slawin, S.E. Thomas and D.J. Williams, *J. Chem. Soc., Chem. Commun.* **1992**, 1778; K.G. Morris, S.P. Saberi and S.E. Thomas, *J. Chem. Soc., Chem. Commun.* **1993**, 209; K.G. Morris, S.P. Saberi, M.M. Salter, S.E. Thomas, M.F. Ward, A.M.Z. Slawin and D.J. Williams, *Tetrahedron* **1993**, *49*, 5617; S.P. Saberi, M.M. Salter, A.M.Z. Slawin, S.E. Thomas and D.J. Williams, *J. Chem. Soc., Perkin Trans. 1* **1994**, 167.
6. S.P. Saberi, A.M.Z. Slawin, S.E. Thomas, D.J. Williams, M.F. Ward and P.A. Worthington, *J. Chem. Soc., Chem. Commun.* **1994**, 2169; S.E. Gibson (née Thomas), S.P. Saberi, A.M.Z. Slawin, P.D. Stanley, M.F. Ward, D.J. Williams and P.A. Worthington, *J. Chem. Soc., Perkin Trans. 1* **1995**, 2147.
7. L. Hill, C.J. Richards and S.E. Thomas, *J. Chem. Soc., Chem. Commun.* **1990**, 1085.
8. A. Marcuzzi, A. Linden, D. Rentsch and W. von Philipsborn, *J. Organomet. Chem.* **1992**, *429*, 87.
9. S.A. Benyunes, S.E. Gibson (née Thomas) and J.A. Stern, *J. Chem. Soc., Perkin Trans 1* **1995**, 1333.
10. F.R. Atherton, H.T. Openshaw and A.R. Todd, *J. Chem. Soc.* **1945**, 660.
11. S.A. Benyunes, S.E. Gibson (née Thomas) and M.F. Ward, *Tetrahedron Asymmetry* **1995**, *6*, 2517.
12. S.A. Benyunes and S.E. Gibson (née Thomas), *J. Chem. Soc., Chem. Commun.* **1996**, 43.
13. J. Klimes and E. Weiss, *Angew. Chem., Int. Ed. Engl.* **1982**, *21*, 205.
14. T. Mitsudo, H. Watanabe, Y. Komiya, Y. Watanabe, Y. Takaegami, K. Nakatsu, K. Kinoshita and Y. Miyagawa, *J. Organomet. Chem.* **1980**, *190*, C39; R. Aumann, *Angew. Chem., Int. Ed. Engl.* **1988**, *27*, 1456.
15. C.J. Richards and S.E. Thomas, *Tetrahedron: Asymmetry* **1992**, *3*, 143.

16. S.A. Benyunes, S.E. Gibson (née Thomas) and M.A. Peplow, *J. Chem. Soc., Chem. Commun.* **1996**, 1757.

17. R.P.C. Cousins and N.S. Simpkins, *Tetrahedron Lett.* **1989**, *30*, 7241.

18. D. Hoppe, F. Hintze and P. Tebben, *Angew. Chem., Int. Ed. Engl.* **1990**, *29*, 1422.

19. D.M. Hodgson and G.P. Lee, *J. Chem. Soc., Chem. Commun.*, **1996**, 1015.

20. J. Blagg, S.G. Davies, N.J. Holman, C.A. Laughton and B.E. Mobbs, *J. Chem. Soc., Perkin Trans. 1*, 1986, 1581.

21. K. Bambridge, M.J. Begley and N.S. Simpkins, *Tetrahedron Lett.*, **1994**, **35**, 3391.

22. E.L.M. Cowton, S.E. Gibson (née Thomas), M.J. Schneider and M.H. Smith, *J. Chem. Soc., Chem. Commun.*, **1996**, 839.

23. S. Top, G. Jaouen and M.J. McGlinchey, *J. Chem. Soc., Chem. Commun.*, **1980**, 1110.

24. C. Derwing and D. Hoppe, *Synthesis*, **1996**, 149.

25. D. Hoppe, A. Carstens and T. Krämer, *Angew. Chem. Int. Ed. Engl.*, **1990**, *29*, 1424.

26. J. Blagg, S.G. Davies, C.L. Goodfellow and K.H. Sutton, *J. Chem. Soc., Perkin Trans. 1*, **1987**, 1805.

27. S.E. Gibson (née Thomas), P.C.V. Potter and M.H. Smith, submitted for publication.

Metal-Centered Coupling Reactions of C_1-Ligands Analogous to Carbon Monoxide

Alexander C. Filippou

Humboldt-Universität zu Berlin, Institut für Chemie, Hessische Str. 1-2, D-10115 Berlin

Abstract

A group of ligand coupling and cleavage reactions of terminal two-faced π-acceptor ligands such as carbon monoxide or isocyanides is discussed, which occur at Group V or Group VI transition metal centres. The common pathway of these reactions involving carbyne complexes as important intermediates is demonstrated by the reductive isocyanide-isocyanide coupling reaction of the seven-coordinate Mo^{II} and W^{II} isocyanide complexes $[M(CNR)_6X]^+$ (R = alkyl; X = Cl-I) to form the bis(alkylamino)acetylene complexes $[X(RNC)_4M[\eta^2-R(H)NC\equiv CN(H)R]]^+$. Factors influencing the consecutive transformation of the starting materials to the coupling products are discussed.

1 Introduction

Transition metal mediated carbon-carbon bond-forming reactions, particularly those involving carbon monoxide or related C_1-ligands, are of continuing interest in organometallic chemistry [1]. Several reports of carbon-carbon coupling reactions of carbon monoxide or isocyanides have been reported in the literature using soluble Group IV transition metal, lanthanide and actinide complexes [2]. The course of these reactions and the nature of the products obtained is determined by the oxophilicity of these metals. Recently a group of closely related ligand coupling and cleavage reactions of terminal two-faced π-bonded ligands to group V and VI transition metal centers has been found, in which carbyne ligands play a central role [3]. The ligands involved in these reactions range from π-acceptor ligands such as carbon monoxide or isocyanides to π-donor ligands such as oxo- or imido ligands [4]. Herein we describe our contributions to the development of this chemistry.

2 Discussion

Our work was stimulated by the report of S. J. Lippard and coworkers that two isocyanide ligands in the Mo^{II} and W^{II} complexes $[M(CNR)_6I]^+$ (R = Cy, tBu) can be coupled to a coordinated bis(alkylamino)acetylene upon reduction with Zn in water-containing THF (equation 1) [5].

$$[M(CNR)_6I]^+ \; I^- \xrightarrow[\text{THF/H}_2\text{O}]{\text{+ Zn}} \quad (1)$$

M = Mo, W; R = Cyclohexyl, t-Bu

Initially these authors suggested a concerted pathway for this C-C coupling reaction, being promoted by the close nonbonded carbon-carbon contact of two isocyanide ligands in the seven-coordinate starting materials, by reduction of the metal center and by coordination of a Lewis acid (such as the Zn^{2+} ion produced during the reaction) to the nitrogen atoms of the coupled isocyanide ligands [5]. Later on these authors studied the reductive carbonyl-carbonyl coupling reaction of the seven-coordinate Nb^I and Ta^I complexes $M(CO)_2(dmpe)_2Cl$ (dmpe = 1,2-bis(dimethylphosphino)ethane) and proposed on the basis of these studies a stepwise pathway for the isocyanide-isocyanide coupling reaction of the $[M(CNR)_6X]^+$ complexes involving aminocarbyne intermediates, but gave no experimental evidence in support of this proposal [6]. In the case of the seven-coordinate Re^{III} isocyanide complexes $[Re(PMePh_2)_2(CNR)_3(Cl)_2]^+$ (R = Me, ^{t}Bu) they succeeded to isolate aminocarbyne complexes under reductive coupling conditions, but failed to demonstrate coupling to give the alkyne complexes [7]. In order to elucidate the mechanism of the reductive isocyanide-isocyanide coupling reaction of the complexes $[M(CNR)_6X]^+$, we decided to investigate thoroughly the reactivity of these compounds and therefore developed first a high yield synthesis of the complexes [M(CNR)$_6$Br]Br (M = Mo, W; R = Et, ^{i}Pr, ^{t}Bu) (**5**) starting from M(CO)$_6$ (**1**) (Scheme 1). This synthesis is distinguished by a sequence of clean, large-scale reactions via the easily isolable and well-characterized isocyanide complexes *fac*-M(CO)$_3$(CNR)$_3$ (**3**) and M(CO)$_2$(CNR)$_3$Br$_2$ (**4**) [8].

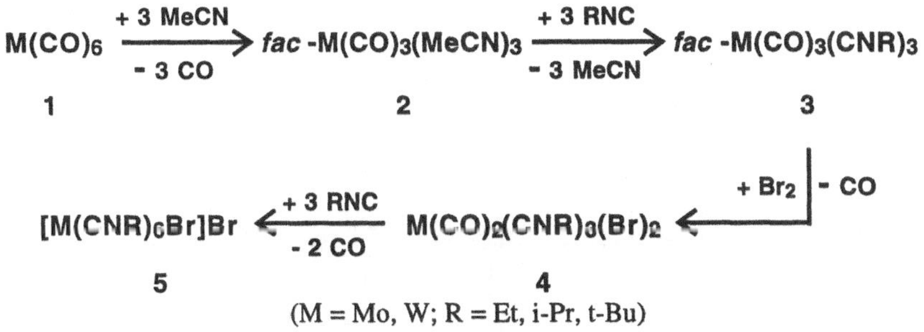

Scheme 1

Reduction of the 18e Mo^{II} and W^{II} isocyanide complexes **5** with an excess of Na/Hg in THF was then carried out and afforded selectively the homoleptic Mo^0 and W^0 isocyanide complexes **6**, indicating that formation of the bis(alkylamino)acetylene coupling products is circumvented, if reduction of **5** is carried out in the absence of water (Scheme 2). This overall two-electron reduction of **5** was found to proceed in two steps. The first step involves an irreversible one-electron transfer to **5** accompanied by the elimination of the bromo ligand from the coordination sphere to afford the 17e Mo^I and W^I isocyanide complexes **7**. This is followed by the reversible one-electron reduction of **7** to yield **6**. The half-wave potential of the complex Mo(CN^{t}Bu)$_6$ in THF was determined by cyclic voltammetry to be -577 mV (reference electrode: Ag/AgCl/3m KCl) indicating in full agreement with the spectroscopic data the presence of an electron-rich metal center in the complexes **6**.

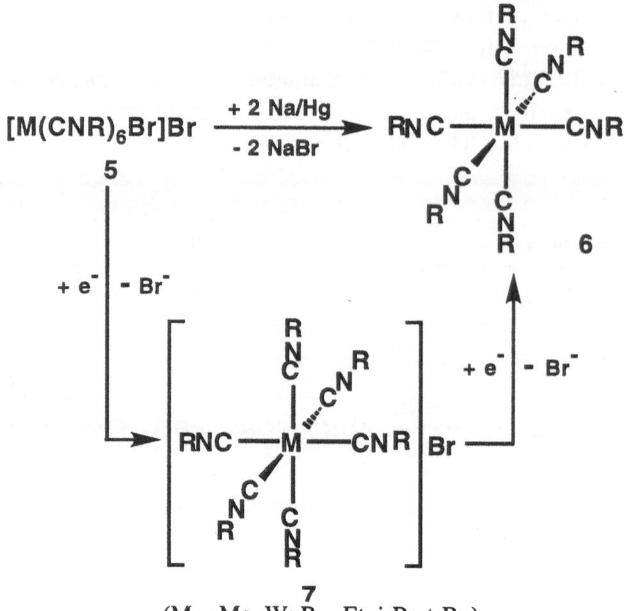

Scheme 2

$(M = Mo, W; R = Et, i\text{-}Pr, t\text{-}Bu)$

We then wanted to demonstrate that the six-coordinate M^0 complexes **6** are important intermediates of the reductive isocyanide-isocyanide coupling reaction of **5** and treated these compounds with two equivalents of a Brönsted acid. In fact, a selective isocyanide-isocyanide coupling reaction was observed resulting in the formation of the bis(alkylamino)acetylene complexes **8** (equation 2) [9]. The same reactivity pattern was independently observed by S. J. Lippard [10].

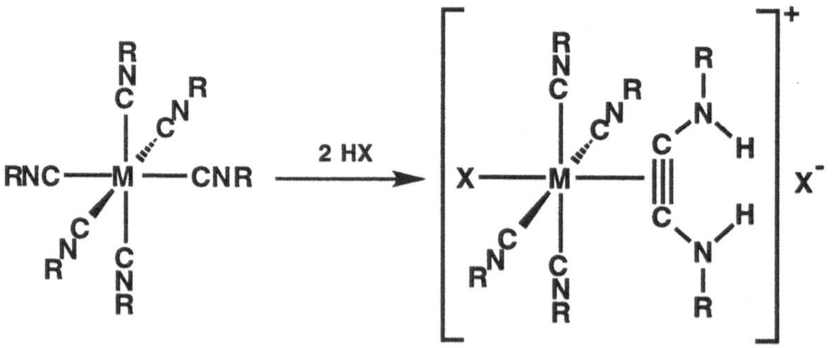

$$M = Mo, W; R = Et, t\text{-}Bu; X = Br, I \qquad (2)$$

100

In this reaction two isocyanide ligands acting as two-electron donors are coupled to a bis(alkylamino)acetylene, which acts as a four-electron donor ligand. Coordination of the bis(alkylamino)acetylene to the metal center inhibits the tautomerization to the corresponding more stable 1,4-diazabutadiene [9].

After we had shown that the complexes **6** were important intermediates of the reductive isocyanide-isocyanide coupling reaction of **5**, we wanted to elucidate the mechanism of the acid-promoted C-C coupling reaction of **6** and treated these compounds with only one equivalent of an electrophile. [Et$_3$O]BF$_4$ or SiMe$_3$OTf were used as electrophiles in order to determine the site of electrophilic attack in the isocyanide complexes **6**. Both electrophiles were found to add selectively to the nitrogen atom of one isocyanide ligand to afford Fischer-type aminocarbyne complexes, as demonstrated by the ethylation of the ethyl isocyanide complexes **6a** to give **9** (equation 3) [8, 9].

6a   9

Similar reactions had been previously reported for other octahedral electron-rich isocyanide complexes such as the Mo^0 and W^0 complexes *trans*-M(CNMe)$_2$(dppe)$_2$ [dppe: 1,2-bis(diphenylphosphino)ethane] [11] or the Re^I complexes *trans*-ReCl(CNR)$_2$(dppe)$_2$ (R = Me, ^{t}Bu) [12].

The aminocarbyne complexes **9** were then found to react with one equivalent of HI to afford the diaminoacetylene complexes **10** (equation 4) [9]. The acid-promoted carbyne-isocyanide reaction of **9** did not only prove that six-coordinate Fischer-type aminocarbyne complexes are important intermediates of the reductive isocyanide-isocyanide coupling reaction of **5** (equation 1), but also allowed the synthesis of diaminoacetylene complexes bearing two different amino groups in the alkyne ligand.

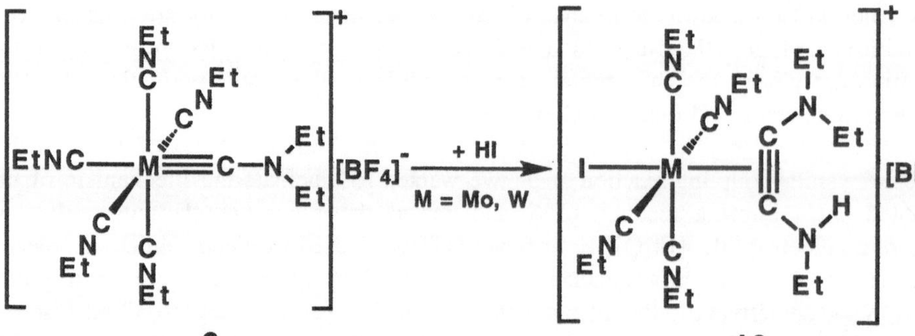

$$\text{(4)}$$

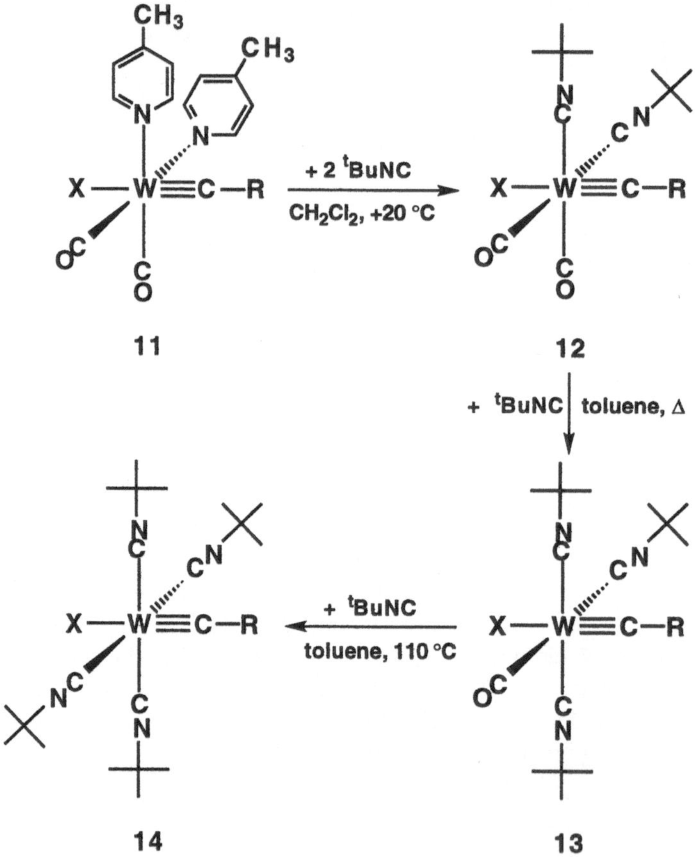

X = Br, I; R = Ph, NEt$_2$

Scheme 3

Our next efforts were to elucidate the mechanism of the acid-promoted carbyne-isocyanide coupling reaction. Various *tert*-butyl isocyanide substituted tungsten carbyne complexes were therefore prepared, structurally characterized and their reactions with Brönsted acids studied. The synthesis of these compounds was achieved starting from the Fischer-type carbyne complexes **11** and involves successive ligand exchange reactions (Scheme 3) [13].

All these compounds were found to undergo with one equivalent of HX a coupling of the carbyne with the tert-butyl isocyanide ligand to give W^{II} (tert-butylamino)alkyne complexes (equation 5) [14]. The back reaction to this C-C coupling reaction also succeeded (equation 5) and was used to prepare electron-rich Fischer-type phenylcarbyne complexes of molybdenum and tungsten, that are not accessible by the route outlined in Scheme 3.

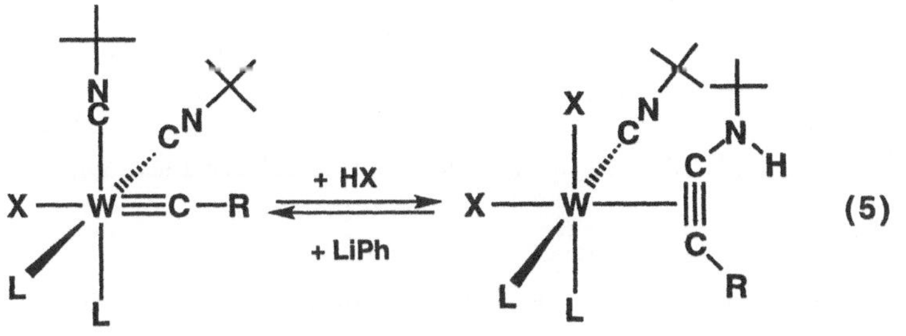

$$X = Br, I; L = CO, t\text{-}BuNC; R = Ph, NEt_2$$

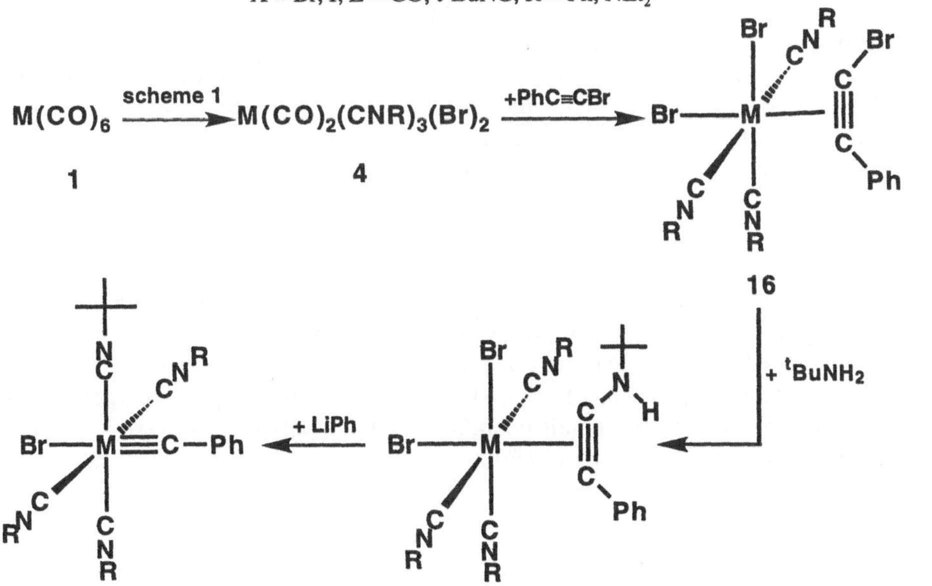

$$M = Mo, W; R = Et, t\text{-}Bu$$

Scheme 4

103

In the essential step of this multistep synthesis starting from M(CO)$_6$ a cleavage of the C-C bond of a coordinated (1-amino-2-phenyl)acetylene ligand occurs with LiPh to afford the desired compounds **14a** (Scheme 4) [15].

Three features of the carbyne-isocyanide coupling reaction of **12-14** let us assume, that a different mechanism was operative here from that suggested for the nucleophile-induced carbyne-carbonyl coupling reaction to give the isolable η^2-ketenyl complexes, which subsequently react with electrophiles to afford alkyne complexes (equation 6) [3, 16].

$$L_nM\equiv C-R \quad \xrightarrow{+ \, Nu^-} \quad \left[L_nM \overset{Nu}{\underset{C-R}{\diagdown}} \overset{C=O}{\diagdown} \right]^- \quad \xrightarrow{+ \, E^+} \quad L_nM-\underset{C}{\overset{C}{\underset{R}{\overset{O-E}{\|\|}}}} \quad (6)$$

M = Mo, W; R = Alkyl, Aryl; Nu = Nukleophil; E = Elektrophil

The first observation was that the carbyne complexes **12-14** did not undergo a nucleophile-induced carbyne-isocyanide coupling reaction to form analogous products (η^2-iminoketenyl complexes) to the η^2-ketenyl complexes. This could be explained, if one would assume that formation of the η^2-iminoketenyl complexes would be the rate-determining step in the carbyne-isocyanide coupling reaction of **12-14**, in which the η^2-iminoketenyl complex either decomposes rapidly to give back the carbyne complex or is trapped rapidly by the electrophile to give the alkyne product [9]. However, this assumption was contradicting to the observation, that enhancement of the electron density at the metal center of the carbyne complex accelerates the carbyne-isocyanide coupling reaction, indicating an interaction of the carbyne complex with the electrophile E$^+$. The second observation was the selective coupling of the carbyne ligand with the isocyanide ligand in **12** and **13**, the carbonyl ligand being only a spectator ligand [14]. Finally, the presence of a π donor substituent at the carbyne-carbon such as an amino group was found to favour the carbyne-isocyanide coupling reaction. In comparison, the nucleophile-induced carbyne-carbonyl coupling reaction of alkyl- and arylcarbyne complexes had been previously reported to be circumvented by the presence of an amino substituent at the carbyne-carbon, as shown by the two reactions in Scheme 5 [17]. All these observations let us suggest a different mechanism for the carbyne-isocyanide coupling reaction involving a bis-carbyne complex as the important intermediate (equation 7) [9, 14].

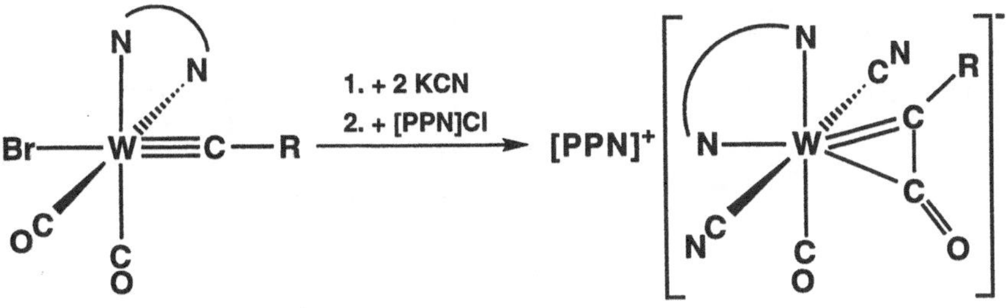

N N : 2.2´-bipy, ophen; R = Me, Ph

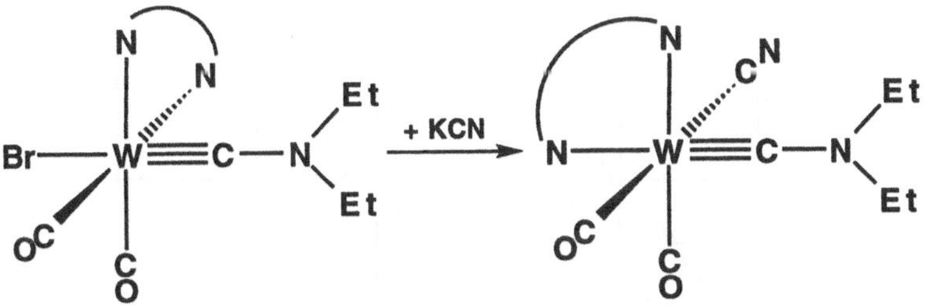

N N : 2.2´-bipy, ophen

Scheme 5

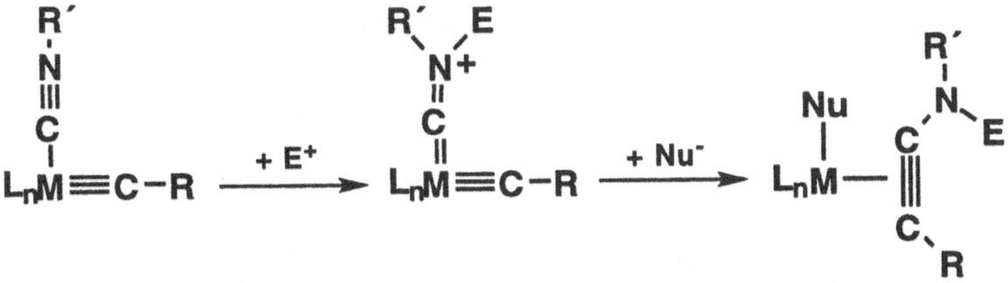

$$\qquad\qquad (7)$$

M = Mo, W; R = Amino, Aryl; R' = Alkyl; Nu = Nukleophil; E = Elektrophil

Searching for a way to verify the intermediate formation of a bis-carbyne complex in the acid-induced carbyne-isocyanide coupling reaction, we treated the diethylaminocarbyne complexes **9** with one equivalent of [Et₃O]BF₄. In fact, addition of the electrophile at the nitrogen-atom of one isocyanide ligand was observed to afford the bis-diethylaminocarbyne complexes **16**. Unfortunately, isolation of the thermolabile bis-carbyne complexes was prevented by the concomitant formation of ionic by-products, which were difficult to remove. However, the complexes **16** were spectroscopically identified and then treated with [NEt₄]Br to afford the bis(diethylamino)acetylene complexes **17** (Scheme 6) [18].

105

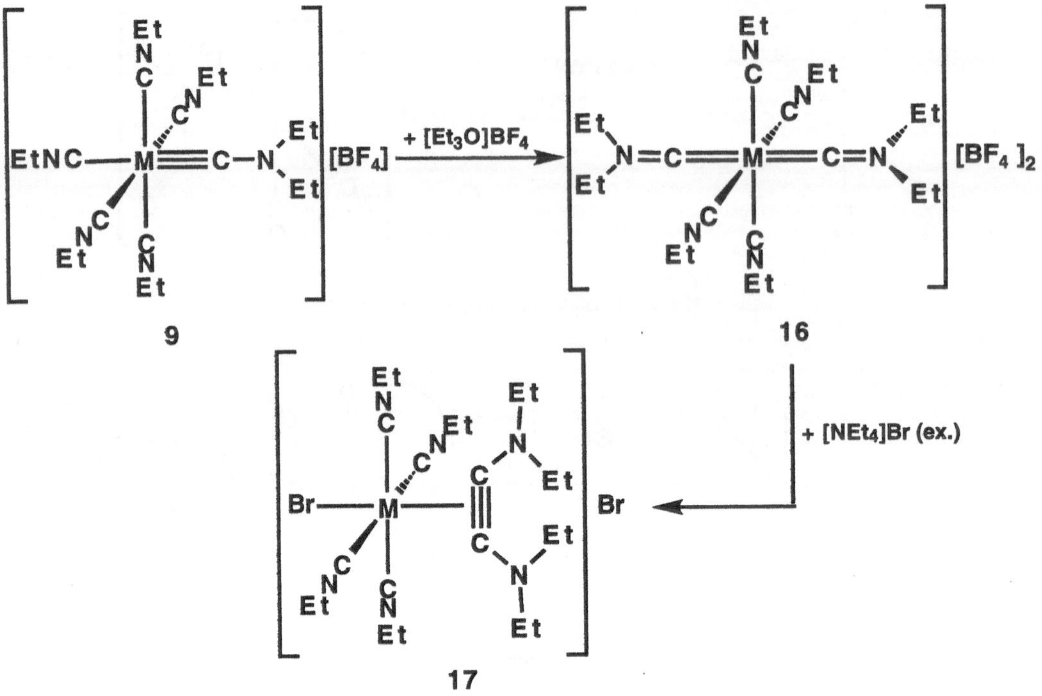

Scheme 6

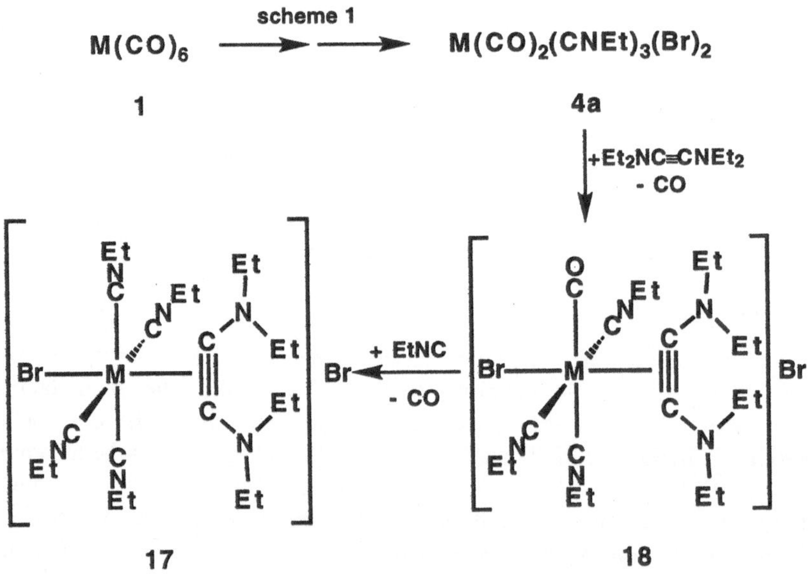

$$M(CO)_6 \xrightarrow{\text{scheme 1}} M(CO)_2(CNEt)_3(Br)_2$$

1 **4a**

M = Mo, W

Scheme 7

Moreover the coupling products **17** were prepared independently from the metal hexacarbonyls **1** following the procedure outlined in Scheme 7. The reaction sequence leading from **9** to **17** shows that the C-C-bond forming step of the reductive isocyanide-isocyanide coupling reaction (equation 1) involves a nucleophile-promoted coupling of two carbyne ligands at a six-coordinate molybdenum or tungsten center. A similar reaction sequence was recently reported by A. J. L. Pombeiro *et al.* for the acid-induced isocyanide-isocyanide coupling reaction of the complexes *trans*-M(CNR)$_2$(dppe)$_2$ (M = Mo, W; R = Me, ^{t}Bu), supporting our work [19].

The electrophile-induced carbyne-isocyanide coupling reactions of **9** and **12-14** (equations 4 and 5) could involve a second intermediate, if H$^+$ is used as the electrophile. Thus, A. Mayr et al. have found that protonation of the carbyne complex Cl(CO)(PMe$_3$)$_2$(^{t}BuNC)W≡CPh occurs first at the carbyne-carbon to give a "distorted" carbene complex. The carbene complex was postulated to rearrange to the bis-carbyne complex after transfer of the proton from the carbene-carbon atom to the isocyanide-nitrogen atom [20].

In order to study the metal-centered carbyne-carbyne coupling reaction in more detail, we had to isolate mononuclear bis-carbyne complexes in pure form and thought to achieve this goal by reducing the electrophilicity of the metal center in the bis-carbyne complexes. Therefore we concentrated our efforts on the synthesis of cyclopentadienyl substituted aminocarbyne complexes (**20**). Two methods were developed to prepare these compounds, the first one involving ethylation of the isocyanide metallate **19** with [Et$_3$O]BF$_4$ [21] and the other treatment of the γ-picoline-substituted aminocarbyne complexes **11a** with KC$_5$R$_5$ (Scheme 8) [22].

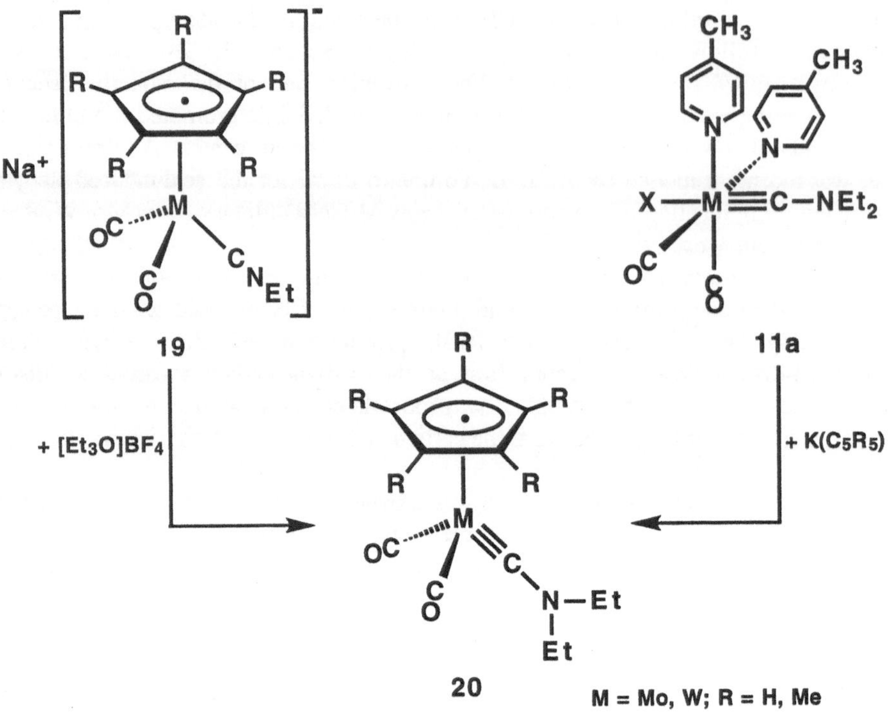

Scheme 8

The aminocarbyne complexes **20** were then converted to the electron-rich isocyanide-substituted derivatives **23** following the procedure depicted in Scheme 9. Thermal or photochemical decarbonylation of **20** in the presence of an excess of isocyanide to afford directly the desired substitution products **23** did not succeed due to the strong metal-carbonyl back-bonding in the starting materials (Scheme 9) [23]. Finally, ethylation of the aminocarbyne complexes **23** with [Et₃O]BF₄ occurs selectively at one of the two isocyanide-nitrogen atoms to give the yellow isolable bis-aminocarbyne complexes **24** (equation 8).

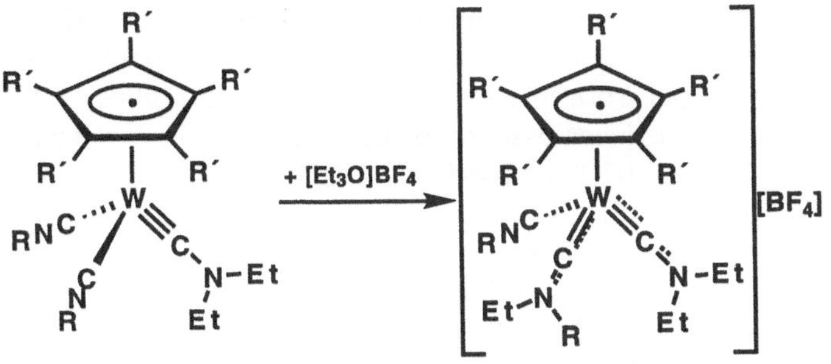

$$R = Et, t\text{-}Bu; \quad R' = H, Me$$

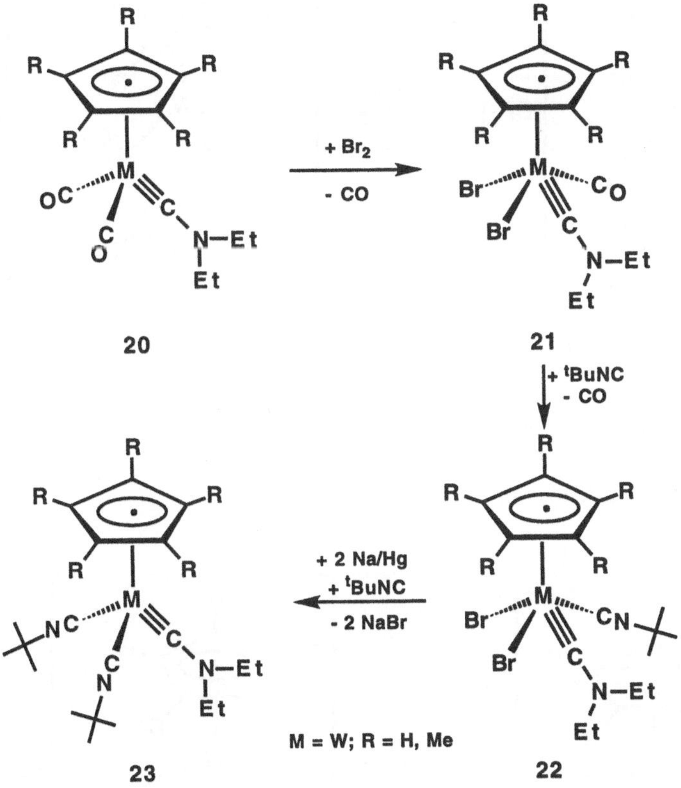

Scheme 9

In order to understand better the electronic structure of the bis-aminocarbyne complexes **24**, Extended-Hückel calculations on the hypothetical bis-carbyne complexes $[CpW(CO)(CH_2)_2]^+$ and $[CpW(CO)(CNH_2)_2]^+$ have been carried out. These predict a spontaneous rearrangement of the 18e bis-methylidyne complex $[CpW(CO)(CH)_2]^+$ to the energetically favoured 16e acetylene complex $[CpW(CO)(\eta^2-HC\equiv CH)]^+$ and show in full agreement with the experimental work that introduction of amino groups on both carbyne-carbons stabilizes the 18e bis-aminocarbyne form $[CpW(CO)(CNH_2)_2]^+$ relative to the 16e diaminoacetylene isomer $[CpW(CO)(\eta^2-H_2NC\equiv CNH_2)]^+$ [24]. However, addition of a nucleophile to $[CpW(CO)(CNH_2)_2]^+$ would be expected to change the overall thermo-dynamics of the aminocarbyne-aminocarbyne coupling reaction stabilizing the 16e diamino-acetylene product $[CpW(CO)(\eta^2-H_2NC\equiv CNH_2)]^+$. Indeed, reaction of **24** with ethyl isocyanide gives the bis(diethylamino)acetylene complex **25** (equation 9).

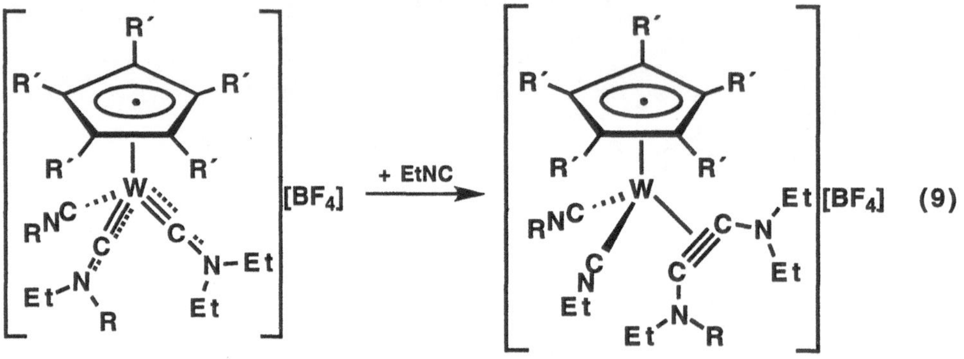

$$\textbf{24} \qquad \qquad \qquad \textbf{25} \qquad \qquad (9)$$

R = Et; R′ = Me

Coupling of the two carbyne ligands of **24** can be induced also by oxidation, as demonstrated by the reaction of **24** with elementary bromine to afford the W^{IV} bis(diethylamino)acetylene complex **26** (equation 10) [23].

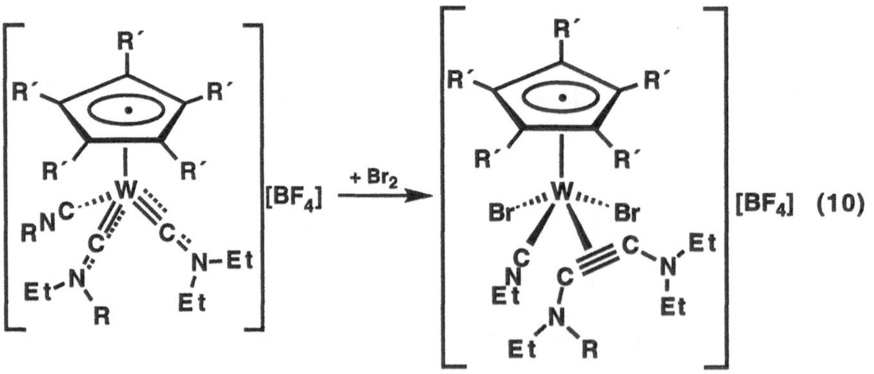

$$\textbf{24} \qquad \qquad \qquad \textbf{26} \qquad \qquad (10)$$

R = Et; R' = Me

All these results show that reductive coupling of two isocyanide ligands in the Mo^{II} and W^{II} complexes $[M(CNR)_6X]^+$ (R = alkyl; X = halogen) (equation 1) is a multi-step process involving aminocarbyne and bis-aminocarbyne intermediates (Scheme 10). This process is initiated by the overall two-electron reduction of the starting materials to afford the 18e Mo^0 and W^0 isocyanide complexes 6, which are activated thereby for an electrophilic attack of H^+ at the isocyanide-nitrogen atom. This occurs successively in the next two steps transforming the two isocyanide ligands to be coupled to two aminocarbyne ligands. Finally a nucleophile-promoted carbyne-carbyne coupling reaction occurs to afford the bis(alkylamino)acetylene complexes 8 (Scheme 10). The course of the reaction and the nature of the products is determined by the ability of the Group VI transition metals to form strong metal-carbon multiple bonds and the stability of the Mo^{II} and W^{II} alkyne coupling products (e.g. 8, 10, 15, 17 and 25).

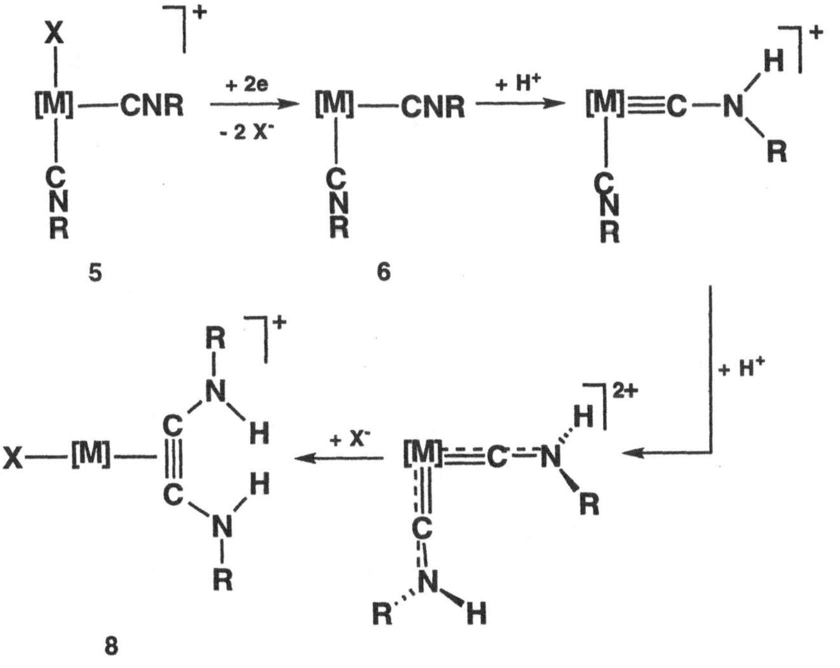

[M] = M(CNR)$_4$; M = Mo, W; R = Alkyl; X = Halogen

Scheme 10

It is noteworthy that reductive coupling of two carbonyl ligands in the seven-coordinate Nb^I and Ta^I complexes $M(CO)_2(dmpe)_2Cl$ (dmpe = 1,2-bis(dimethylphosphino)ethane), and reductive coupling of an isocyanide with a carbonyl ligand in the Nb^I and Ta^I complexes $M(CO)(CNMe)(dmpe)_2Cl$, which are isoelectronic with the Mo^{II} and W^{II} isocyanide complexes $[M(CNR)_6X]^+$ (R = alkyl; X = halogen) follow the same pathway [3c, 3e, 6, 25].

The reaction sequence illustrated in Scheme 10 implies that factors influencing the selectivity of the reactions of the isocyanide complexes 6 and the aminocarbyne complexes with electrophiles will also influence the outcome of the overall reaction. Therefore we studied the influence of the electrophile, the isocyanide ligand and the metal centre on the course of these reactions [26]. Only the effect of the metal centre is described here, which has been elucidated by studying the reactions of analogous chromium complexes with electrophiles.

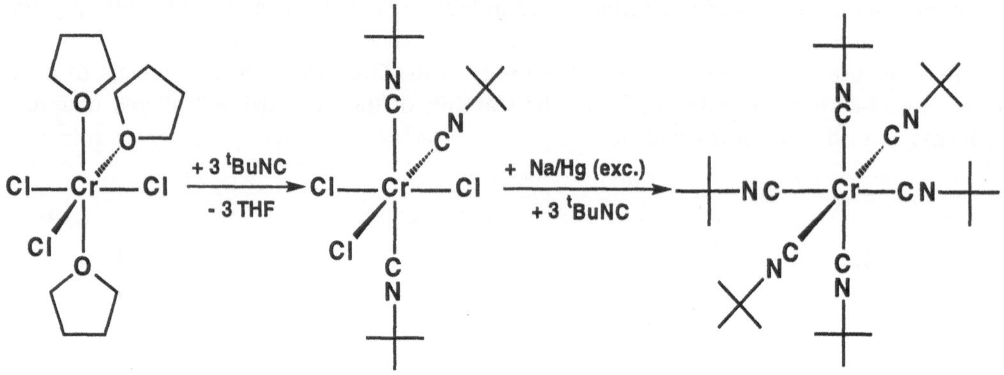

Scheme 11

For this purpose the Cr^{III} complex **27** was first treated with tert-butyl isocyanide to give selectively the complex **28**, which was then reduced with an excess of Na/Hg to afford the desired Cr^0 complex Cr(CN^{t}Bu)$_6$ (**29**) in high yield (Scheme 11). This procedure failed to give Cr^0 isocyanide complexes bearing smaller alkyl substituents at the isocyanide-nitrogen atom. However, the synthesis of Cr(CN^{i}Pr)$_6$ (**31**) has been achieved starting from CrCl$_2$ and following the procedure outlined in Scheme 12.

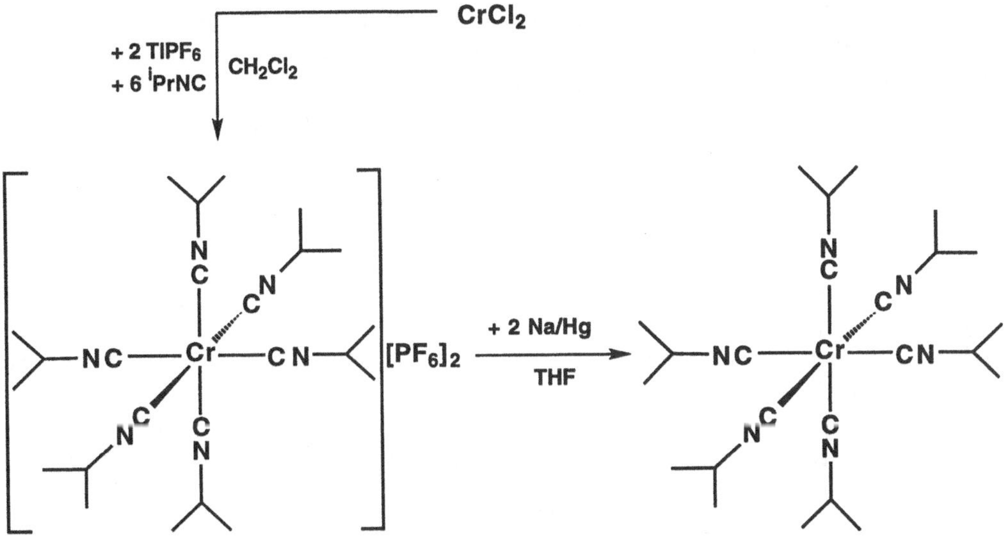

30 31

Scheme 12

Complexes **29** and **31** show the same reactivity pattern as the analogous molybdenum and tungsten compounds **6** reacting with one equivalent of $[Et_3O]BF_4$ to give the aminocarbyne complexes **32** (equation 11).

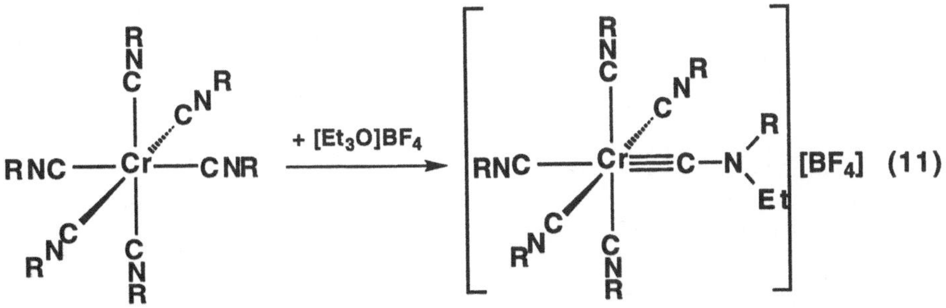

29, 31 32

R = i-Pr, t-Bu

However, no coupling of the carbyne with the isocyanide ligand was observed in the reaction of **29** and **31** with one equivalent of HI. This let us assume, that the isocyanide-iso-cyanide coupling reaction of **29** with two equivalents of an aqueous HI solution to give the Cr^{II} bis(tert-butylamino)acetylene complex **33** (equation 12) [27], follows a different pathway from that of the analogous molybdenum and tungsten complexes **6** (equation 2, Scheme 10).

In order to get an insight into this C-C bond forming reaction, the reactivity of **29** towards HI was studied. Surprisingly, treatment of **29** with one equivalent of an aqueous HI solution affords in an electron-transfer reaction the 17e Cr^I isocyanide complex **34** (Scheme 13). This can be also prepared selectively by oxidation of **29** with half an equivalent of I_2. Reaction of **34** with one equivalent of an aqueous HI solution gives then the Cr^{II} bis(tert-butylamino)acetylene complex **33** (Scheme 13), showing unequivocally that complex **34** is an intermediate of the HI-induced isocyanide-isocyanide coupling reaction of **29** (equation 12).

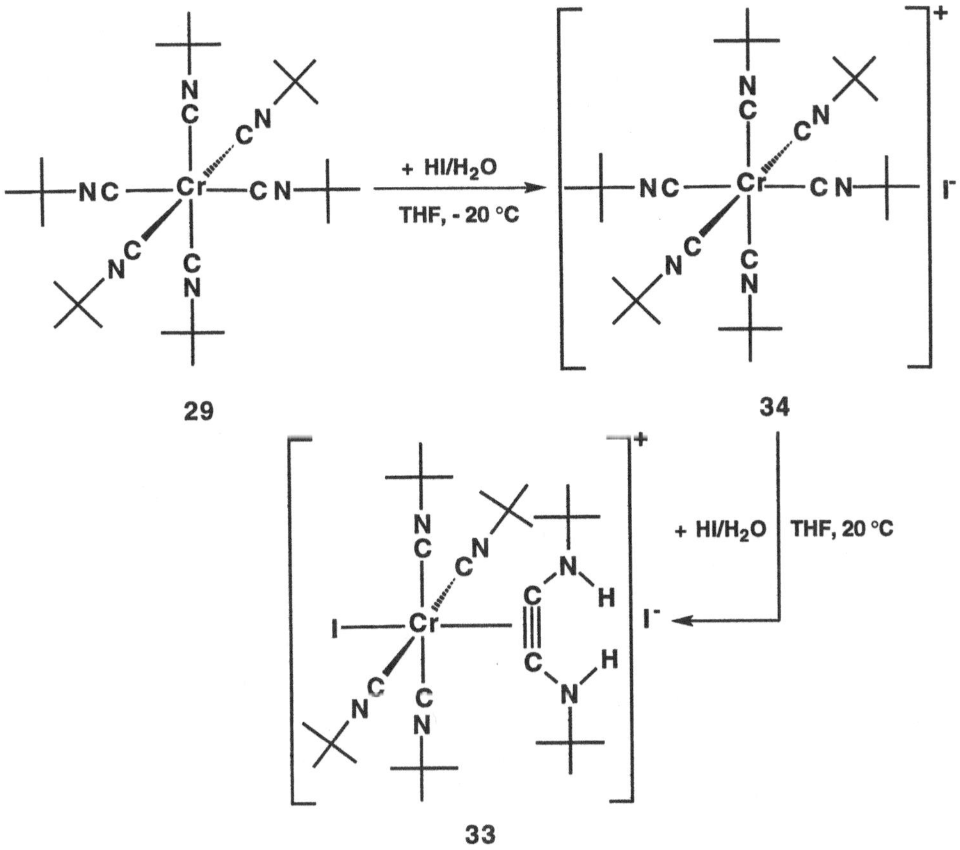

Scheme 13

Similarly, the reaction of **29** with one equivalent of HCl or HBr afforded the analogous Cr^I isocyanide complexes [Cr(CN^{t}Bu)$_6$][X] (X = Cl, Br) (**34a**).

In the reaction of **29** with two equivalents of HI and of **34** with one equivalent of HI, a by-product is formed in small amount, which precipitates out of the solution. This was identified to be the Cr^{II} isocyanide complex [Cr(CN^{t}Bu)$_6$][I]$_2$ (**35**) and can be independently prepared by the oxidation of **29** with one equivalent of I$_2$ in THF (Scheme 14). Complex **35** disproportionates in solution to give the Cr^I isocyanide complex **34** and the Cr^{III} isocyanide complex *mer*-Cr(CN^{t}Bu)$_3$I$_3$ (**36**), the latter being selectively formed in the oxidation reaction of **29** with 1.5 equivalents of I$_2$ (Scheme 14).

Treatment of **29** with two equivalents of HI, prepared in situ from [NBu$_4$]I and CF$_3$SO$_3$H gave a mixture of **33** and **34**. This suggests, that water is necessary for the complete conversion of **29** to **33** in equation 12, and is the source of the second amino-hydrogen atom in the coupling product **33** obtained from **34** and HI (Scheme 13). No isocyanide-isocyanide coupling reaction was observed, when **29** was treated with two equivalents of an aqueous solution of HCl or HBr, indicating that the counter anion is crucial for the observed isocyanide-isocyanide coupling reaction of **29** (equation 12). Studies are in progress to elucidate the mechanism of this chromium-mediated C-C bond-forming reaction.

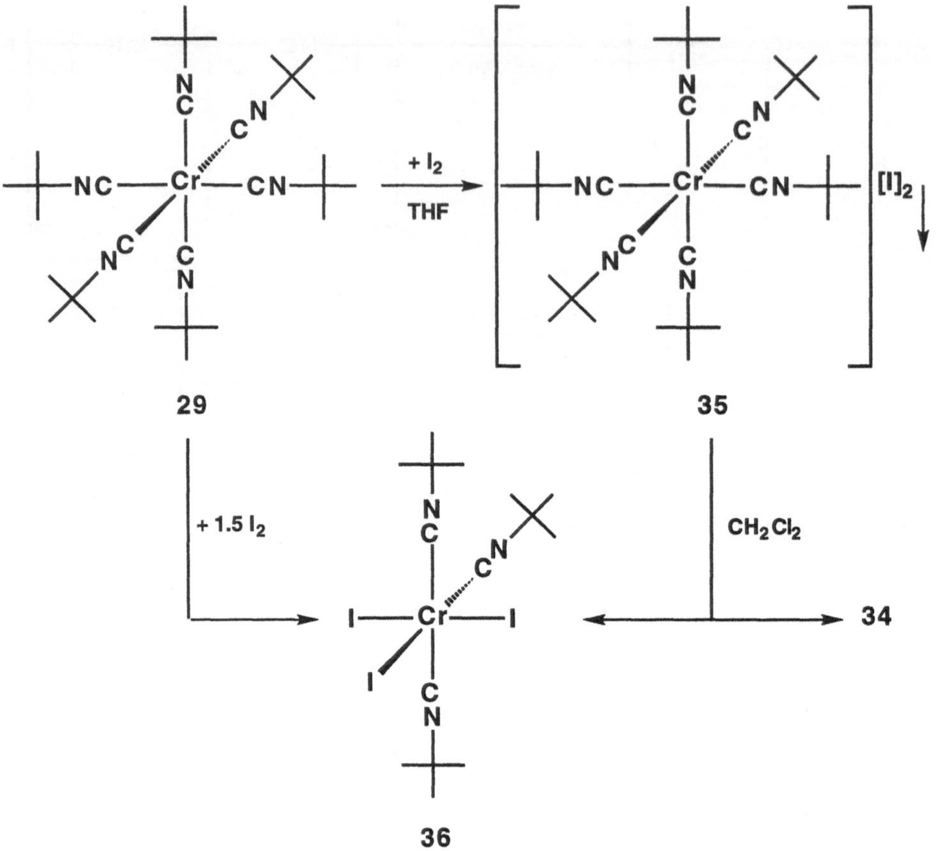

Scheme 14

116

Acknowledgements

I appreciate greatly the contributions of my coworkers, whose names appear in the references. I would like also to thank Professor Dr. W. A. Herrmann and Professor Dr. E. O. Fischer for the continuous interest of my work and their support. Financial support of this work by the VW-foundation, the Deutsche Forschungsgemeinschaft, the Fonds der Chemischen Industrie, the Technische Universität München and the Humboldt Universität zu Berlin is highly acknowledged.

References

1. a) Comprehensive Organometallic Chemistry, (G. Wilkinson, Ed.), Vol. 8, Pergamon, New York 1982; b) J. P. Collman, L. S. Hegedus, J. R. Norton, R. G. Finke, Principles and Applications of Organotransition Metal Chemistry, University Science Books, Mill Valley, USA, 1987.
2. a) P. T. Wolczanski, J. E. Bercaw, *Acc. Chem. Res.* **1980**, *13*, 121; b) P. J. Fagan, J. M. Manriquez, T. J. Marks, V. W. Day, S. H. Vollmer, C. S. Day, *J. Am. Chem. Soc.* **1980**, *102*, 5393; c) W. J. Evans, A. L. Wayda, W. E. Hunter, J. L. Atwood, *J. Chem. Soc. Chem. Comm.* **1981**, 706; d) S. Gambarotta, C. Floriani, A. Chiesi-Villa, C. Guastini, *J. Am. Chem. Soc.* **1983**, *105*, 1690; e) G. Erker, *Acc. Chem. Res.* **1984**, *17*, 103; f) W. J. Evans, J. W. Grate, L. A. Hughes, H. Zhang, J. L. Atwood, *J. Am. Chem. Soc.* **1985**, *107*, 3728; g) K. G. Moloy, P. J. Fagan, J. M. Manriquez, T. J. Marks, *J. Am. Chem. Soc.* **1986**, *108*, 56; h) K. Tatsumi, A. Nakamura, P. Hofmann, R. Hoffmann, K. G. Moloy, T. J. Marks, *J. Am. Chem. Soc.* **1986**, *108*, 4467; i) L. D. Durfee, I. P. Rothwell, *Chem. Rev.* **1988**, *88*, 1059; j) D. Roddick, J. E. Bercaw, *Chem. Ber.* **1989**, *122*, 1579.
3. a) H. Fischer, P. Hofmann, F. R. Kreißl, R. R. Schrock, U. Schubert, K. Weiss, Carbyne Complexes, VCH, Weinheim, 1988; b) A. Mayr, *Comments Inorg. Chem.* **1990**, *10*, 227; c) R. N. Vrtis, S. J. Lippard, *Isr. J. Chem.* **1990**, *30*, 331; d) A. Mayr, C. M. Bastos, *Prog. Inorg. Chem.* **1992**, *40*, 1; e) E. M. Carnahan, J. D. Protasiewicz, S. J. Lippard, *Acc. Chem. Res.* **1993**, *26*, 90.
4. W. A. Nugent, J. M. Mayer, Metal-Ligand Multiple Bonds, Wiley, New York, 1988.
5. a) C. T. Lam, P. W. R. Corfield, S. J. Lippard, *J. Am. Chem. Soc.* **1977**, *99*, 617; b) C. M. Giandomenico, C. T. Lam, S. J. Lippard, *ibid.* **1982**, *104*, 1263; c) C. Caravana, C. M. Giandomenico, S. J. Lippard, *Inorg. Chem.* **1982**, *21*, 1860; d) R. Hoffmann, C. N. Wilker, S. J. Lippard, J. L. Templeton, D. C. Brower, *J. Am. Chem. Soc.* **1983**, *105*, 146.
6. a) P. A. Bianconi, I. D. Williams, M. P. Engeler, S. J. Lippard, *J. Am. Chem. Soc.* **1986**, *108*, 311; b) P. A. Bianconi, R. N. Vrtis, C. P. Rao, I. D. Williams, M. P. Engeler, S. J. Lippard, *Organometallics* **1987**, *6*, 1968; c) R. N. Vrtis, C. P. Rao, S. Warner, S. J. Lippard, *J. Am. Chem. Soc.* **1988**, *110*, 2669.
7. S. Warner, S. J. Lippard, *Organometallics* **1989**, *8*, 228.

8. A. C. Filippou, W. Grünleitner, *J. Organomet. Chem.* **1990**, *398*, 99.

9. a) A. C. Filippou, W. Grünleitner, *J. Organomet. Chem.* **1990**, *393*, C10; b) A. C. Filippou, W. Grünleitner, *Z. Naturforsch.* **1991**, *46b*, 216.

10. a) E. M. Carnahan, S. J. Lippard, *J. Chem. Soc. Dalton Trans.* **1991**, 699.

11. a) J. Chatt, A. J. L. Pombeiro, R. L. Richards, *J. Chem. Soc. Dalton Trans.* **1980**, 492; b) J. Chatt, A. J. L. Pombeiro, R. L. Richards, *J. Organomet. Chem.* **1980**, *184*, 357; c) A. J. L. Pombeiro, R. L. Richards, Coord. *Chem. Rev.* **1990**, *104*, 13.

12. A. J. L. Pombeiro, M. F. N. N. Carvalho, P. B. Hitchcock, R. L. Richards, *J. Chem. Soc. Dalton Trans.* **1981**, 1629.

13. a) A. C. Filippou, E. O. Fischer, R. Paciello, *J. Organomet. Chem.* **1988**, *347*, 127; b) A. C. Filippou, E. O. Fischer, *ibid.* **1988**, *352*, 141; c) A. C. Filippou, E. O. Fischer, *ibid.* **1989**, *365*, 317; d) A. C. Filippou, C. Mehnert, K. M. A. Wanninger, M. Kleine, *ibid.* **1995**, *491*, 47; e) B. Lungwitz, A. C. Filippou, *ibid.* **1995**, *498*, 91.

14. a) A. C. Filippou, W. Grünleitner, *Z. Naturforsch.* **1989**, *44b*, 1023; b) A. C. Filippou, *Polyhedron* **1990**, *9*, 727.

15. a) A. C. Filippou, C. Völkl, W. Grünleitner, P. Kiprof, *Angew. Chem.* **1990**, *102*, 224; *Angew. Chem. Int. Ed. Engl.* **1990**, *29*, 207; b) A. C. Filippou, C. Völkl, W. Grünleitner, P. Kiprof, *Z. Naturforsch.* **1990**, *45b*, 351.

16. a) F. R. Kreißl, in Organometallics in Organic Synthesis, A. deMeijere, H. tom Dieck (Eds.), Springer Verlag Berlin, 1987, 105; b) A. Mayr, S. M. Holmes, C. M. Bastos, *Organometallics* **1992**, *11*, 4358; c) A. Mayr, H. Hoffmeister, *Adv. Organomet. Chem.* **1992**, *32*, 227.

17. a) E. O. Fischer, A. C. Filippou, H. G. Alt, *J. Organomet. Chem.* **1984**, *276*, 377; b) E. O. Fischer, A. C. Filippou, H. G. Alt, *ibid.* **1985**, *296*, 69.

18. A. C. Filippou, C. Völkl, W. Grünleitner, P. Kiprof, *J. Organomet. Chem.* **1992**, *434*, 201.

19. a) J. J. R. F. da Silva, M. A. Pellinghelli, A. J. L. Pombeiro, R. L. Richards, A. Tiripicchio, Y. Wang, *J. Organomet. Chem.* **1993**, *454*, C8; b) R. A. Henderson, A. J. L. Pombeiro, R. L. Richards, J. J. R. F. da Silva, Y. Wang, *J. Chem. Soc., Dalton Trans.* **1995**, 1193.

20. A. Mayr, C. M. Bastos, *J. Am. Chem. Soc.* **1990**, *112*, 7797.

21. a) A. C. Filippou, W. Grünleitner, *Z. Naturforsch.* **1989**, *44b*, 572; b) A. C. Filippou, W. Grünleitner, *J. Organomet. Chem.* **1991**, *407*, 61; c) A. C. Filippou, W. Grünleitner, E. O. Fischer, W. Imhof, G. Huttner, *ibid.* **1991**, *413*, 165.

22. A. C. Filippou, E. O. Fischer, *J. Organomet. Chem.* **1988**, *349*, 367.

23. A. C. Filippou, W. Grünleitner, C. Völkl, P. Kiprof, *Angew. Chem.* **1991**, *103*, 1188; *Angew. Chem. Int. Ed. Engl.* **1991**, *30*, 1167.

24. A. C. Filippou, P. Hofmann, P. Kiprof, H. R. Schmidt, C. Wagner, *J. Organomet. Chem.* **1993**, *459*, 233.

25. a) E. M. Carnahan, S. J. Lippard, *J. Am. Chem. Soc.* **1990**, *112*, 3230; b) E. Carnahan, S. J. Lippard, *ibid.* **1992**, *114*, 4166; c) J. D. Protasiewicz, B. S. Bronk, A. Masschelein, S. J. Lippard, *Organometallics* **1994**, *13*, 1300.

26. a) A. C. Filippou, W. Grünleitner, P. Kiprof, *J. Organomet. Chem.* **1991**, *410*, 175;
b) A. C. Filippou, W. Grünleitner, C. Völkl, P. Kiprof, *ibid.* **1991**, *413*, 181; c) A.
M. Martins, M. J. Calhorda, C. C. Romao, C. Völkl, P. Kiprof, A. C. Filippou, *ibid.*
1992, *423*, 367; d) A. C. Filippou, W. Grünleitner, E. O. Fischer, *ibid.* **1992**, *428*,
C37; e) A. C. Filippou, C. Wagner, E. O. Fischer, C. Völkl, *ibid.* **1992**, *438*, C15.
27. J. A. Acho, S. J. Lippard, *Organometallics* **1994**, *13*, 1294.

The Synthesis of Indoles by Transition Metal Catalyzed Addition of Alkynes to 1,2-Diazenes

Horst Kisch* and Uwe Dürr

Institut für Anorganische Chemie der Universität Erlangen-Nürnberg, Egerlandstr.1, D-91058 Erlangen.

Although the cycloaddition of a 1,2-diazene derivative, namely bis(ethoxycarbonyl)-1,2-diazene, to cyclopentadiene represents one of the very first Diels-Alder reactions[1], this and other C-N coupling transformations with unsaturated substrates occur only when the N=N bond is activated by electron withdrawing substituents. However, activation may be also achieved through coordination to a transition metal, as amply documented for the isoelectronic olefins.

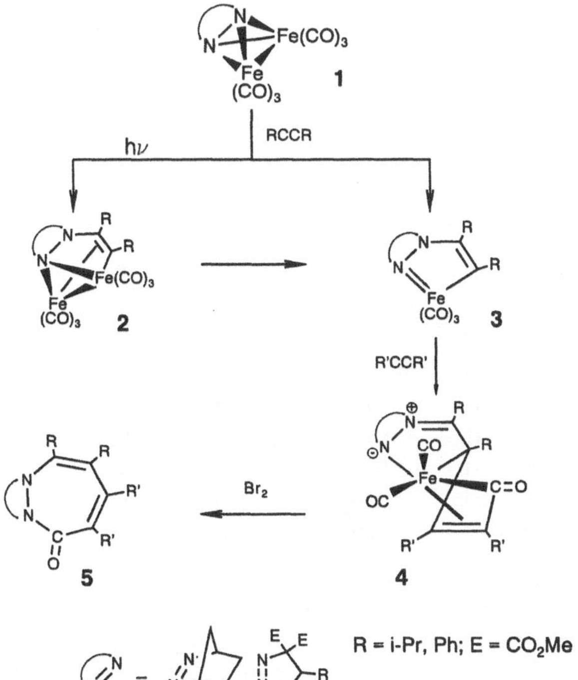

Scheme 1

121

In previous work we isolated a great manifold of different metal carbonyl complexes of cyclic 1,2-diazenes and tested whether they may undergo C-N coupling with alkynes or olefins[2]. Such types of transformations are related to the direct conversion of dinitrogen to organic nitrogen compounds, since 1,2-diazenes may be obtained from dinitrogen complexes[3].

The rather long N-N distance of 149 pm in the binuclear complex 1[4] suggested that complexation by the hexacarbonyldiiron fragment formally resulted in a reductive activation of the 1,2-diazene. Irradiation in the presence of an alkyne indeed induced C-N coupling as indicated by the structure of 2. This compound is rather labile in solution and decomposes to the diazaferrole 3 which in turn can react with a second alkyne, either thermally or photochemically, to afford the double-addition product 4[5, 6]. The latter can be prepared also in one step by heating of 1 in the presence of excess alkyne. Oxidative degradation by bromine in glacial acetic acids produced the hitherto unknown 1,2,3-diazepinones 5[7]. Thus, complexation of the cyclic 1,2-diazene by iron carbonyl assists a series of C-N and C-C coupling reactions to finally produce a novel organic heterocycle. The same type of reaction could not be performed with acyclic 1,2-diaryldiazenes like azobenzene, since in this case N=N cleavage produces unreactive orthosemidine and orthometalated complexes [8].

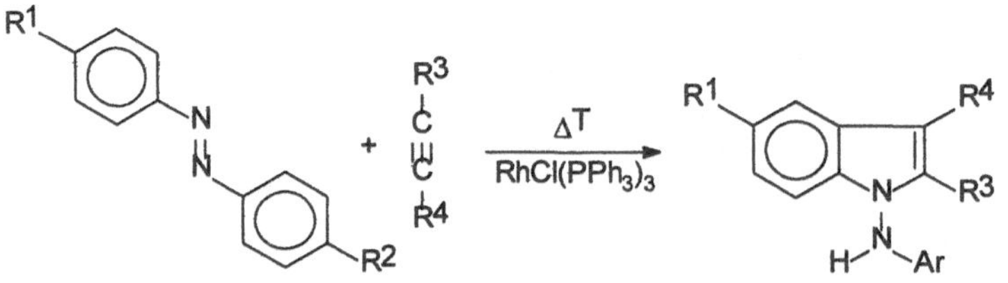

6	R1	R2	R3	R4	yield [%]
a	H	H	C_6H_5	C_6H_5	90
b	H	H	p-CH$_3$-C$_6$H$_4$	p-CH$_3$-C$_6$H$_4$	86
c	H	H	p-CH$_3$O-C$_6$H$_4$	p-CH$_3$O-C$_6$H$_4$	78
d	H	H	C_3H_7	C_3H_7	52
e	H	H	adamantyl	C_6H_5	29
f	H	H	p-CH$_3$-C$_6$H$_4$	C_6H_5	42
g	H	H	C_6H_5	p-CH$_3$-C$_6$H$_4$	31
h	CH$_3$	H	C_6H_5	C_6H_5	52
i	H	CH$_3$	C_6H_5	C_6H_5	26
k	Cl	Cl	C_6H_5	C_6H_5	21

Scheme 2

Attempts to use carbonyl-free metal complexes like CpCo(COD) (COD = trans-1,5-cyclooctadiene) which are known to catalyze pyridine formation from alkynes and nitriles through a related sequence of C-C and C-N coupling steps[9, 10], were not successful. But

CoCl(PPh$_3$)$_3$ and CpRu(Cl)L$_2$ at about 200 °C assisted formation 2,3-diphenylindole from azobenzene and tolan[11]. With the former complex, traces of a sensitive, red compound were obtained[12]. The unknown mechanism of this novel transformation should contain as a key step the insertion of the alkyne into the M-C bond. An analogous reaction was reported for an orthopalladated azobenzene complex affording the corresponding cinnolinium salt and metallic palladium as products[13].

Although it was known that RhCl(PPh$_3$)$_3$ does not orthometalate azobenzene[14, 15], we tested whether this may occur in the presence of an alkyne and indeed observed a catalytic formation of the hitherto unknown 1-(arylamino)indole derivatives **6** (Scheme 2)[16].

Thus, this novel addition reaction includes C-C and C-N coupling steps. Although some transition metal catalyzed indole syntheses are known from the literature[17, 18] none of them affords a 1-arylsubstituted derivative. Best results were obtained in refluxing 1-BuOH or 1-PrOH in the presence of 0.06 vol% of HOAc while at higher acid concentration the rate was slowed down considerably. Isolated yields were in the range of 85 - 55 % for tolan derivatives substituted symmetrically by electron-donating substituents in the para-position and also for 4-CH$_3$- and 4-CH$_3$O-monosubstituted derivatives; maximum turnover numbers amounted to about 100. Yields decreased to 30 - 6 % when electron-withdrawing groups were present or when one of the two alkyne substituents was aliphatic. Changing the electronic properties of the azo compounds had only minor consequences, however, no indole was formed when the meta-positions were substituted like in 3,3',5,5'-tetramethylazobenzene. As a secondary reaction a reductive N-N cleavage of **6** afforded the corresponding N-unsubstituted indoles in low yields of 1 - 7%.

When the reaction was made heterogeneous by introducing silica and water-soluble phosphines under supported-aqueous-phase conditions, the catalyst could be filtered off and used again[19, 20]. Kinetic investigations of the homogeneous reaction in BuOH/HOAc revealed a first order dependence of rate on complex and azobenzene but a broken negative order on tolan concentration[16]. The Arrhenius activation energy was determined as 75 ± 6 kJ/mol. UV-Vis reaction spectroscopy revealed formation of a red intermediate which could be isolated as an impure material in small amounts by HPLC. It was converted to the indole **6a** upon heating in BuOH/HOAc in the *absence* of RhCl(PPh$_3$)$_3$.

In attempts to obtain this intermediate on a preparative scale, we went back to the stoichiometric CoCl(PPh$_3$)$_3$ assisted reaction of azobenzene and tolan but used now the more labile CoH(N$_2$)(PPh$_3$)$_3$. When the latter was added at 85 °C to a melt of various 1,2-diazenes and tolan, no 1:1 adduct but the double addition products **7** or their isomeric 2,3-dihydrocinnolines **8** (Scheme 3) were catalytically formed. Usually either **7** or **8** were produced except in the case of the difluoro derivative where both isomers (**7e**, **8e**) were obtained. Isolated yields reached 70 - 80% but maximum turnover numbers were not higher than 15. The structure of these new compounds was resolved by X-ray analysis[21, 22]. In no case evidence could be obtained for products in which the second aryl ring was vinylated also. Thus, the distilbenyl azobenzene derivatives **7** result from a regioselective insertion of two alkyne molecules into the two ortho-positions of one azobenzene aryl ring. Note, that only trans-stilbenyl groups are present. To our knowledge, no similar type of aromatic C-H activation has been reported in the literature.

123

When the meta-positions were unsubstituted like in diazenes **a** and **b**, the ortho-quinoide dihydrocinnolines **8** were obtained. These are formed through a thermal electrocyclic ring closure from **7** as demonstrated for the difluoro derivative **e** while the reverse ring-opening takes place only photochemically. Since the corresponding methyl and chloro derivatives **c** and **d** afforded only the distilbenyl isomers, it is likely that these larger substituents (R^3) prevent the formation of a planar transition state required for ring-closure to **8**. Thus, different to the previous examples, C-N bond formation occurs without involvement of the transition metal which is however responsible for the preceding activation of the ortho-CH bonds.

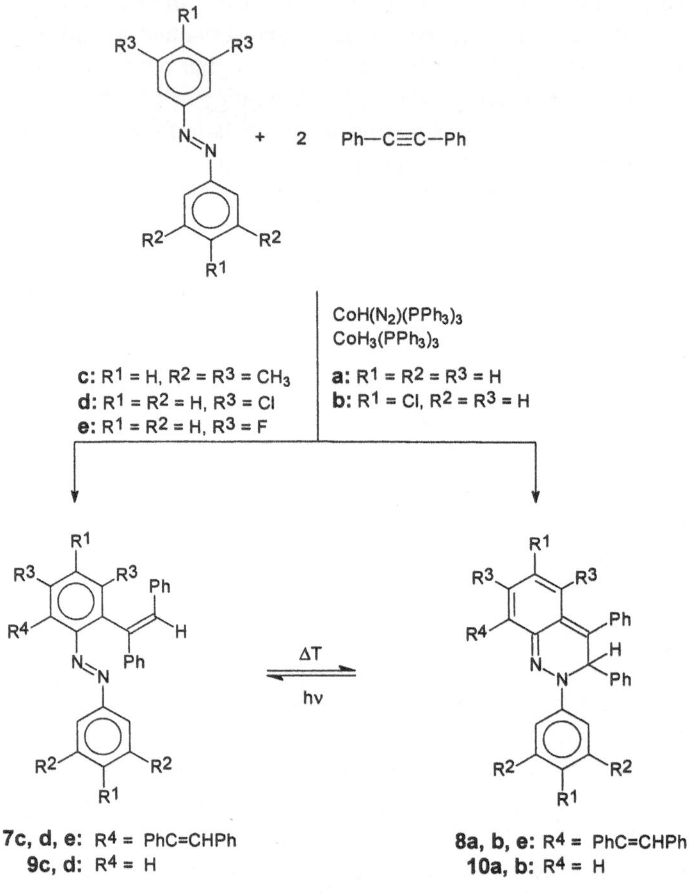

R1
R3
R3
N
N
+ 2 Ph—C≡C—Ph
R2
R2
R1
CoH(N2)(PPh3)3
CoH3(PPh3)3
c: R1 = H, R2 = R3 = CH3
d: R1 = R2 = H, R3 = Cl
e: R1 = R2 = H, R3 = F
a: R1 = R2 = R3 = H
b: R1 = Cl, R2 = R3 = H
Ph
H
Ph
R4
ΔT
hν
R4
Ph
H
Ph
7c, d, e: R4 = PhC=CHPh
9c, d: R4 = H
8a, b, e: R4 = PhC=CHPh
10a, b: R4 = H

Scheme 3

When the cobalt catalyzed reaction of azobenzene and tolan was performed at room temperature in THF or ether solution, in addition to the double addition product **8a** the mono insertion product **10a** could be isolated[23]. The latter undergoes a fast rearrangement to the indole **6a** when refluxed in BuOH/HOAc, which are the experimental conditions of the rhodium catalyzed indole synthesis. UV-Vis reaction spectroscopy and HPLC analysis revealed that **10a** is the red intermediate observed previously in the reaction solution (*vide supra*). While the activation energies of 56 ± 9 and 54 ± 3 kJ/mol are identical within experimental error, the activation entropy of -136 ± 7 J/Kmol, measured for the HOAc catalyzed reaction, is more positive as compared to -191 ± 14 J/Kmol obtained for the uncatalyzed rearrangement[23].

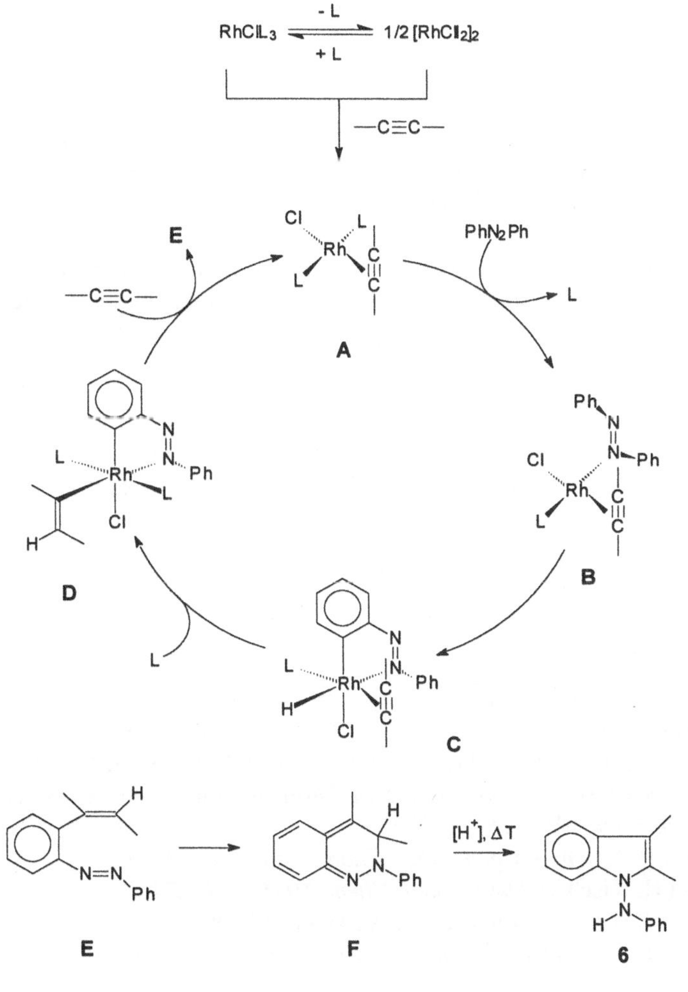

Scheme 4

From the results presented above, a mechanism for the rhodium catalyzed indole synthesis is schematically summarized in Scheme 4. In the first step the well known alkyne complex **A** is formed, followed by substitution of one phosphine ligand by azobenzene. *Ab initio* model calculations suggest that σ-coordination is favored over π-coordination[24]. From the rate law and the fact that cyclometalations of azobenzene usually require rather high temperatures, it is likely that the orthometalation step **B** to **C** is rate-determining. Successive cis-insertion of the alkyne leads to a cis-stilbenyl complex, which is assumed to isomerize to the trans-stilbenyl compound **D** before undergoing reductive elimination to the 2-trans-stilbenylazobenzene **E**. In most cases described in the literature, the formation of trans-insertion products is formulated as initial cis-insertion followed by isomerization[25]. In the case of the tolan-azobenzene system the intermediate appearance of this compound could be demonstrated by UV reaction spectroscopy and HPLC experiments.

As described above the same product (**10a**) was obtained in the cobalt catalyzed reaction and therefore it could be shown that the further conversion to the indole *via* ring closure to **F** occurs only when no sterically demanding substituents are present in meta-positions.

The final step, an acid catalyzed rearrangement to the 1-(arylamino)indole **6**, proceeds without involvement of the rhodium complex. Thus, as in the cobalt catalysis, C-N bond formation does not require the activation by a transition metal, while that is necessary of course for the C-H activation step. The main difference to the cobalt system is, that due to the absence of a hydride ligand in the intermediate **C**, insertion of a second alkyne is disfavored.

Acknowledgement

This work was supported by *Volkswagenstiftung*, *Degussa AG*, and *Fonds der Chemischen Industrie*.

References

1. O. Diels, J. H. Blom and W. Koll, *Liebigs Ann.Chem.* **1925**, *443*, 242.
2. H. Kisch and P. Holzmeier, *Adv. Organomet. Chem.* **1992**, *34*, 67.
3. D. Sellmann, *Angew.Chem.* **1974**, *86*, 692; *Angew. Chem., Int. Ed. Engl.* **1974**, *13*, 639; D. Sellmann, W. Soglowek, F. Knoch and M. Moll, *Angew. Chem.* **1989**, *101*, 1244; *Angew. Chem., Int. Ed. Engl.* **1989**, *28*, 1271.
4. C. Krüger and H. Kisch, *J. Chem. Soc., Chem. Comm.* **1975**, 65.
5. A. Albini and H. Kisch, *J. Organomet. Chem.* **1975**, *101*, 231.
6. A. Albini and H. Kisch, *J. Am. Chem. Soc.* **1976**, *98*, 3869.
7. R. Millini, H. Kisch and C. Krüger, *Z. Naturforsch, B: Anorg. Chem. Org. Chem.* **1985**, *40b*, 187.
8. P. Reißer, F. Knoch and H. Kisch, *Chem. Ber.* **1991**, *124*, 1143.
9. H. Bönnemann, *Angew. Chem.* 1985, 97 264; *Angew. Chem., Int. Ed. Engl.* **1985**, *24*, 248.
10. H Yamazaki and Y. Wakatsuki, *Tetrahedron Lett.* **1973**, 3383.
11. D. Garn and H. Kisch, *J. Organomet. Chem.* **1991**, *409*, 347.
12. H. Gstach and H. Kisch, *Z. Naturforsch, B: Anorg. Chem. Org. Chem.* **1983**, *38b*, 251.

13. G. Wu, A. L. Rheingold and R. F. Heck, *Organometallics* **1987**, *6*, 2386.
14. M. I. Bruce, M. Z. Iqbal and F. G. A. Stone, *J. Organomet. Chem.* **1972**, *40*, 393.
15. J. F. van Baar, K. Vrieze and D. J. Stufkens, *J. Organomet. Chem.* **1975**,*85*, 249.
16. U. R. Aulwurm, J. U. Melchinger and H. Kisch, *Organometallics* **1995**, *14*, 3385.
17. B. Robinson, *The Fischer Indole Synthesis*: J. Wiley & Sons: Chichester, **1982**.
18. L. S. Hegedus, *Angew. Chem.* 1988, *100*, 1147; *Angew. Chem., Int. Ed. Engl.* **1988**, *27*, 1113.
19. S. Westernacher and H. Kisch, *Monatsh. Chemie* **1996**, *127*, 469.
20. S. Westernacher, J. Blümel and H. Kisch, *J. Mol. Catal. A: Chemical* **1996**, *109*, 169.
21. G. Halbritter, F. Knoch, A. Wolski and H. Kisch, *Angew. Chem.* **1994**, *106*, 1676; *Angew. Chem., Int. Ed. Engl.* **1994**, *33*, 1603.
22. G. Halbritter, F, Knoch and H. Kisch, *J. Organomet. Chem.* **1995**, *492*, 87.
23. U. Dürr, F. Heinemann and H. Kisch, to be published.
24. P. Reißer, Y. Wakatsuki and H. Kisch, *Monatsh. Chemie* **1995**, *126*,1.
25. See e.g. alkyne insertions into Ni-C bonds: J. M. Huggins and R. G. Bergman, *J. Am. Chem. Soc.* **1981**, *103*, 3002.

Polycyclic Ring Systems *via* Palladacycles as Reactive Intermediates

Gerald Dyker

FB 6, Organische/Metallorganische Chemie der Gerhard-Mercator-Universität-GH Duisburg, Lotharstr. 1, D-47048 Duisburg, FRG

Summary

Five-membered palladacycles, formed by cyclometallation, are of special interest as key intermediates of palladium-catalyzed processes: in general, the reductive elimination of Pd^0 to give a four-membered ring is unfavourable; therefore the addition of a second equivalent of starting material successfully competes, resulting in domino processes that lead from simple starting materials to complex ring systems. Annelated fulvenes, dibenzopyranes, benzo-furans, and even propellanes are efficiently accessible.

1 Introduction

Palladium-catalyzed reactions can open up short and efficient pathways from simple starting materials to complex target molecules. Especially the Heck reaction is well known as a powerful tool for C-C bond formation[1]. The intramolecular Heck reaction of suitable functionalized substrates leads to carbo- and heterocyclic rings of various ring size[2]. For substrates of type **1** (with X = halogen atom or triflate) the mechanism is briefly outlined in Scheme 1 (Additional ligands are ommited for clarity). The sequence consists of an oxidative addition of a generally sp^2-centered C-X bond to the Pd^0-catalyst to give a Pd^{II}-intermediate **2** that undergoes a carbopalladation of a C-C double bond; if the β-hydrogen elimination is possible in a syn-manner, then the product **4** with the exocyclic double bond is finally produced. As the result of this sequence C-C bond formation has taken place by elimination of HX.

Ring closure via intramolecular Heck reaction:

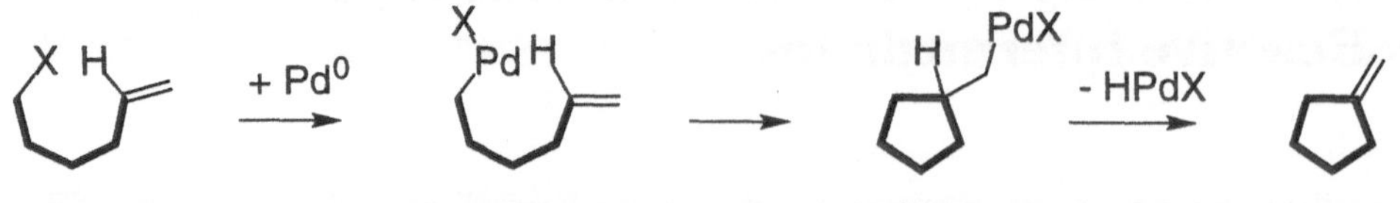

1 +Pd⁰→ **2** → **3** -HPdX→ **4**

Ring closure via palladacycles:

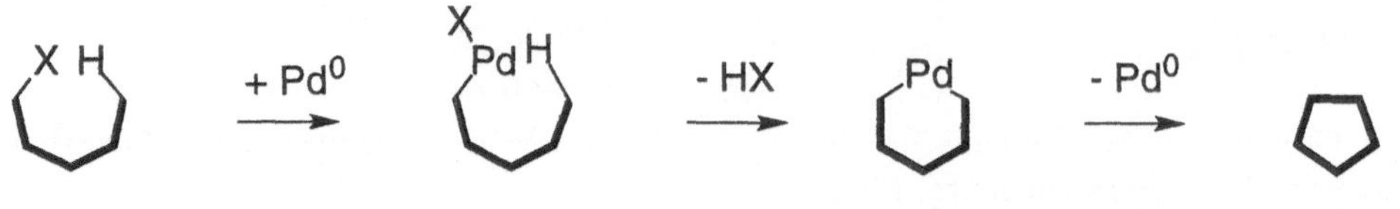

5 +Pd⁰→ **6** -HX→ **7** -Pd⁰→ **8**

Scheme 1

Most interestingly, a similar result can be achieved by a different mechanistic pathway: for substrates of type **5**, without a suitable double bond for the carbopalladation, the ring closure to give **8** proceeds via a cyclopalladation/reductive elimination sequence with the palladacycle **7** as reactive intermediate. The formation of the annelated fluoranthene **10** (Scheme 2), reported by Rice et al.[3], presumably proceeds according to this mechanism.

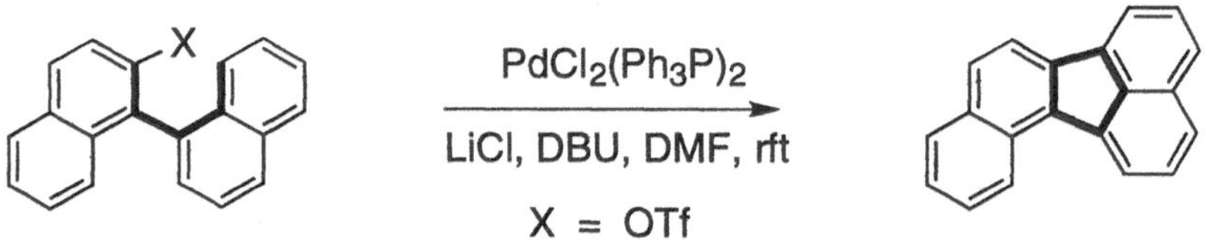

$$\underset{\textbf{9}}{} \xrightarrow[\substack{\text{LiCl, DBU, DMF, rft} \\ X = \text{OTf}}]{\text{PdCl}_2(\text{Ph}_3\text{P})_2} \underset{\textbf{10} \ (93\,\%)}{}$$

Scheme 2

The cyclopalladation is probably achieved by an intramolecular electrophilic attack of a Pd^{II}-species of type **6**. Obviously, 6-membered palladacycles **7** rapidly transform to 5-membered rings **8** by reductive elimination of Pd⁰, even if considerable ring strain is introduced like in the case of the fluoranthene **10**.

In contrast, 5-membered palladacycles **13** behave differently (Scheme 3): the formation of 4-membered rings **14** via reductive elimination is unfavourable and restricted to special cases. Instead, these 5-membered palladacycles **13** can react with a second equivalent starting material **11** or with added reagents, resulting in domino processes[4].

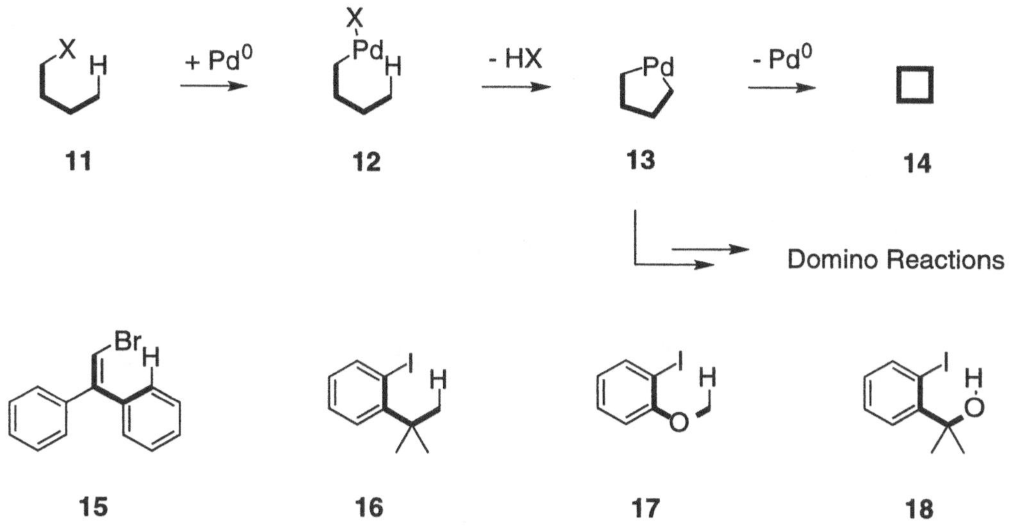

Scheme 3

The bold lines of the structures **11-13** emphasize the generality of this scheme. In principle, the centres connected by bold lines can be sp^2- hybridized as well as sp^3. Both, carbon atoms and heteroatoms are allowed. We have studied Pd-catalyzed transformations of the model compounds **15-18**, which fulfill the structural requirements of this general mechanistic scheme.

2 Discussion

The phenyl-substituted olefinic bromide **15** as well as the corresponding iodide is efficiently transformed by a Pd-catalyzed domino coupling process to the annelated fulvene **21**. One mol-% palladium acetate is sufficient. The active Pd^0-catalyst is formed in situ under the reaction conditions (K_2CO_3, n-Bu$_4$NBr, DMF, N_2, 3 d, 100 °C)[5]. Scheme 4 briefly summarizes the mechanistic interpretation. With the halide and a hydrogen atom in a suitable distance the cyclometallation step leads to the 5-membered palladacycle **19**. The reductive elimination of Pd^0 is inhibited in this case, because the result would be a highly strained and antiaromatic benzocyclobutadiene. Instead, the palladacycle **19** adds another equivalent starting material **15**, presumably to give the Pd^{II}-intermediate **20** (or its regioisomer with the palladium atom at the vinylic position). The final ring closure proceeds in the sense of an intramolecular Heck reaction.

We have tested scope and limitations of this new synthesis for annelated fulvenes by introducing a bridge between the phenyl substituents: substrates of type **22** with a central six-membered ring and also with a central seven-membered ring give polycyclic hydrocarbons like **23** and **24** in excellent yields. With a central 5-membered ring the cyclometallation at substrate **22** is not feasible due to geometric constraints. In this case a Pd-catalyzed Ullmann coupling reaction takes place instead.

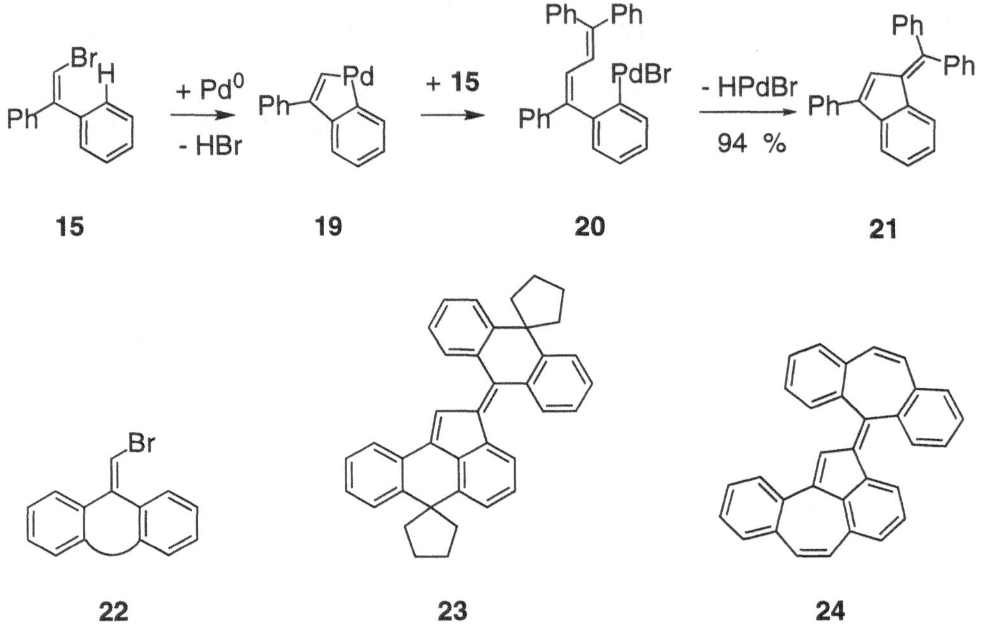

Scheme 4

The coupling reaction of the iodoanisole **17** is an especially interesting case, because the C-H group placed in a suitable distance from the iodide is sp³-hybridized and therefore regarded to be unactivated. Nevertheless, a palladium-catalyzed domino process efficiently leads to the substituted dibenzopyrane **26**, derived from three equivalents starting material **17** (Scheme 5)[6]. Obviously, a CH-activation has taken place at an sp³-hybridized center, because one methoxy group has been arylated. Again palladacycles of type **25** are assumed as key intermediates. The C-H activation is not restricted to methoxy groups. In comparison, the reaction of the substrate **27** with the *ortho*-benzyloxy group took the same regiochemical course. Does the heteroatom have an influence on the C-H activation? The reaction of the *tert*-butyl substituted substrate **16** shows that it does not! Even at a *tert*-butyl group the palladium-catalyzed C-H activation can occur. The 9,10-dihydrophenanthrene derivative **29**, which corresponds to **26** and **28**, is a minor byproduct. Surprisingly the benzocyclobutene **30** is isolated as the main product[7]. The formation of the four-membered carbocycle by reductive elimination at an intermediary five-membered palladacycle might be favoured in this case either by the steric interaction of the aryl substituent with the dimethylmethylene group or by the slightly increased temperature, which is required for the C-H activation at the *tert*-butyl group (110 ^{0}C instead of 100 ^{0}C).

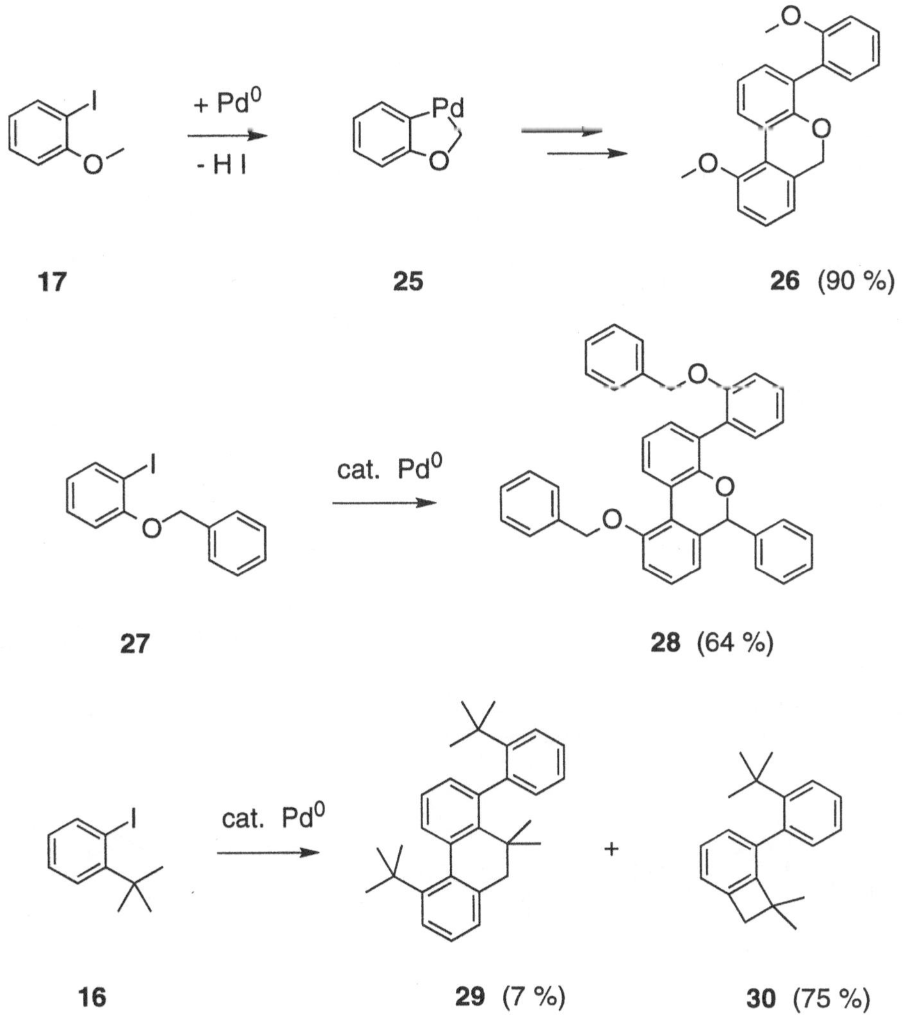
+ Pd0
- H I
17
25
26 (90 %)
cat. Pd0
27
28 (64 %)
cat. Pd0
16
29 (7 %)
30 (75 %)

Scheme 5

The preparative value of this type of C-H activation is further enhanced by cross-coupling reactions with vinylic bromides like **31** as added reagents. The *tert*-butyl substituted substrate **16** leads to the isomeric indene derivatives **32** and **33**, whereas starting from the methoxy substituted substrate **34**, the corresponding benzofurans **35** and **36** are obtained (Scheme 6)[8].

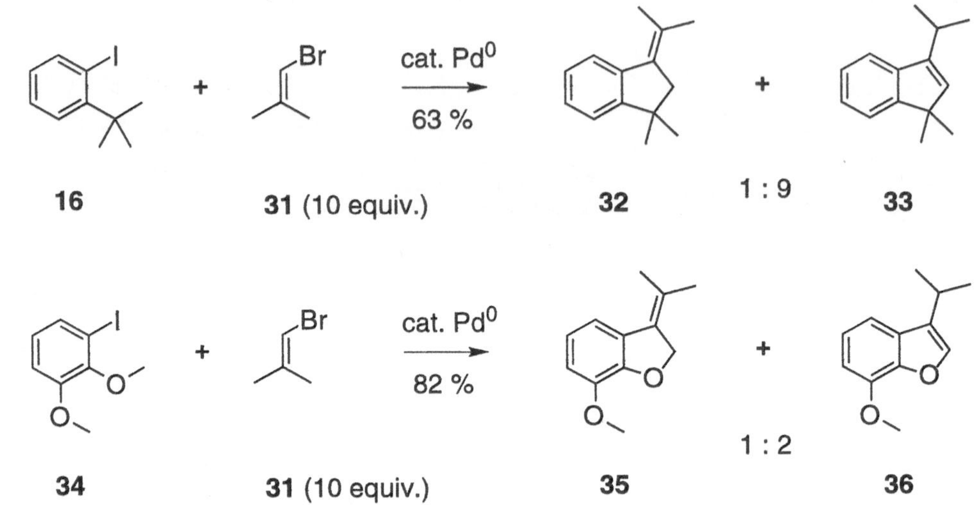
Br
cat. Pd0
63 %
1 : 9
16
31 (10 equiv.)
32
33
Br
cat. Pd0
82 %
1 : 2
34
31 (10 equiv.)
35
36

Scheme 6

The reaction of the hydroxy-functionalized substrate **18** is a special case, in that the anticipated palladacycle **37** contains a palladium-oxygen bond (Scheme 7). Although the details of the mechanism remain unclear, it is obvious, that the formation of the isolated product **38** requires the elimination of acetone involving a C-C bond cleavage, and a ring closure involving a C-O bond formation.

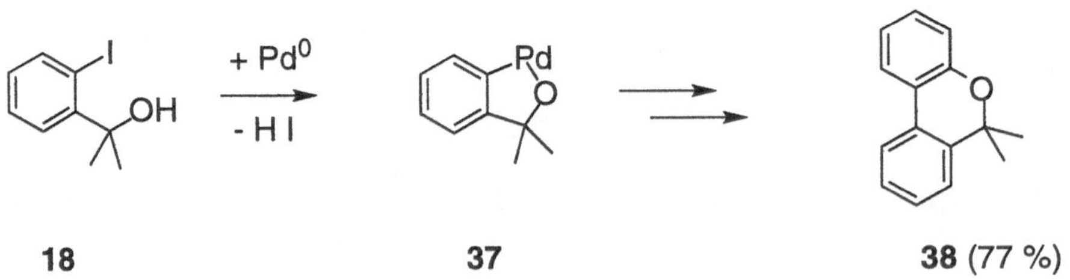
OH
+ Pd0
- HI
Pd
18
37
38 (77 %)

Scheme 7

In the examples discussed so far the substrates are well suited for a direct cyclopalladation. In the following examples a suitable structural unit has to be assembled first: a carbopalladation is preceding the cyclopalladation step. The well known palladium-catalyzed annelation reaction of norbornene (**39**) with iodobenzene (**40**) illustrates this (Scheme 8).

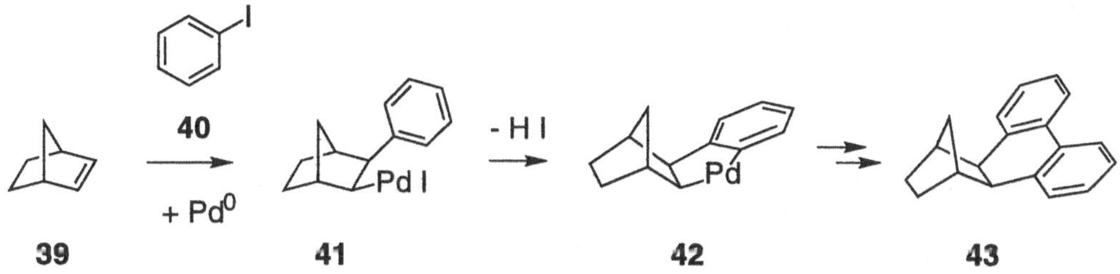

Scheme 8

The intermediary five-membered palladacycle **42** finally leads to the 1:2 product **43**[9] or to 1:3 products[10] depending on the reaction conditions. We found that a similar annelation reaction takes place with the hexacyclic hydrocarbon **44** as a strained tetrasubstituted olefinic coupling component (Scheme 9)[11]; propellanes with a hexaarylethane moiety are the result. Iodobenzene (**40**) leads to the [4.3.3]propellane **46** via the palladapropellane **45**. The reaction with iodoprehnitene (**47**) took a surprizing course; as the only isolated product we obtained the [3.3.2]propellane **49**. Presumably because of steric hindrance the addition of a second molecule aryl iodide **47** to the palladapropellane **48** is inhibited and therefore the reductive elimination to the four-membered ring can compete.

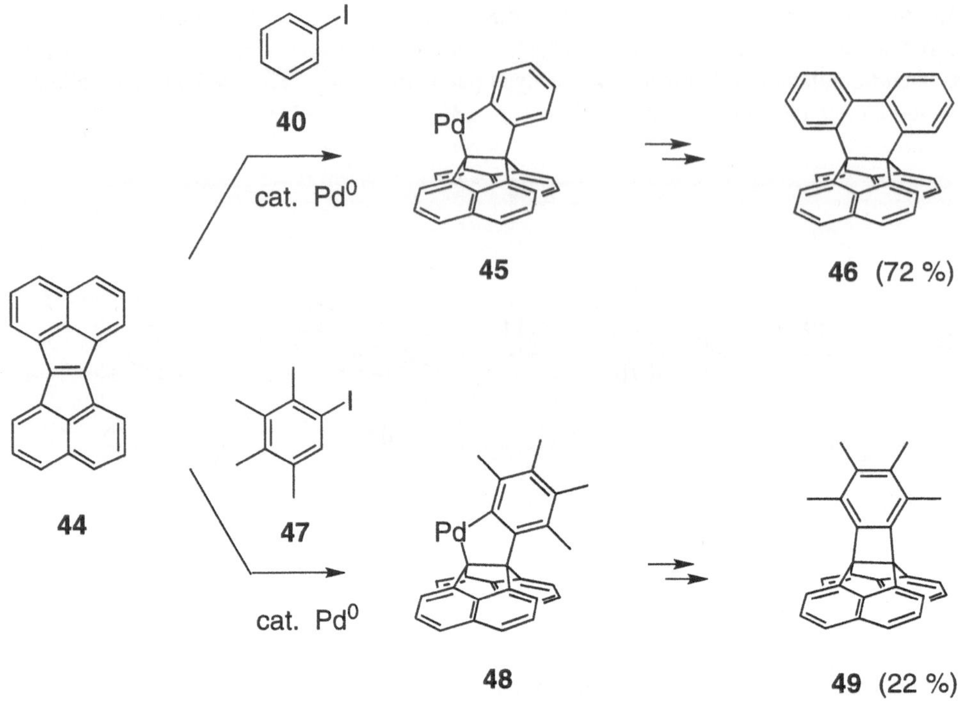

Scheme 9

3 Conclusion

The studies presented here shed some light on palladacycles as intermediates of palladium-catalyzed processes. The ring size of the palladacycles is of special importance: Whereas six-membered palladacycles rapidly transform to five-membered rings via reductive elimination of Pd⁰, the corresponding cyclobutane formation from five-membered palladacycles is limited to special cases. Instead, five-membered palladacycles as reactive intermediates offer opportunities to develop domino processes that lead from simple starting materials to complex ring systems.

Acknowledgement

These studies were supported by the Volkswagenstiftung and by the Fonds der Chemischen Industrie. I wish to express my sincere appreciation to my coworkers, whose names appear in the references, for their efforts in exploring this field of organometallic chemistry.

References

1. a) J. Tsuji, *Palladium Reagents and Catalysts: Innovations in Organic Synthesis*, Wiley, Chichester, **1995**. b) A. de Meijere, F. E. Meyer, *Angew. Chem.* **1994**, *106*, 2473-2506; *Angew. Chem. Int. Ed. Engl.* **1994**, *33*, 2379. c) R. F. Heck, *Org. React.* **1982**, *27*, 345-390.
2. a) N. E. Carpenter, D. J. Kucera, L. E. Overman, *J. Org. Chem.* **1989**, *54*, 5846-5848. b) K. Nagasawa, H. Ishihara, Y. Zako, I. Shimizu, *J. Org. Chem.* **1993**, *58*, 2523-2529. c) L. F. Tietze, R. Schimpf, *Synthesis* **1993**, 876-880.
3. J. E. Rice, Z. W. Cai, *J. Org. Chem.* **1993**, *58*, 1415-1424.
4. L. F. Tietze, U. Beifuss, *Angew. Chem.* **1993**, *105*, 137-170; *Angew. Chem. Int. Ed. Engl.* **1993**, *32*, 131-163.
5. G. Dyker, J. Körning, F. Nerenz, P. Siemsen, S. Sostmann, A. Wiegand, P. G. Jones, P. Bubenitschek, *Pure Appl. Chem.* **1996**, *68*, 323-326.
6. a) G. Dyker, *Angew. Chem.* **1992**, *104*, 1079-1081; *Angew. Chem. Int. Ed. Engl.* **1992**, *31*, 1023-1025. b) G. Dyker, *Chem. Ber.* **1994**, *127*, 739-742.
7. G. Dyker, *Angew. Chem.* **1994**, *106*, 117-119; *Angew. Chem. Int. Ed. Engl.* **1994**, *33*, 103-105.
8. G. Dyker, *J. Org. Chem.* **1993**, *58*, 6426-6428.
9. M. Catellani, G. P. Chiusoli, *J. Organomet. Chem.* **1985**, *286*, C13-C16.
10. a) O. Reiser, M. Weber, A. de Meijere, *Angew. Chem.* **1989**, *101*, 1071-1072; *Angew. Chem. Int. Ed. Engl.* **1989**, *28*, 1037. b) K. Albrecht, O. Reiser, M. Weber, A. de Meijere, *Tetrahedron* **1994**, *50*, 383-401.
11. G. Dyker, J. Körning, P. G. Jones, P. Bubenitschek, *Angew. Chem.* **1993**, *105*, 1805-1807; *Angew. Chem. Int. Ed. Engl.* **1993**, *32*, 1733-1735.

Substrate Directed Diastereoselective Hydroformylation of Methallylic Alcohols

Bernhard Breit

Fachbereich Chemie der Philipps-Universität Marburg, Hans-Meerwein Str.,
D 35043 Marburg

1 Introduction

One of the strategies often employed for the construction of new stereocenters in the course of a synthesis, is the use of substrate directed diastereoselective reactions. Although substrate directed oxidation, reduction and cyclopropanation methods are well-known and used routinely, there is a lack of knowledge of transition-metal catalyzed C/C-bond forming reactions of this type [1]. If one could develop such reactions to become diastereoselective this could be a major improvement in synthetic efficiency, because one could simul-taneously build the carbon core of a given target and set the right stereochemistry.

The acyclic fragment **A** (Scheme 1.1.), with two adjacent stereocenters, is a widespread structural motif found especially in polyketide natural products. The stereoselective synthesis of this unit has been a probe for the efficiency of current synthetic methods to control stereoselectivity on reactions of acyclic substrates. Successful approaches known to this fragment via C/C-bond forming reactions include addition of organometallic reagents to chiral, α-substituted aldehydes (path **1**) [2], homo-aldol type processes developed by *Hoppe* and *Thomas* (path **2**)[3], as well as the recently discovered zirconium-catalyzed carbomagnesiation reaction (path **3**) [4]. Although the latter method belongs to the few known cases of transition metal catalyzed, substrate directed diastereoselective C/C-bond forming reactions, it is restricted to the introduction of an ethyl substituent. All three types of methods suffer from the fact, that they need stoichiometric amounts of organometallic reagents.

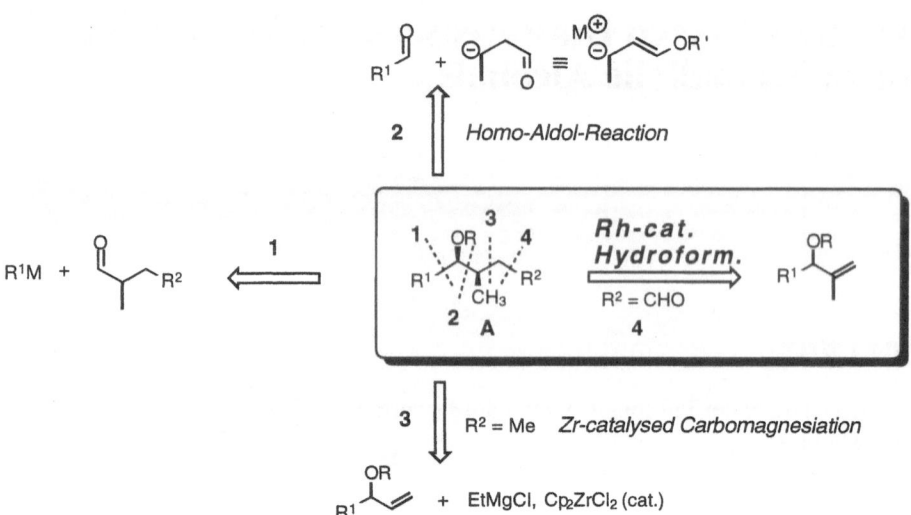

Scheme 1.1. Known (path 1-3) and hypothetical (path 4) C/C-bond forming reactions for the diastereoselective generation of substructure A

We wanted to explore, whether the hydroformylation reaction of methallylic alcohols can be used to construct fragment **A** (path **4**). Although hydroformylation is known since almost 60 years as an industrially important transition metal catalyzed C/C-bond-forming reaction, no efficient variants to control diastereoselectivity for acyclic substrates are known[5]. Thus we had to develop a new method for such a diastereoselective hydroformylation. Methallylic alcohols seemed to be ideal substrates for that purpose for several reasons:

1. The 1,2-relation of the directing stereocenter to the one formed in the course of hydroformylation reaction might guarantee a maximum of asymmetric induction.
2. The control of regioselectivity would not be a problem for methallylic alcohols as substrates. Because of the rule of *Keulemans*, the formyl group will be attached exclusively at the less substituted olefin terminus [6].
3. Methallylic alcohols are readily available in enantiomerically pure form via the *Sharpless* epoxidation method.
4. The hydroxyl-functionality allows the easy introduction of a *catalyst directing group*.

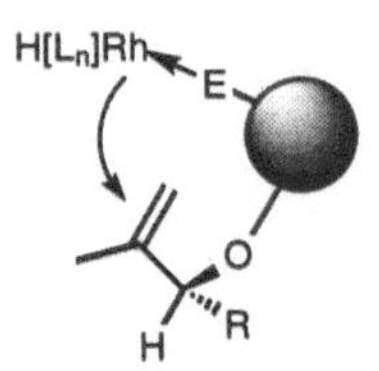

Scheme 1.2. Concept for diastereoselective hydro-formylation of methallylic alcohols with the aid of a *catalyst directing group* (E = element able to coordinate to rhodium)

As a concept for the development of a diastereoselective hydroformylation of methallylic alcohols we wanted to employ a catalyst directing group (Scheme 1.2.). Such a directing group should precoordinate the catalytically active rhodium-species and position it intramolecularly to hopefully only one of the two diastereotopic olefin faces.

2 Development of Diastereoselective Hydroformylation of Methallylic Alcohols

2.1 Development of an Efficient Catalyst Directing Group

Since exploratory experiments on the hydroformylation of unmodified methallylic alcohols had shown to be non-diastereoselective an effective catalyst directing group had to be developed, which meets the following requirements:

First, this catalyst directing group should successfully compete with the carbon monoxide present in large excess as a ligand for the rhodium. At the same time it should not coordinate irreversibly to the catalytically active transition metal center. If this would be the case, this would require stoichiometric amounts of rhodium, i.e. catalysis would not be possible. A ligand with such a property could be a triarylphosphane, because the rapid exchange of monodentate triarylphosphanes at rhodium complexes under hydroformylation conditions is documented [7]. Furthermore this coordinating group should be readily attached to and removed from the methallylic alcohol, and also should have sufficient stability against oxidation and hydrolysis. A system fulfilling all these requirements is the *ortho*-diphenylphosphinobenzoic acid (*o*-DPPBA), readily available from triphenyl-phosphane, sodium and *ortho*-chlorobenzoic acid [8]. The corresponding methallylic *ortho*-diphenylphosphino benzoates **2** (Figure 2.1.1.) had been obtained using the dicyclo-hexylcarbodiimide/ DMAP coupling method [9].

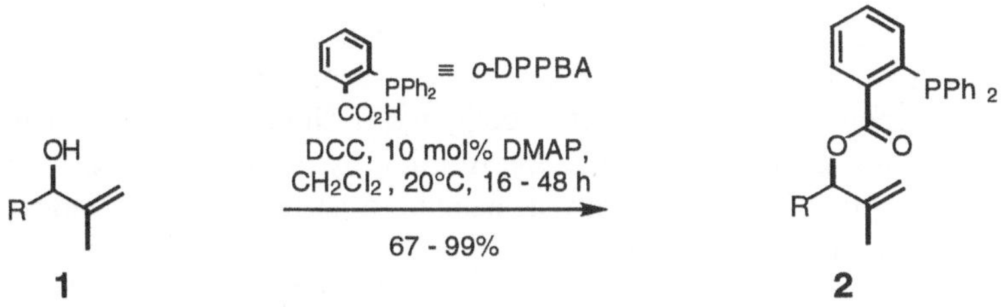

Figure 2.1.1. Synthesis of methallylic *o*-DPPB-ester **2**

We were pleased, that hydroformylation of the methallylic *o*-DPPB-ester **2a** led in almost quantitative yield to the *syn*-aldehyde **3a** in a diastereomer ratio of 96 : 4 (Figure 2.1.2.).

141

2 *syn* 96 : 4 *anti*

3

(**2a** , **3a**: R = *i*-Pr) (*o*-DPPB = *ortho*-diphenylphosphinobenzoate)

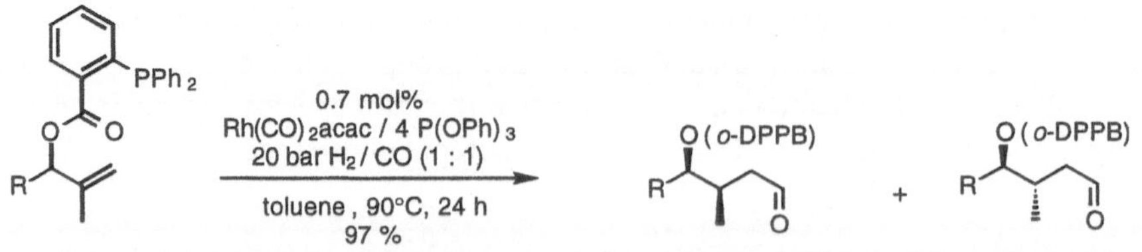

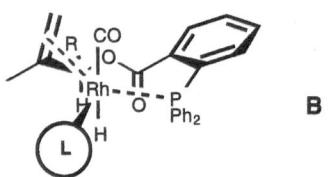

Figure 2.1.2. Diastereoselective hydroformylation of *o*-DPPB-ester **2a**

The choice of $Rh(CO)_2acac/P(OPh)_3$ as the catalyst system for this experiment was based on careful considerations. Thus as a working hypothesis we envisioned complex **B** (Figure 2.1.3.), as a plausible intermediate in the hydroformylation reaction of an *o*-DPPB-ester **2**. If we assume that the ester conformation is preferred as drawn, the substrate should adopt a chelating binding mode. Such a conformational restriction might result in a highly ordered transition state for the stereochemistry defining step.

B

Figure 2.1.3. Trigonal bipyramidal rhodium complex B - a proposed intermediate in the hydroformylation of *o*-DPPB esters **2**

Furthermore catalytically active rhodium/phosphane hydroformylation catalysts are supposed to possess two P-donor ligands [7]. Therefore a second P-donor besides the coordinating *o*-DPPB unit, becomes necessary. Such a coligand **L** should simultaneously stabilize the catalyst and enhance stereoselectivity of the reaction by increased steric interactions. Both might be achieved by using a sterically demanding π-acceptor ligand, which induced us to use triphenylphosphite as a coligand.

2.2 Probing the Role of the *o*-DPPB Group as a Catalyst Directing Substituent via Reversible Coordination

Even though the *ortho*-diphenylphosphinobenzoate unit represents an intrinsically coordinating functionality, this role had not yet been proven as the basis for the observed diastereoselection. An experiment which could give further mechanistic insight was the hydroformylation of the methallylic benzoate **4** (Figure 2.2.1.), which differs from **2b** only by the replacement of the phosphorus atom by a CH-moiety. The benzoate-function in **4** should have almost the same steric features as the *o*-DPPB unit in **2b**, although **4** lacks the ability to coordinate temporarily to the catalytically active transition metal center. If simply the steric demand of the *o*-DPPB group would account for the observed diastereoselection we would expect to get at least some diastereoselection with the derivative **4**.

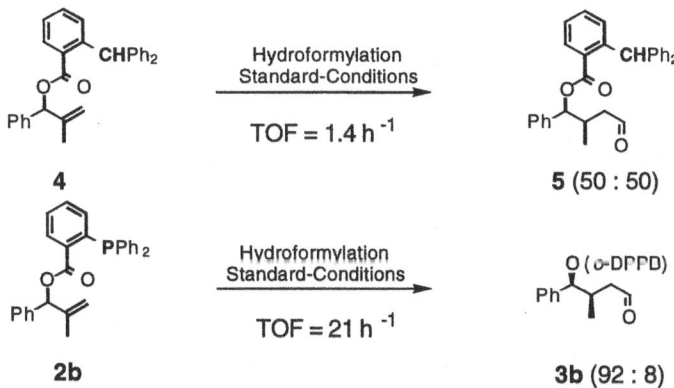

Figure 2.2.1. Probing the role of the *o*-DPPB unit in the course of hydroformylation

Hydroformylation of **4** under identical conditions as for **2b** led to a 1 : 1 *syn/anti* mixture of the two diastereomers **5**, i.e. *no diastereoselectivity*. In contrast, hydroformylation of **2b** provided **3b** in a *syn/anti*-ratio of 92 : 8. Furthermore, determination of turnover frequencies showed the reaction with the *o*-DPPB-ester **2b** to be 15 times faster than for derivative **4**. Both the diastereoselection as well as the rate acceleration for **2b** in comparison to **4** strongly support the role of the *o*-DPPB unit as a catalyst directing group via reversible catalyst coordination.

2.3 Limit and Scope of Diastereoselective Hydroformylation of Methallylic *o*-DPPB-Esters

To explore limit and scope of the substrate directed hydroformylation of acyclic methallylic *o*-DPPB-esters we subjected the derivatives **2a-i** to the same hydroformylation conditions. In all cases the *syn*-aldehydes **3a-i** were formed as the major diastereomers usually in high yields and with predominantly excellent diastereoselectivities (Figure 2.3.1.).

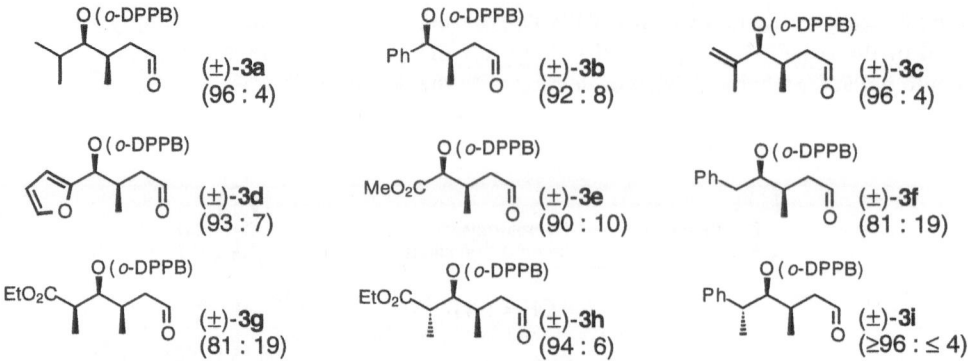

Figure 2.3.1. Diastereoselectivities for aldehydes **3a-i**

On hydroformylation of methallylic alcohols with a branched substituent in the α−position, especially when attached either via an sp2 or sp3 carbon atom, diastereomer ratios from 90 : 10 up to 96 : 4 (compare **3a-e**) were obtained. Only in the case of primary alkyl substituents at the stereogenic center of the methallylic alcohol (**3f**) was the diastereoselectivity inferior. Evidently, a specific steric demand of this α-substituent is necessary for an effective differentiation of the two diastereotopic olefin faces. The influence of a second stereocenter in the methallylic alcohol was investigated using the two diastereomeric substrates **2g** and **2h**. Hydroformylation proceeded with excellent diastereoselectivity (*syn/anti*, 94 : 6) only for **2h** (→**3h**). The phenyl-derivative **2i** (→**3i**) reacted with even better diastereoselectivity (≥ 96 : ≤ 4). The resulting products **3g-i** with three adjacent stereocenters represent stereotriads, which are considered to be central structural units of polyketide natural products [10].

Up to that point all substrates investigated were racemic. But the reaction is not restricted to the synthesis of racemic products. We can easily use this methodology for the preparation of enantiomerically pure compounds. Thus the Sharpless epoxidation method provides access to the enantiomerically pure methallylic alcohol (+)-**1b** (ee ≥ 95 %), which could be hydroformylated as its *o*-DPPB-ester (+)-**2b** to give almost quantitatively the *syn*-aldehyde (+)-**3b** (Figure 2.3.2.). Formation of the cyclic ketal **6** with an enantiomerically pure 1,2-diol demonstrated that the reaction occurred without racemization.

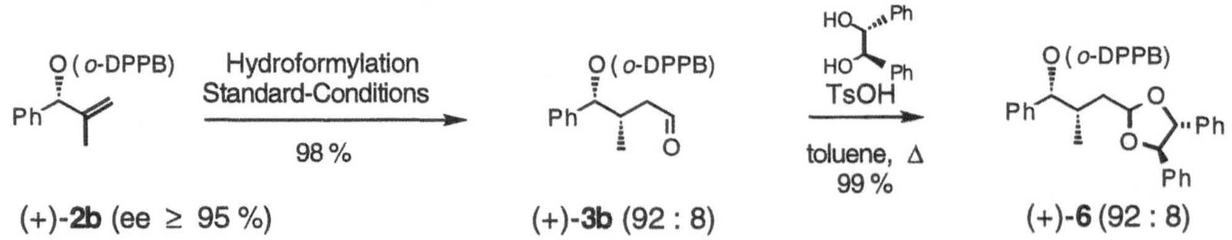

Figure 2.3.2. Route to enantiomerically pure products

2.4 Removal of the *o*-DPPB Unit

To become a valuable preparative method the directing substituent should be removed without problems (Figure 2.4.1.). As tested for **3b**, alkaline hydrolysis gave in quantitative yields both the lactols **7** and the phosphino-carboxylic acid. Oxidation of the lactols **7** provided the known lactones *syn*-**8** and *anti*-**8** [11].

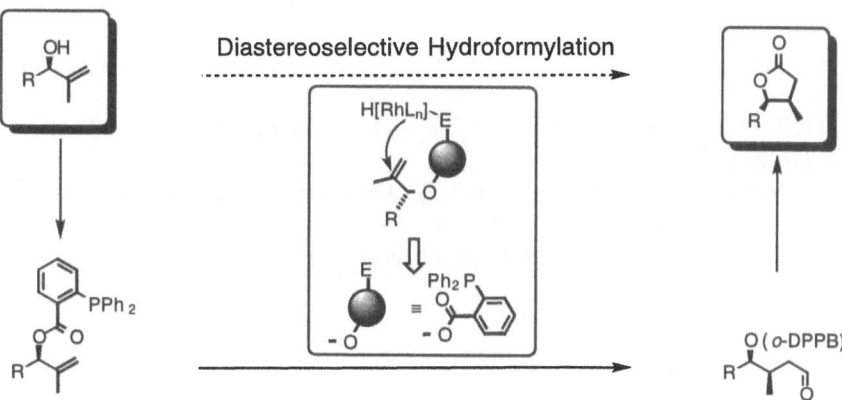

Figure 2.4.1. Removal of the *o*-DPPB group

3 Summary

We have developed an efficient catalyst directing group for the substrate directed diastereoselective hydroformylation of acyclic methallylic alcohols, based on the concept of reversible catalyst coordination (Figure 3.1.). Methallylic alcohols could be hydroformylated in high diastereoselectivities to give the corresponding *syn*-aldehydes. Removal of the directing substituent and oxidation leads in three steps overall to the corresponding γ-lactones. The reaction could be used efficiently for the construction of stereotriads which are central structural units of polyketide natural products.

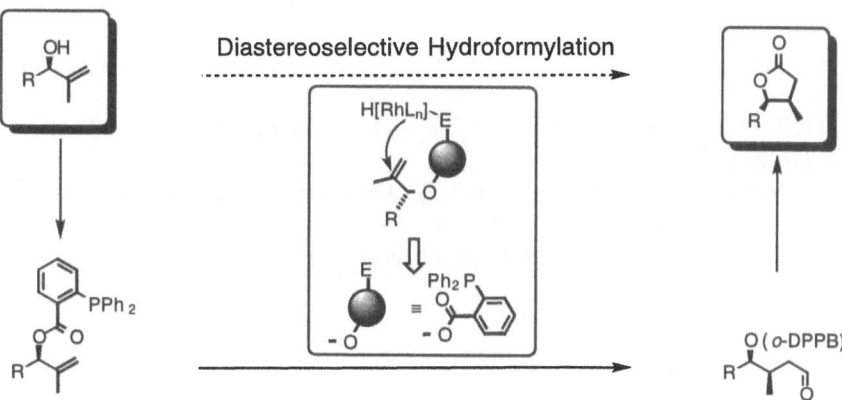

Scheme 3.1. Summary of substrate directed hydroformylation of methallylic alcohols

Acknowledgments

This work was supported by the Fonds der Chemischen Industrie (Liebig-fellowship for B. B.) and by the Deutsche Forschungsgemeinschaft (SFB 260). The author thanks Prof. R. W. Hoffmann for his generous support as well as BASF and Degussa for their gifts of chemicals.

References

1. A. H. Hoveyda, D. A. Evans, G. C. Fu, *Chem. Rev.* **1993**, *93*, 1307.
2. For a review see Houben-Weyl, "Methods of Organic Chemistry" (4th Ed.) *"Stereoselective Synthesis"*, Ed. G. Helmchen, R.W. Hoffmann, J. Mulzer, E. Schaumann, Vol. E 21b, Thieme Stuttgart **1995**, pp 1172.
3. a) For a review see D. Hoppe *Angew. Chem.* **1984**, *96*, 930; *Angew. Chem. Int. Ed. Engl.* **1984**, 23, 932; b) A. J. Pratt, E. J. Thomas, *J. Chem. Soc., Chem. Commun.* **1982**, 1115.
4. a) A. H. Hoveyda, Z. Xu, *J. Am. Chem. Soc.* **1991**, *113*, 5079; b) A. H. Hoveyda, Z. Xu, P. P. Morken, A. F. Houri, *ibid.* **1991**, *113*, 8950; A. F. Houri, M. T. Didiuk, Z. Yu, N. R. Horan, A. H. Hoveyda, *ibid.* **1993**, *115*, 6614.
5. For substrate-directed diastereoselective hydroformylation of cyclic systems see a) S. D. Burke, J. E. Cobb, *Tetrahedron Lett.* **1986**, 27, 4237; b) W. R. Jackson, P. Perlmutter, E. E. Tasdelen, *J. Chem. Soc., Chem. Commun.* **1990**, 763.
6. According to the rule of Keulemans CO-insertion does not occur at a quarternary carbon atom. A. J. M. Keulemans, A. Kwantes, T. Van Bavel, *Recl. Trav. Chim. Pays-Bas* **1948**, 67, 298.
7. J. M. Brown, A. G. Kent, *J. Chem. Soc., Perkin Trans. II* **1987**, 1597. A bidentate ligand would not be suited here, because of the chelate effect the resulting binding constant would be to big.
8. J. E. Hootes, T. B. Rauchfuss, D. A. Wrobleski, H. C. Knachel, *Inorg. Synth.* **1982**, *21*, 175.
9. G. Höfle, W. Steglich, H. Vorbrüggen, *Angew. Chem.* **1978**, *17*, 602; *Angew. Chem. Int. Ed. Engl.* **1978**, *17*, 569.
10. For a definition of stereotriads see R. W. Hoffmann, *Angew. Chem.* **1987**, *99*, 503; *Angew. Chem. Int. Ed. Engl.* **1987**, *26*, 489.
11. J.-M. Fang, B.-C. Hong, L.-F. Liao, *J. Org. Chem.* **1987**, *52*, 855.

Parahydrogen and *in situ* NMR-Spectroscopy as Mechanistic Tools in Catalysis

J. Bargon, R. Giernoth, A. Harthun, and C. Ulrich

Institute of Physical and Theoretical Chemistry, University of Bonn, Wegelerstr. 12, 53115 Bonn, Germany

Abstract

Parahydrogen (p-H_2), one of the two spin isomers of molecular hydrogen, easily accessible from regular H_2 at cryogenic temperatures using a suitable catalyst (like activated charcoal), is totally diamagnetic; therefore, it possesses no ^{1}H-NMR spectrum. If p-H_2 is used in homogeneous hydrogenation reactions, its original high symmetry is broken, and transitions, which were totally forbidden in p-H_2, typically become allowed in the hydrogenation products. This together with the fact that only some of the energy levels of these products, between which the transitions occur, become selectively populated due to the usage of the p-H_2 spin isomer only, gives rise to strong signal enhancement of the NMR lines of both the reaction products and of short-lived reaction intermediates. The associated enhancement of the sensitivity of the NMR detection method can reach values of a few powers of ten, but typically ranges around a factor of a few thousand. Both asymmetric and symmetric substrates give rise to strong nuclear spin polarization upon hydrogenation, and not only of their protons, but also of their hetero-nuclei. ^{13}C-isotopes in natural abundance cause asymmetry even in seemingly symmetric substrates, and permit a very sensitive detection of the reaction products by ^{13}C-NMR-spectroscopy. The ^{13}C- or ^{29}Si-NMR spectra of the hydrogenation products can be recorded after only one pulse despite their low natural abundance without using electronic enhancement based upon polarization transfer techniques. The spectra so obtained reveal a very good signal-to-noise ratio. If in addition nuclear spin polarizations is observed in the starting material as well, this reveals reversibility of the hydrogenation step and proves the fact that the corresponding catalysts promote both the oxidative addition and the reductive elimination of H_2. Even chiral products can be differentiated, whereby this is most convenient, if diastereoisomers result as the products.

Even though this method is primarily useful to investigate hydrogenations, it represents a technique which is of general utility to boost the sensitivity of in situ NMR studies of almost any reaction, especially in organometallic chemistry and in catalysis: Using a suitable precursor and parahydrogen to guarantee a high degree of nuclear polarization in an initially formed hydrogenation product, renders it possible to transfer it on to other can be investigated when initially hydrogenating acetylene and using prepolarized ethylene as the starting material for the subsequent studies. Even the kinetic constants can be obtained in this fashion. Attractive applications of the phenomenon also include photochemically induced reactions.

1 Introduction

Catalytic reactions are gaining in importance due to a variety of reasons. Homogeneous catalysis, in particular asymmetric hydrogenation, is an especially attractive method for the synthesis of chiral drugs. During such hydrogenations catalyzed by transition metal catalysts, nuclear spin polarization has initially been observed by accident [1], but its origin has only been explained much later[2]. It is the consequence of the fact that homogeneous hydrogenation catalysts break the symmetry of H_2 and transfer the H-atoms pairwise, i. e., simultaneously and while maintaining some degree of (magnetic) coupling between them. The nuclear spin polarization so obtained can be detected via *in situ* NMR studies, if parahydrogen is used for the hydrogenation instead of ordinary H_2. Some confusion has arisen concerning the name of this nuclear polarization phenomenon: Bowers and Weitekamp[3] initially assigned the acronym PASADENA (Parahydrogen and Synthesis Allow Dramatically Enhanced Nuclear Alignment) to the polarization obtained when conducting the experiments within the high magnetic field of the NMR spectrometer, and ALTADENA (Adiabatic Longitudinal Transport After Dissociation Engenders Net Alignment), if the reaction itself is performed and completed in the low magnetic field of the earth or the lab, upon which the sample is transferred into the high field of the NMR spectrometer for subsequent analysis[4]. In the following we will use these two acronyms, i.e., PASADENA or ALTADENA, in order to differentiate these distinctively different experimental boundary conditions, since they cause characteristically different polarization patterns. In order to refer to this type of polarization phenomenon in general, in particular to the variety obtained when using parahydrogen, we adopt the nomenclature of Eisenberg et al[5], who used PHIP (Para-Hydrogen Induced Polarization) as the corresponding acronym. This usage corresponds of the various acronyms corresponds to a recent suggestion by Buntkowsky et al.[6].

As such the PHIP spectra resemble those resulting from the well known CIDNP (Chemically Induced Dynamic Nuclear Polarization) phenomenon[7], and indeed the accidentally observed emission lines have been initially mistaken to provide evidence for the occurrence of free radicals during transition metal catalyzed reactions[8]. This conclusion was erroneous, however. The true origin of such nuclear polarization phenomena can indeed be easily identified: Whether the polarization stems from either free radicals or from a perhaps unsuspected formation of parahydrogen during the storage of samples in, for example, liquid nitrogen (as it was the case in Bryndza's pioneering studies[1]) can be tested using both enriched fractions of orthohydrogen (o-H_2) and of parahydrogen, one at a time: using o-H_2, namely, gives rise to an inverted polarization pattern relative to the more convenient case where p - H_2 is used instead[9].

2 Experimental Details

A convenient procedure to obtain fractions of ortho-enriched fractions H_2 is to use Al_2O_3 free of paramagnetic impurities for adsorption of regular H_2 at cryogenic (typically that of liquid nitrogen) temperature. Fractionating desorption upon warming the adsorption cell allows to obtain o-H_2, which desorbs much more slowly, i.e., *after* the parahydrogen. This is because of the difference in symmetry between the two spin isomers, which demands different modes for their rotation. Therefore, a regular mix of the two spin isomers of H_2 can be split into enriched fractions of the individual spin isomers, whereas activated charcoal acts as an efficient catalyst to enrich only p-H_2, the thermodynamically favored component, since such catalysts speed up the rearrangement, thereby shortening the time requires to achieve thermal equilibrium between the two spin isomers at the low temperature of liquid nitrogen considerably. Accordingly, a U-shaped cell filled with charcoal immersed into liquid nitrogen provides for a continuous flow of hydrogen consisting of 51% p-H_2 and 49% o-H_2 at pressures up to 20 bar, whereas at room temperature regular hydrogen consists of 25% p-H_2 and 75% o-H_2. The actual composition of individual fractions (i.e., of "enriched p-H_2") can conveniently be determined taking advantage of their characteristically different thermal conductivity. A suitable apparatus for this purpose has been outlined elsewhere[10]. Specialized probes for investigating reactions of air-sensitive compounds have been developed[11]. When recording spin polarized NMR spectra, care has to be taken with respect to appropriate flip angles, and if polarization transfer routines are used, those have to be modified and optimized for PHIP conditions[12,13].

3 Additional information content

The polarization patterns of the final hydrogenation products, especially those of the ^{1}H-NMR spectra, very sensitively depend on the lifetimes of intermediates, for example of intermediate dihydride complexes of the catalysts. This is due to a process called singlet/triplet mixing[14,15], which occurs in the intermediates. This phenomenon yields even more detailed information about organometallic and catalytic reaction mechanisms, and it provides data to characterize the dihydrogen or dihydride complexes of homogeneous catalysts and to clock there lifetimes, even in those cases, where these intermediates cannot directly be observed by means of NMR spectroscopy.

4 Calculation of the theoretical PHIP spectra

A computer program has been written, which allows to calculate and to simulate graphically the expected PHIP spectra both for the PASADENA and for the ALTADENA mode[16]. A previous version restricted to PASADENA conditions has been available since some time. The latter has only recently been achieved, using different algorithms to solve the corresponding differential equations numerically for essentially an unlimited number of coupled nuclei with spins of either 1/2 or 1, for example ^{1}H, ^{2}D, ^{13}C, ^{14}N, ^{15}N, ^{29}Si, ^{31}P, etc. This program runs both on IBM-compatible PC's under OS2 as the operating system or alternatively under UNIX or its equivalents on workstation, for example on an IBM RS 6000. The time required for a calculation depends on the speed of the processor and the computer architecture, but ranges for 10 spin system around 5 hours.

5 Applications

A very characteristic application of PHIP is the in situ investigation of catalytic hydrogenation reactions mediated by soluble complexes of transition metals[17]. An obvious example is the homogeneous hydrogenation of alkynes or alkenes using homogeneous catalysts containing Rh, Ir, Ru, Pt, and others. Various examples using different transition metals have recently been described[18]. The typical steps associated with catalytic hydrogenations have been outlined by Huey[19]. It has to be kept in mind, however, that the mechanism of homogeneous hydrogenation may proceed according to different routes, whereby the sequence of complexation of H_2 (X = H) and of the unsaturated substrate are interchanged. According to Halpern[20], the initial complexation of molecular hydrogen may result in so-called *"classical"* dihydrides, a pathway which is referred to as the *hydride route*. According to this concept, the unsaturated substrate is complexed subsequent to the molecular hydrogen, namely by the dihydride complex of the initial transition metal complex.

In the alternate *unsaturate route*[20] the substrate is complexed directly by the transition metal complex, i.e., at first, and subsequently the molecular hydrogen becomes activated by the resulting substrate / transition metal chelate complex. In reality, both processes may happen simultaneously, according to their relative values of the corresponding kinetic constants. The ratio of these constants is characteristically different for individual cases following the two alternates routes.

6 Asymmetric Hydrogenation

Of particular interest are PHIP investigations of asymmetric hydrogenations, especially of those which yield diastereoisomers as the products. This is due to two facts:

a) On the one hand, chiral forms of homogeneous catalysts form specific (i. e., chiral) complexes with the simultaneously occurring individual enantiomers of the substrate, yielding chiral products. This renders them superior to heterogeneous catalysts, in particular for asymmetric hydrogenations and hydroformylation, which are gaining in significance as synthetic pathways to chiral drugs. In order to investigate such mechanisms, PHIP studies turn out to be very informative.

b) If in particular diastereoisomers result as the products of the asymmetric hydrogenation, they can conveniently by differentiated directly by NMR without the need of additional chiral aids, such as for example either shift-reagents or chiral solvents. Accordingly, PHIP studies have been conducted during the hydrogenation of a variety of suitable starting materials using a number of different asymmetric catalysts[21]. Already the PHIP spectra of a variety of intermediate dihydrides yield vital information about the geometrical preferences, which cause the difference in the rates of their reactivities, and therefore of the preference of the formation of the diastereoisomers, according to a so-called *major* and *minor* pathway[22]. Figure 1 shows spectra obtained during the hydrogenation of acetamidoacrylic acid methyl ester in D_2O in the presence of the detergent sodium dodecyl sulfate (SDS)[23].

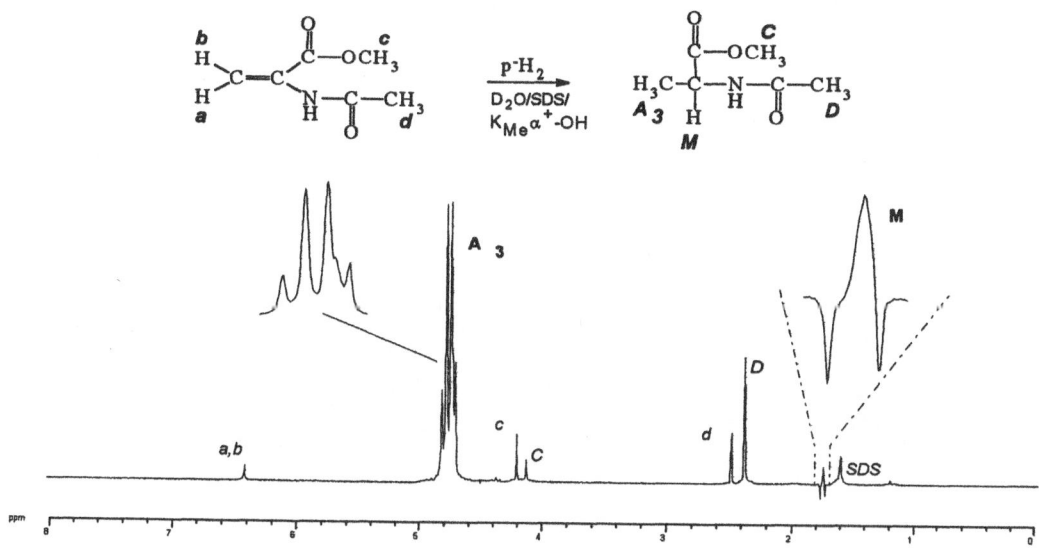

Figure 1. Hydrogenation of acetamidoacrylic acid methyl ester in D_2O/SDS

7 Reversible Reactions

Using p-H_2 in homogeneous hydrogenations even allows to provide evidence for the reversibility of the hydrogenation step, since in this case proton polarization is carried into the starting material, i.e., into the educt. This has been successfully demonstrated recently for the hydrogenation of various a, b- and b, g-unsaturated carboxylic acids and of their derivatives as well as for styrene and its substitued forms[24]. Figure 2 shows the 200 MHz ^{1}H-NMR spectrum recorded during the hydrogenation of styrene using p-H_2 and Wilkinson's catalyst $[(PPh)_3RhCl]$[25].

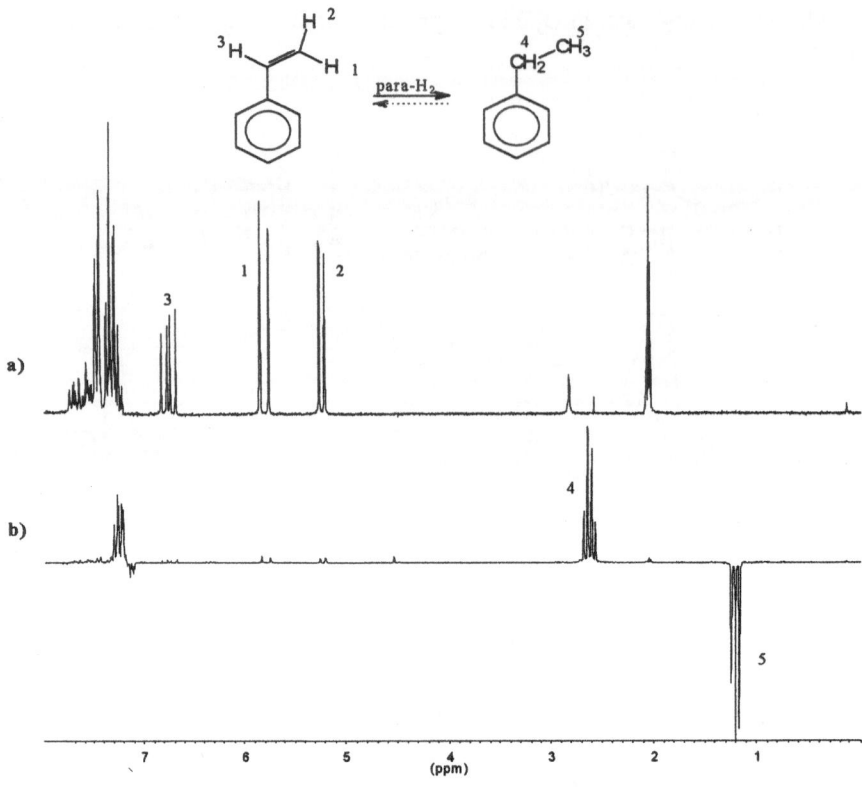

Figure 2. Hydrogenation of styrene with parahydrogen using Wilkinson's catalyst [Rh(PPh₃)₃Cl]: a) ¹H-NMR -spectrum of the educt styrene in acetone-d₆, b) PHIP-NMR-spectrum detected after only few seconds of hydrogenation.

8 Photochemically induced reactions

In order to activate the catalyst, UV light has been used successfully. This has been demonstrated successfully using complexes of Cr, Mo, and W. Using an excimer laser and synchronizing the laser with the pulsed FT NMR spectrometer allows for an attractive determination of kinetic constants of organometallic reactions[26].

9 Hydrogenation with *p*-H₂ as an auxiliary step to introduce polarization into educts

Using an initial hydrogenation with p - H$_2$ as an auxiliary step to introduce nuclear spin polarization into a starting material for *any* subsequent reaction allows to take advantage of the PHIP phenomenon when investigating a big variety of reactions other than just hydrogenations. This concept renders it possible to investigate the subsequent reactions of prehydrogenated components by in situ NMR taking advantage of the associated signal enhancement. Even acetylene is a suitable precursor for spin polarized ethylene, because during its hydrogenation, the few acetylene molecules which contain ^{13}C isotopes in natural abundance cause asymmetry and, therefore, they give rise to sufficient polarization. This is because the ^{13}C isotopes occur individually in the starting acetylene, and, therefore, they break the symmetry in the product ethylene, which causes PHIP, both of the ^{1}H as well as of the ^{13}C nuclei of ethylene[27]. Even though the efficiency of this carbon isotope induced built up of PHIP is modest, it suffices to study a variety of reactions of ethylene via PHIP, other than just hydrogenation. Accordingly, polymerizations, addition of various molecules of the type X-Y, or substitution reactions can thus attractively be investigated via the PHIP method.

Acknowledgements

This work has been supported by the Volkswagen-Stiftung, the Bundesministerium für Bildung, Wissenschaft, Forschung und Technologie (BMBF), the Deutsche Forschungsgemeinschaft (DFG-SFB 334), and the Fonds der Chemischen Industrie.

References

1. H. Bryndza, J. Bargon, R. Bergman, UC Berkeley, in H.E. Bryndza, Thesis, University of California, Berkeley, 1981, Chapter V.
2. C. R. Bowers, D. P. Weitekamp, *Phys. Rev. Lett.* **1986**, *57*, 2645-2648.
3. C. R. Bowers, D. P. Weitekamp, *J. Am. Chem. Soc.* **1987**, *109*, 5541-5542.
4. M. G. Pravica, D. P. Weitekamp, *Chem. Phys. Lett.* **1988**, *145*, 255-258.
5. T. C. Eisenschmidt, R. U. Kirrs, P. P. Deutsch, S. I. Hommeltoft, R. Eisenberg, J. Bargon, R. G. Lawler, A. L. Balch, *J. Am. Chem. Soc.* **1987**, *109*, 8089-8091.
6. G. Buntkowsky, J. Bargon, H.-H. Limbach, *J. Am. Chem. Soc.*, in press.
7. J. Bargon, H. Fischer, U. Johnsen, *Z. Naturfschg 22a*, **1967**, 1551-1555.
8. P.F. Seidler, H.E. Bryndza, J.E. Frommer, L.S. Stuhl, and R.G. Bergman, *Organometallics* **1983**, *2*, 1701-1705.
9. J. Bargon, J. Kandels, K. Woelk, *Angew. Chem. Int. Ed. Engl.* **1990**, *29*, 58-59.
10. J. Bargon, J. Kandels, H. Woelk, *Z. Phys. Chem.* **1993**, *180*, 65-93.
11. K. Woelk, J. Bargon, *Rev. Sci. Instrum.* **1992**, *636*, 3307-3310.

12. J. B Arkemeyer, J. Bargon, H. Sengstschmid, R. Freeman,
    J. Magn. Reson. A **1996,** *120*, 129-132.
13. H. Sengstschmid, R. Freeman, J. Barkemeyer, J. Bargon,
    J. Magn. Reson., Series A **1996,** *120*, 249-257.
14. J. Bargon, J. Kandels, P. Kating, *J. Chem. Phys.* **1993,** *98*, 6150-6153.
15. P. Kating, A. Wandelt, R. Selke, J. Bargon, *J. Phys. Chem.* 1993, *97*, 13313-13317.
16. T. Greve, PhD Thesis, University of Bonn, 1996, unpublished.
17. A. Harthun, J. Barkemeyer, R. Selke, J. Bargon,
    Tetrahedron Lett. **1994,** *35,* 7755-7758; *ibid.* 36 1995 7423.
18. J. Bargon "Recent Developments in Homogeneous Catalysis", in "Applied Homogeneous
    Catalysis by Organometallic Complexes", Eds.: B. Cornils, W. A. Herrmann, VCH
    Publishers, Weinheim, Vol. 2, **1996,** 672-683.
19. J. E. Huheey, Inorganic Chemistry, 3rd. ed., Harper & Row, Publishers, New York, **1983.**
20. J. Halpern, *Trans. Am. Crystallogr. Assoc.,* **1978,** *14*, 59-70.
21. A. Harthun, J. Barkemeyer, R. Selke, J. Bargon, *Tetrahedron Lett.* **1995,** *36,* 7423-7426.
22. A. Harthun, R. Selke, J. Bargon, *Angew. Chem. Int. Ed. Engl.,* in press.
23. A. Harthun, R. Selke, J. Bargon, to be published.
24. A. Harthun, R. Giernoth, J. Bargon, *J. Chem. Soc., Chem. Commun.,* in press.
25. J. A. Osborn, F. H. Jardine, J. F. Young, G. Wilkinson,
    J. Chem. Soc., Chem. Comm. **1965,** 131-132.
26. A. Thomas, M. Haake, F.-W. Grevels, J. Bargon,
    Angew. Chem., Int. Ed. Engl. **1994,** *33,* 755-757.
27. (a) M. Haake, J. Barkemeyer, J. Bargon, *J. Phys. Chem.* **1995,** *99,* 17539-17543;
    (b) J. Barkemeyer, M. Haake, J. Bargon, *J.Am.Chem.Soc.* **1995,** *117,* 2927-2928.

Chiral Ligands and Chiral Pockets. Some Thoughts on Enantioselective Homogeneous Catalysis

P. S. Pregosin

Laboratorium für Anorganische Chemie der ETH-Zürich,
ETH-Zentrum, CH-8092 Zürich, Switzerland

1 Introduction

There are now numerous applications of chiral transition metal complexes in enantioselective organic synthesis. Many of these reactions concern the use of atropisomeric bidentate phosphines, such as BINAP[1] or the interesting MeO-BIPHEP, derivative **1a**, (= 6,6'-dimethoxybiphenyl-2,2'-diyl) bis (3,5-di-t-butylphenylphosphine)[2].

In the area of enantioselective hydrogenation[3,4], the metal of choice is often ruthenium. In addition to the complexed bidentate, the catalyst precursor often contains either chloride or acetate ligands bound to the Ru(II). Although relatively little is known with respect to the complexes which are generated during the hydrogenation, in situ, it is generally assumed that many of the intermediates are six-coordinate[5-7] and that the hydrogen activation is heterolytic[6,7].

2 Results

In the course of hydrogenation studies using the complex $Ru(OAc)_2(1a)^{[2]}$, we isolated[8] the novel bis-solvento five-coordinate hydride $[RuH(i\text{-}PrOH)_2(1a)]^+$, **2**, from the hydrogenation autoclave, in good yield as a red crystalline solid:

$$Ru(OAc)_2(\mathbf{1a}) + 2\,HBF_4 \quad \xrightarrow[333K,\ i\text{-}PrOH]{H_2\ (60\ atm)} \quad [RuH(i\text{-}PrOH)_2(\mathbf{1a})]\,BF_4,\ \mathbf{2} + 2HOAc + HBF_4 \qquad (1)$$

The solid-state structure of **2** has been determined by X-ray diffraction methods and a view of the cation is shown in Fig. 1. It is clear that, although the space associated with the coordinated iso-propanol molecules is relatively open, coordinated **1a** takes up a significant amount of space about the ruthenium ion. This cationic metal complex is interesting in that it represents the first structure of a "stripped-down" hydrogenation catalyst precursor, i.e., it contains only the metal, chiral auxiliary and a hydride[8].

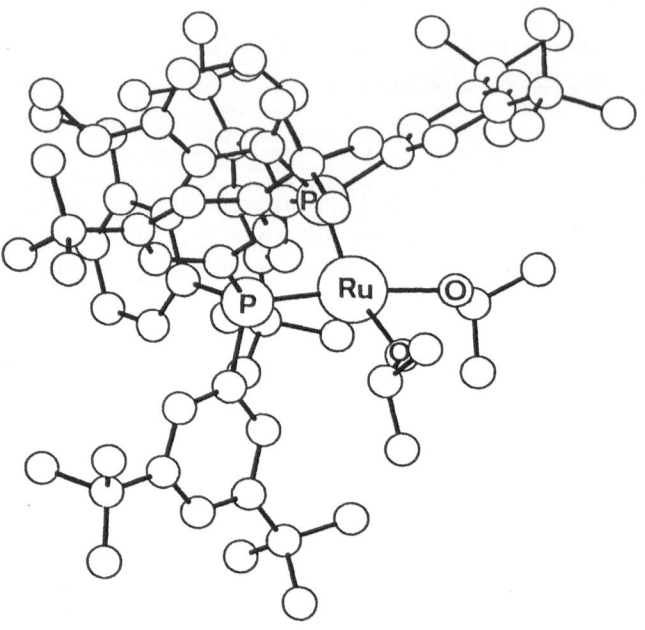

Fig. 1. ORTEP view of the cation of 2

In order to estimate how coordinated **1a** would affect a complexed olefin, we prepared and studied the arene complex [RuH(p-cymene)(**1**)]BF$_4$, **3**, which, we arbitrarily take as model for the interaction of coordinated **1a** with an organic substrate.

3

A detailed NMR analysis[8] of **3** shows unexpected rigidity of coordinated **1a**, in that, at room temperature, there is restricted rotation about one of the four P-C(ipso) bonds as shown below:

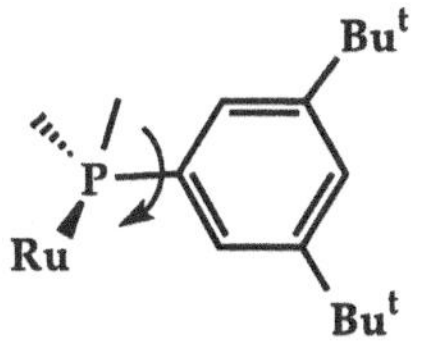

fragment showing the site of restricted rotation

This is the first report of such a relatively high rotational barrier in a coordinated tertiary aryl phosphine not having ortho substituents. Indeed there are four distinct rotational barriers which can be detected via a series of variable temperature proton NMR experiments. As **3** was not typical of an olefinic substrate we attempted to prepare a 1,5-COD complex by reacting $Ru(OAc)_2$(**1a**) with 1,5-COD in the presence of HBF_4, which acid is commonly used to remove coordinated acetate[2]. To our surprise, the product was $[Ru(\eta^5\text{-}C_8H_{11})$(**1a**)$]BF_4$, **4a**,[9] in which:

a) the 1,5-COD has lost a proton, to give a six-electron donor $\eta^5\text{-}C_8H_{11}$ ligand and

b) ligand **1a** functions as a 6e-donor to Ru(II) via an unexpected coordination of one of the biaryl double bonds. Consequently **4a** is coordinatively saturated.

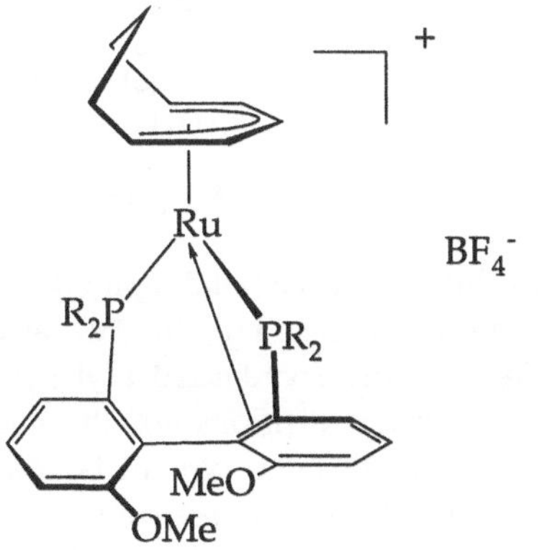

4

An iso-propyl analog $[Ru(\eta^5\text{-}C_8H_{11})$(**1b**)$]CF_3CO_2$, **4b**, **1b** = (6,6'-dimethoxybiphenyl-2,2'-diyl) bis (iso-propylphosphine) was prepared starting from $[Ru(CF_3CO_2)_2(1,5\text{-}COD)]_2$

157

plus, **1b** and reveals the same η^4-bonding mode. Both complexes were characterized by detailed multi-dimensional NMR studies and the X-ray structure[9] for **4b** is given in Fig. 2.

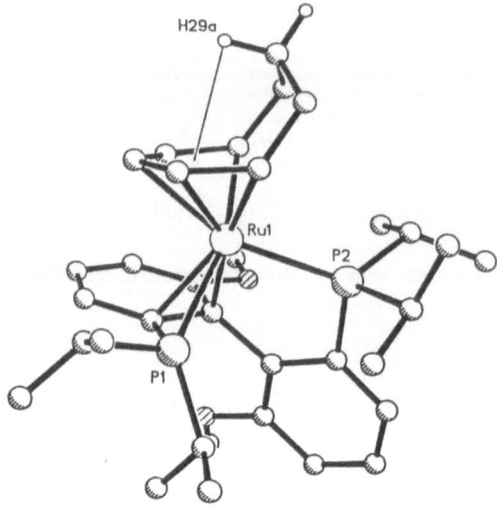

Fig. 2. ORTEP view of the cation of **4b**

Perhaps not surprisingly, the coordinated biaryl ring now has localized double bonds, as illustrated from the bond length data shown:

$$C(1)-C(2) = 1.453(7)$$
$$C(2)-C(3) = 1.430(7)$$
$$C(3)-C(4) = 1.362(8)$$
$$C(4)-C(5) = 1.410(9)$$
$$C(5)-C(6) = 1.354(9)$$
$$C(1)-C(6) = 1.409(7)$$

Ru-C(1), 2.299(5)Å and Ru-C(2), 2.366(5)Å,

The ^{31}P chemical shifts for this new η^4-6e bonding mode of **1** are unusual, in that there are two widely separated phosphorus signals, i.e., for **1a,** at 69.6 ppm and - 11.2 ppm (!) with the phosphorus resonance for the coordinated aryl ring appearing at relatively low frequency. However, the ^{13}C resonance positions for the coordinated biaryl carbons are a more reliable criterion for recognizing this new type of bonding. Unfortunately, the carbon signals for C(1), 74.5 ppm and C(6), 95.1 ppm, are sometimes difficult to locate due to the relatively long T_1 's for these carbons, combined with the splitting due to the ^{31}P spins. Consequently, a long-range ^{13}C,^{1}H-correlation using proton detection is recommended as the method of choice[9].

The complex Ru(Cp)(BINAP) has been recently shown[10], by X-ray crystallography, to have a similar bonding mode for the BINAP ligand.

Complex **4a** exhibits dynamic behavior in solution as shown by 2-D exchange spectroscopy[11], with a section of the exchange spectrum given in Fig. 3.

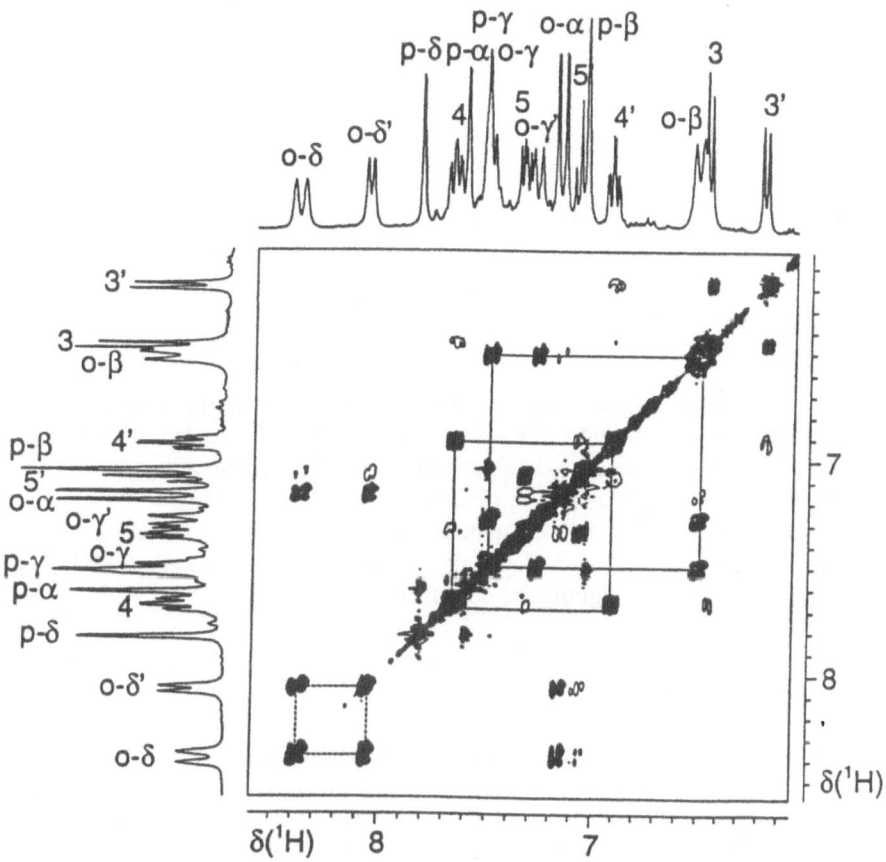

Fig. 3. Aromatic section of the 2-D NOESY for 4 showing the exchange cross-peaks. The two halves of the ligand **1a** are in equilibrium. The "open" cross-peaks arise from NOE, the filled in cross-peaks from exchange (300 MHz, CD_2Cl_2, room temperature, mixing time: 0.6 sec.).

Since the isomerization product has exactly the same structure, one observes only one set of resonances. This intramolecular process, which exchanges the two "halves" of coordinated **1a**, can be rationalized by assuming that the double bond dissociates and that the η^5-C_8H_{11} ligand can rotate.

Interestingly, there is no exchange involving any of the η^5-C_8H_{11}- protons, i.e., the allyl protons do not exchange amongst themselves and the CH_2 protons do not exchange with the allyl protons, so that we can exclude a process involving sequential migration of H^+(or H^-) during the isomerization. We observe a second type of relatively slow dynamic process in **4a** occurring simultaneously at room temperature. This involves restricted rotation around the P-C(ipso) bonds for two of the four 3,5-disubstituted t-butyl phenyl rings, again suggesting some interesting rigidity for coordinated **1a**.

3 Discussion

Ligand **1a** is obviously interesting in that it can stabilise a five-coordinate hydrido-solvent complex. In addition, it would seem that the 3,5-t-butyl groups contribute additional rigidity to complexed **1a** as demonstrated by relatively slow rotation about the P-C bonds on the NMR time scale. This relatively slow phenyl ring rotation is valid for both complexes **3** and **4** and it is worth noting that there is no restricted rotation in uncoordinated **1a** at ambient temperature.

It is known[2] that introducing the 3,5-di-t-butyl groups into MeO-BIPHEP, improves the observed enantioselectivity in the Ru(II) catalysed hydrogenation of a pyrone, by ca 7%, relative to the parent MeO-BIPHEP. We have suggested previously[12], based on results in the Pd(II) catalysed enantioselective allylic alkylation reaction, with BINAP, BIPHEP, JOSIPHOS and other auxiliaries, that the intrusiveness and rigidity of the chiral pocket are important contributors to the success of a given chiral auxiliary. As it seemed that **1a** was an improvement over the parent ligand, we carried out the "standard" enantioselective allylic alkylation using **1a**[13] and find an ee of 91%, ca 4% higher than our reported result[9] for the unsubstituted auxiliary. We attribute this structural "tuning" (which admittedly involves only a few tenths of a Kcal in energy) in both the Ru(II) and Pd(II) reactions, to the combined slightly larger and more rigid chiral pocket offered by **1a** relative to the conventional MeO-BIPHEP.

The unexpected *bonding flexibility* of coordinated **1** is noteworthy in that this ability to act as a 6e-donor might well help this ligand to stabilise reactive intermediates. Equally important in this connection, is the relative ease with which **1a** can resume its "normal" 4e bis-phosphine coordination mode via double bond dissociation, as shown by the exchange spectroscopy. Naturally, the chiral pocket of **1a** in its 6e-donor form will be even more rigid; however, it is not proven that this interesting donor mode has any relevance for catalytic chemistry.

References

1. Ohta, T.; Tonomura, Y.; Nozaki, K.; Takaya, H.; Mashima, K. *Organometallics* **1996**, *15* , 1521; Zhang, X.; Uemura, T.; Matsumura, K.; Sayo, N.; Kumobayashi, H.; Takaya, H. *Synlett* **1994**, 501; Kitamura, M.; Tokunaga, M.; Noyori, R. *J. Am. Chem. Soc.* **1993**, *115*, 144; Ohta, T.; Miyake, T.; Seido, N.; Kumobayashi, H.; Akutagawa, S.; Takaya, H. *Tetrahedron Letters* **1992**, *33* , 635; Mashima, K.; Hino, T.; Takaya, H. *J. Chem. Soc. Dalton Trans* **1992**, 2099; Ohta, T.; Takaya, H.; Noyori, R. *Inorg. Chem.* **1988**, *27*, 566.
2. Schmid, R.; Broger, E. A.; Cereghetti, M.; Crameri, Y.; Foricher, J.; Lalonde, M.; Mueller, R. K.; Scalone, M.; Schoettel, G.; Zutter, U. *Pure & Appl. Chem.* **1996**, *68* , 131; Heiser, B.; Broger, E. A.; Crameri, Y. *Tetrahedron Asymmetry* **1991**, *2* , 51. BIPHEP is the generic name suggested.
3. Kitamura, M.; Noyori, R. in *Modern Synthetic Methods* (Ed. R. Scheffold), **1989**, *5*, 116; Genet, J. P.; *Acros. Organics Acta* **1995**, *1*, 4.
4. Brown, J. M. *Chem. Soc. Rev.*, **1993**, 25.
5. a) Ashby, M. T.; Khan, M. A.; Halpern, J.*Organometallics.* **1991**, *10*, 2011.; (b) Ashby, M. T.; Halpern, J. *J. Am. Chem. Soc.* **1991**, *113* , 589.
6. Mezzetti, A.; Tschumper, A.; Consiglio, G. *J. Chem. Soc. Dalton Trans* **1995**, 49; Mezzetti, A.; Costella, L.; Del Zotto, A.; Rigo, P.; Consiglio, G. *Gazzetta Chimica Italiana* **1993**, *123* , 155.
7. D. C. Mudalige, S. J. Rettig, B. R. James, W. R. Cullen, *J. Chem. Soc. Chem. Commun.* **1993**, 830; R. S. M. Cashman, I. R. Butler, W. R. Cullen, B. R. James, BJ. P. Charland, J. Simpson, *Inorg. Chem.* **1992**, *31* , 5509.
8. Currao, A.; Feiken, N.; Macchioni, A.; Nesper, R.; Pregosin P. S.; Trabesinger, G *Helv. Chim. Acta* **1996**, *79* , 1587.
9. Feiken, N.; Pregosin, P. S.; Scalone, M.; Trabesinger, G. submitted to *Organometallics*.
10. Pathak, D. D.; Adams, H.; Bailey, N. A.; King, P. J.; White, C. *J. Organomet. Chem.* **1994**, *479* , 237.
11. For applications of 2-D exchange spectroscopy see: Barbaro, P.; Currao, A.; Herrmann, J.; Nesper, R.; Pregosin, P. S.; Salzmann, R. *Organometallics* **1996**, *15* , 1879; Herrmann, J.; Pregosin, P. S.; Salzmann, R.; Albinati, A. *Organometallics* **1995**, *14* , 3311; Pregosin, P. S.; Salzmann, R.; Togni, A. *Organometallics* **1995**, *14* , 842; Breutel, C.; Pregosin, P. S.; Salzmann, R.; Togni, A. *J. Am. Chem S oc.* **1994** , *116*,

4067; *Magn. Reson. Chem.* **1994**, *32* , 297; Lianza, F.; Macchioni, A.; Pregosin, P. S.; Rüegger, H. *Inorg. Chem.* **1994**, *33* , 4999.

12. Barbaro, P. Pregosin, P. S.; Salzmann, R.; Albinati, A.; Kunz, R., W. *Organometallics* **1995**, *14* , 5160.

13. Tschoerner, M.; Pregosin, P. S. unpublished results.

Application of High Pressure in Palladium Catalysis

Stephan Hillers, Anne Mengel, Kerstin Bodmann and Oliver Reiser*

Inst. f. Organische Chemie der Georg-August-Universität, Tammannstr. 2,
D-37077 Göttingen, Germany

1 Introduction

The development of efficient chemical transformations is one of the most important challenges to date. In organic synthesis reactions are required to be highly selective as well as economically and ecologically benign, especially in large scale productions. Since catalysts, consisting of a transition metal modified by ligands, are at least in theory indestructible vehicles for carrying out reactions under mild conditions with high selectivity, the development of catalytic processes has become a major research goal in recent decades[1].

The catalyst performance can be accessed by its *selectivity* and *reactivity* for a given reaction. Great progress and high standards have been achieved for the control of selectivity under all aspects – chemo-, regio-, diastereo-, and enantioselectivity – mainly by developing ligands which effectively force a well defined orientation of substrates and reagents by coordination at the metal. The reactivity of a catalyst, reflected by its turnover numbers (TON) and turnover rates (TOR), leaves in general much room for improvement. Heteroatom transfer reactions and especially hydrogenations are notable exceptions (TON up to 10^6), while in carbon carbon bond forming reactions a much lower turnover is reached on a routine basis (TON ≤ 100).

A possible reason for the break down of a catalyst might be the lack of stabilization, which is usually in the responsibility of the ligands. However, ligands are sometimes sensitive to the reaction conditions and might degenerate during the course of the reaction[2]. Consequently, an obvious but potentially expensive solution to obtain high TON's might be to use a large excess of ligands[3]. Another possibility is to use particularly stable and strongly coordinating ligands[4], however, this might effectively block the metal for coordination to substrates, thus requiring high reaction temperatures to obtain a catalytically active system. Moreover, it is desirable to maintain flexibility in the choice of ligands, since not only stability but also selectivity of a reaction is controlled by them. A third option for catalyst stabilization might be to enhance coordination of weak ligands to the metal by applying pressure. This hypothesis was our starting point to systematically investigate the influence of high pressure in liquid phases for transition metal catalysis, both under selectivity as well as under reactivity aspects.

2 Principles

High pressure in liquid phases has been traditionally used in organic synthesis to accelerate reactions[5]. A rate acceleration can be expected if the volume of activation $\Delta V^{\ddagger}$, i.e. the difference of the volume of the transition state and the starting materials, is negative.

This is generally the case for addition reactions, and consequently pressure has been proven to be most beneficial in cycloadditions. Dissociation reactions can also be greatly favored by pressure if ions are formed in that process. This results in an ordering of charged and uncharged species based on electrostatic effects (electrostriction), and thus in an substantial volume decrease.

Transition metal catalyzed reactions consist of many reaction steps, each could be favored or disfavored by pressure. Consequently, to predict the net effect of pressure is much more difficult. Most critical of course is whether the catalytic cycle of a reaction can still take place at all if pressure is applied to the system, therefore, ligand / substrate exchange at the metal must remain facile. The substitution of a ligand can occur either by a dissociative or an associative mechanism (Scheme 1), and the latter should have a negative activation volume. Indeed, it has been shown for many inorganic coordination compounds that associative substitution, which is accelerated by pressure ($\Delta V^{\ddagger} \approx -5$ to $-12 \text{ cm}^3 \text{ mol}^{-1}$), is operating[6]. Taking this hypothesis as the guiding principle of this investigation, we began to look into some palladium catalyzed reactions at high pressure.

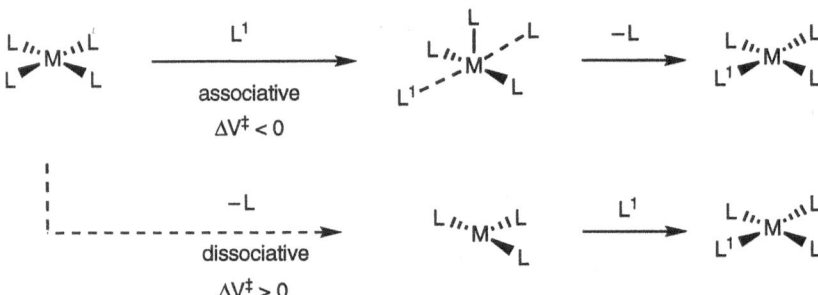

Scheme 1. Possible mechanisms for ligand displacement

3 Heck-type Reactions at High Pressure

Palladium(0) has been proven to be one of the most versatile catalysts for the construction of new carbon carbon bonds. In particular, coupling reactions of the Heck type have flourished recently in that this reaction can be used for very selective and complicated carbon carbon bond constructions[7]. Nevertheless, a serious drawback for large scale applications is the generally low reactivity of the catalyst, making high reaction temperatures and the use of palladium amounts in the order of 1 to 20 mol% necessary. Therefore, we were especially eager to test our hypotheses for the application of high pressure in this area of catalysis.

As a model we have chosen the palladium catalyzed cross coupling of iodobenzene and 2,3-dihydrofuran (**2**, R = H)[8], which has attracted considerable interest in recent years since it offers the possibility to carry out this reaction asymmetrically[9], as well as providing an indirect access to aldol products[10]. In particular, it is possible to convert allylic alcohols to *anti*-aldol products using only transition metal catalyzed bond forming reactions: Rhodium(I)-catalyzed hydroformylation of **1** and subsequent acid catalyzed elimination of water leads directly to 4-substituted 2,3-dihydrofurans **2** in a one pot procedure. Palladium(0)-catalyzed arylation or vinylation of **2** proceeds highly diastereoselectively to give the *trans*-disubstituted dihydrofurans **4**, which can be converted by ozonolysis to *anti*-aldol products **3**[11]. The weak step in this reaction sequence is the palladium catalyzed transformation of **2** to **4**, especially if R is different from hydrogen: Alkyl-substituted dihydrofurans **2** are sluggish substrates, while aryl substituted derivatives of **2** did not undergo coupling reactions under any standard conditions tried.

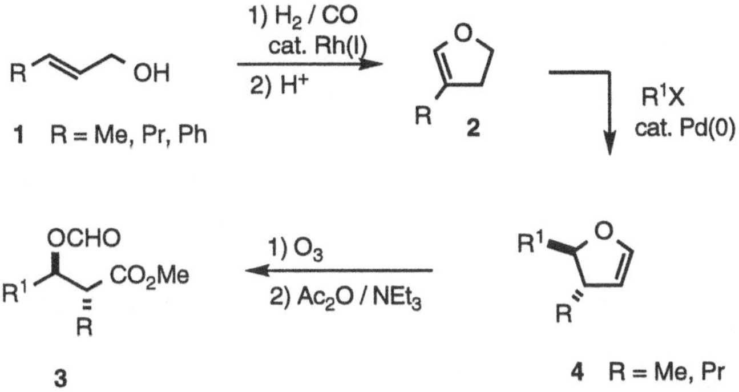

Scheme 2. A catalytic approach to *anti*-aldol products starting from a llyl alcohols

Carrying out the reaction of 2,3-dihydrofuran (**2**, R = H) and iodobenzene at pressures between 1 bar and 8 kbar revealed that this reaction is indeed accelerated by pressure, however, the increase in rate is rather modest (k_{rel} (1 bar) = 1; k_{rel} (8 kbar) = 23; $\Delta V^{\ddagger} = -12$ cm^3 mol^{-1}) compared with that of Diels-Alder reactions ($\Delta V^{\ddagger} = -25$ to -45 cm^3 mol^{-1})[12]. Taking into consideration that ligand exchange via an associative mechanism as well as ion formation should be favored by pressure (Scheme 3) all reaction steps except metal migration (**10** → **9** → **8**) should benefit. This analysis is further corroborated by the finding that an increase of the concentration of triphenylphosphine, which acts as a good nucleophile, favors the substitution **9** → **12** rather than metal migration **9** → **8** (no PPh$_3$: **11/12** = 95:5 (1 bar) and 90:10 (10 kbar); Pd:PPh$_3$ = 1:60: **11/12** = 90:10 (1 bar) and 25:75 (10 kbar))[13].

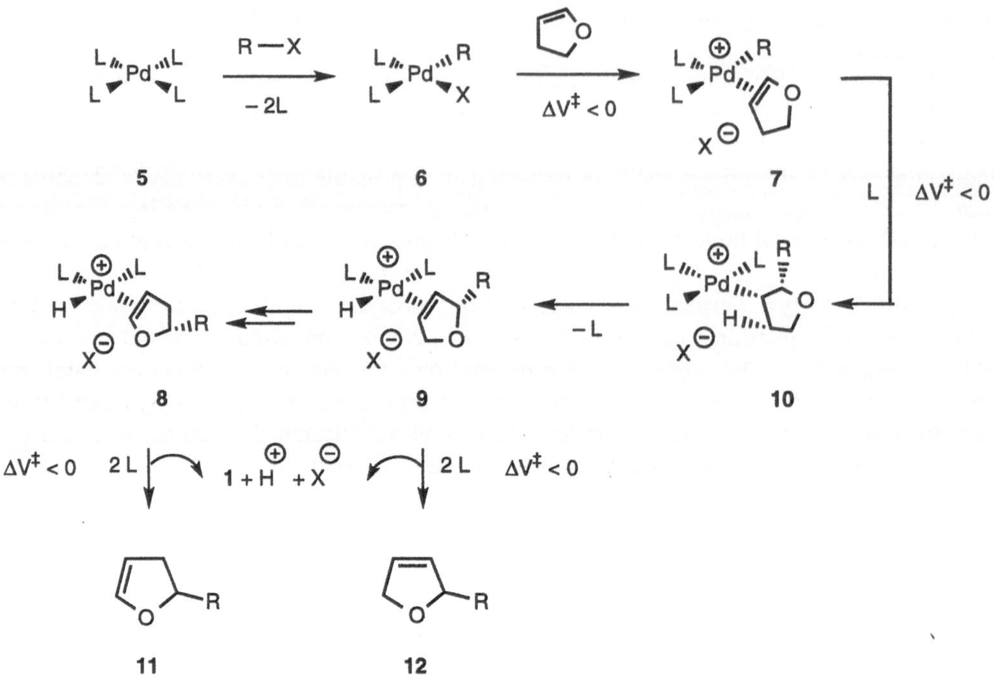

Scheme 3. Effect of pressure on the coupling between 2,3-dihydrofuran and iodobenzene

Most exciting, however, was the discovery that the lifetime of the catalyst seems to be greatly enhanced by pressure. Coupling reactions carried out at 60°C between iodobenzene and 2,3-dihydrofuran, 2,3-dihydropyrrol or cyclopentene, respectively, revealed that the maximum TON was about 270-280 at 1 bar, while at 2 to 8 kbar the TON was found to be up to 20000. Even in the absence of triphenylphosphine a TON of up to 7500 was reached, indicating that the solvent (THF / acetonitrile) can act as an effective ligand at high pressure. Raising the reaction temperature to 100°C the catalyst concentration could be further lowered. With 2,3-dihydrofuran or 2,3-dihydropyrrol TON's of around 140000 were observed, while cyclopentene proved to be even more reactive (TON 770000; TOR 5500 cycles / h[14].

Classical Heck reactions between aryl iodides or bromides and alkenes such as styrene also could be carried out with considerably increased TON at high pressure. Also, in palladium catalyzed allylations higher TON's and TOR's were found, while the Suzuki Cross Coupling seems to be slowed down at high pressure.

4 Palladium-catalyzed Acylations

The application of high pressure also opens up the possibility of developing new catalytic reactions. The Heck reaction is so far limited to the arylation and vinylation of alkenes. Negishi has demonstrated that under carbon monoxide pressure (40 atm) palladium acyl complexes are formed which can subsequently acylate an alkene in an *intramolecular* cyclization reaction[15]. However, in these reactions the potentially competitive intramolecular vinylation seemed to be unlikely due to an unfavorable ring closure which would have been occured. Spencer has shown, that benzoyl chlorides **13** are suitable substrates in *intermolecular* Heck reactions, however, the products obtained are exclusively the arylated alkenes **15** which arise from loss of CO during the reaction[16]. Reinvestigation of these reactions revealed that even under CO gas pressure only the arylated products are obtained as the sole products.

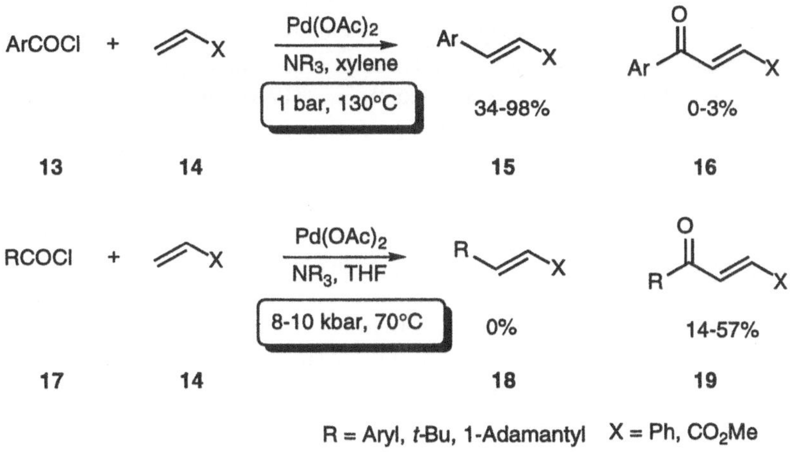

R = Aryl, *t*-Bu, 1-Adamantyl   X = Ph, CO$_2$Me

Scheme 4. Heck reactions with acid chlorides

Nevertheless, the extrusion of CO should have a positive activation volume, so we began to investigate the Heck reaction of benzoyl chlorides **17** (R = Ar) at high pressure in liquid phases. Up to 6 kbar, mixtures of acylated and arylated products were obtained, however, at a pressure range of 7 to 10 kbar the acylated alkenes **19** were obtained as the only products. Also, certain alkyl substituted acid chlorides **17** (R = *t*-Butyl, 1-Adamantyl) were successfully employed as substrates as long as no β-hydrogens were present which would react to form ketenes under the reaction conditions. Gratifyingly, a reaction temperature of 70°C was sufficient for the reaction to proceed compared to 130°C needed in the normal pressure process. However, the reaction turned out to be highly sensitive towards the variation of reaction parameters. The presence of triphenylphosphine inhibits the reaction completely, and an amine base seems to be absolutely necessary. The role of the amine must be more than that of a acid scavenger, since numerous attempts to replace the amine by other bases, molecular sieves or 2-methyl-2-butene as an HCl trap failed to give the desired product. Moreover, other activated carboxylic acid derivates could not be employed as substrates. Nevertheless, these studies show that acid chlorides in principle can be used in Heck reactions opening the way to catalytic acylation reactions of alkenes. Further studies to optimize this progress are currently in progress.

5 Domino-HWE-Heck Reaction

The alkylation of aldehydes with phosphonates (Horner-Wadsworth-Emmons (HWE) reaction) is usually carried out with a stochiometric amount of a strong base such as KHMDS or with the combination of a stochiometric amount of LiCl and triethylamine or DBU as the base[17]. Schmidt, however, already demonstrated in the case of (Z)-protected amino substituted phosphonates that the use of LiCl is superfluous if DBU is used as the base[18]. This can most likely be attributed to the greater acidity of the particular phosphonate used. In our studies of asymmetric olefination reactions[19], we recently found that under a pressure of 8 kbar aldehydes can be cleanly alkenylated at room temperature just in the presence of triethylamine[20]. These mild conditions open the possibility for the development of new domino reactions. For example, a HWE reaction can be combined with a Heck coupling, allowing the direct synthesis of trisubstituted alkenes such as 21 from an aldehyde, an aryl halide and the phosphonate 20.

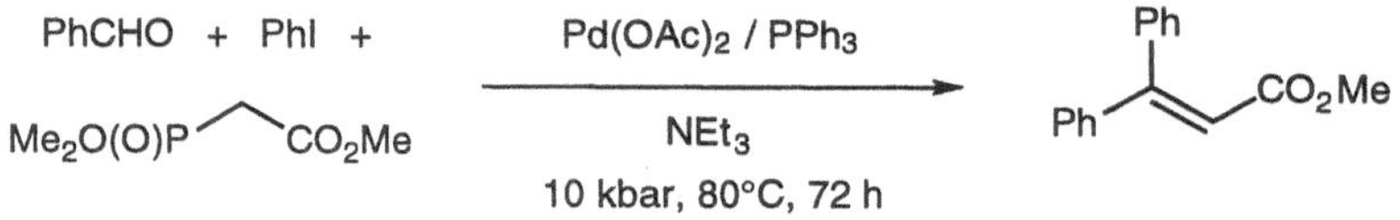

PhCHO + PhI +

$Me_2O(O)P\diagup\diagdown CO_2Me$

20

$\xrightarrow[\substack{NEt_3 \\ 10\ kbar,\ 80°C,\ 72\ h}]{Pd(OAc)_2\ /\ PPh_3}$

Ph

$Ph\diagup C=CH\diagdown CO_2Me$

21 (78%)

Scheme 5. Domino-HWE-Heck Reaction

In conclusion, we believe we have demonstrated that pressure in liquid phases should be one parameter to be considered in transition metal mediated synthesis. Since our initial study[13], there have been a few other reports dealing with different aspects of palladium catalyzed reactions at high pressure[21], giving us confidence that there is considerable potential in this technique for catalysis.

Acknowledgement

We thank the Volkswagenstiftung (AZ 68/870), the Deutsche Forschungsgemeinschaft (Graduiertenkolleg Hochdruck), Fonds der Chemischen Industrie, Hoechst AG and Degussa AG for their generous support of our work.

References

1. *Applied Homogeneous Catalysis with Organometallic Compounds*, B. Cornils; W. Herrmann (Eds.); VCH: Weinheim, 1996.
2. W. A. Herrmann, C. Broßmer, K. Öfele, M. Beller, H. Fischer, *J. Organomet. Chem.* **1995**, *491*, C1.
3. B. A. Patel, C. B. Ziegler, N. A. Cortese, J. E. Plevyak, T. C. Zebovitz, M. Terpko, R. F. Heck, *J. Org. Chem.* **1977**, *42*, 3903.
4. (a) A. Spencer, *J. Organomet. Chem.* **1983**, *258*, 101. (b) W. A. Herrmann, C. Broßmer, K. Öfele, C.-P. Reisinger, T. Priermeier, M. Beller, H. Fischer, *Angew. Chem.* **1995**, *107*, 1989; *Angew. Chem. Int. Ed. Engl.* **1995**, *34*, 1844.
5. (a) *Organic Synthesis at High Pressure*, K. Matsumoto, R. M. Acheson (Eds.) *New York* **1991**. (b) F.-G. Klärner, *Chemie in unserer Zeit* **1989**, *23*, 53.
6. R. v. Eldik, T. Asano, W. J. l. Noble, *Chem. Rev.* **1989**, *89*, 549.
7. A. de Meijere, F. E. Meyer, *Angew. Chem.* **1994**, *106*, 2473; Angew. Chem. Int. Ed. Engl. **1994**, *33*, 2379.
8. (a) R. C. Larock, W. H. Gong, *J. Org. Chem.* **1990**, *55*, 407. (b) R. C. Larock, W. H. Gong, B. E. Baker, *Tetrahedron Lett* **1989**, *30*, 2603.
9. (a) F. Ozawa, Y. Kobatake, T. Hayashi, *Tetrahedron Lett* **1993**, *34*, 2505. (b) F. Ozawa, A. Kubo, Y. Matsumoto, T. Hayashi, E. Nishioka, K. Yanagi, K. Moriguchi, *Organometallics* **1993**, *12*, 4188. (c) F. Ozawa, A. Kubo, T. Hayashi, *J. Am. Chem. Soc.* **1991**, *113*, 1419.
10. S. Hillers, A. Niklaus, O. Reiser, *J. Org. Chem.* **1993**, *58*, 3169.
11. S. Hillers, O. Reiser, *Synlett* **1995**, 153.
12. S. Hillers, O. Reiser, *J. Chem. Soc. Chem. Commun.* **1996**, in press.
13. S. Hillers, O. Reiser, *Tetrahedron Lett* **1993**, *34*, 5265.
14. S. Hillers, S. Saratori, O. Reiser, *J. Am. Chem. Soc.* **1996**, *118*, 2077.
15. J. M. Tour, E. Negishi, *J. Am. Chem. Soc.* **1985**, *107*, 8289.
16. H.-U. Blaser, A. Spencer, *J. Organomet. Chem.* **1982**, *233*, 267.
17. (a) M. A. Blanchette, W. Choy, J. T. Davis, A. P. Essenfeld, S. Masamune, W. R. Roush, T. Sakai, *Tetrahedron Lett.* **1984**, *25*, 2183. (b) M. W. Rathke, M. Nowak, *J. Org. Chem.* **1985**, *50*, 2624-26.
18. U. Schmidt, H. Griesser, V. Leitenberger, A. Lieberknecht, R. Mangold, R. Meyer, *Synthesis* **1992**, 487.
19. (a) T. Rein, J. Anvelt, A. Soone, R. Kreuder, C. Wulff, O. Reiser, *Tetrahedron Lett.* **1995**, *36*, 2302. (b) T. Rein, R. Kreuder, P. v. Zezschwitz, C. Wulff, O. Reiser, *Angew. Chem.* **1995**, *107*, 1099; *Angew. Chem. Int. Ed. Engl.* **1995**, *34*, 1023. (c) T. Rein, N. Kann, R. Kreuder, B. Gangloff, O. Reiser, *Angew. Chem.* **1994**, *106*, 597; *Angew. Chem. Int. Ed. Engl.* **1994**, *33*, 556.
20. K. Bodmann, R. Kreuder, G. Santoni, O. Reiser, *unpublished results*.
21. (a) B. M. Trost, J. R. Parquette, A. L. Marquart, *J. Am. Chem. Soc.* **1995**, *117*, 3284. (b) T. Sugihara, M. Takebayashi, C. Kaneko, *Tetrahedron Lett.* **1995**, *36*, 5547. (c) K. Voigt, U. Schick, F. E. Meyer, A. de. Meijere, *Synlett* **1994**, 189.

Asymmetric Reactions Catalyzed by Palladium-MOP Complexes

Tamio Hayashi

Department of Chemistry, Faculty of Science, Kyoto Univ., Sakyo, Kyoto 606-01, Japan

1 Introduction

Asymmetric reactions catalyzed by transition metal complexes containing optically active phosphine ligands have attracted significant interest due to their synthetic utility[1]. One of the most exciting and challenging subjects in the research of catalytic asymmetric synthesis is the development of a chiral ligand which will influence reaction efficiency in terms of catalytic activity and enantioselectivity. Most of the chiral phosphine ligands prepared and used for catalytic asymmetric reactions hitherto are the bisphosphines which are in the general case anticipated to be effective in constructing a chiral environment by the chelate coordination to a metal. The representatives are BINAP[2] and ferrocenylbisphosphines[3,4]. These chelating bisphosphines have been demonstrated to be effective for several types of asymmetric reactions including rhodium- or ruthenium-catalyzed hydrogenation[2], palladium- or nickel-catalyzed allylic substitution reactions[3], and gold- or silver-catalyzed aldol reactions[4]. On the other hand, there have been reported only a limited number of monodentate chiral phosphine ligands, probably because with the exception of some they have been exhorted as being of little practical use as a bisphosphine ligand[1,5]. However, there exist transition metal-catalyzed reactions where the bisphosphine-metal complexes can not be used because of their low catalytic activity and/or low selectivity toward a desired reaction pathway and therefore chiral monodentate phosphine ligands are required for the catalytic asymmetric synthesis to be viable. We have previously reported a nickel-catalyzed asymmetric cross-coupling forming axially chiral binaphthyls which was realized for the first time by use of a monophosphine ligand containing ferrocene planar chirality[6], and we have continued our efforts to develop new enantioselective chiral monodentate phosphine ligands for the transition metal-catalyzed reactions where only monodentate phosphine ligand can be used. We found that high enantioselectivity and high catalytic activity can be achieved in the palladium-catalyzed asymmetric reactions, including hydrosilylation of olefins (Scheme 1), 1,4-hydroboration of 1,3-enynes (Scheme 2), and reduction of allylic esters with formic acid (Scheme 3), by use of palladium complexes coordinated with an axially chiral monodentate phosphine ligand, 2-(diphenylphosphino)-2'-methoxy-1,1'-binaphthyl (MeO-MOP (**1a**)) or its derivatives such as H-MOP-Ar or MOP-Phen.

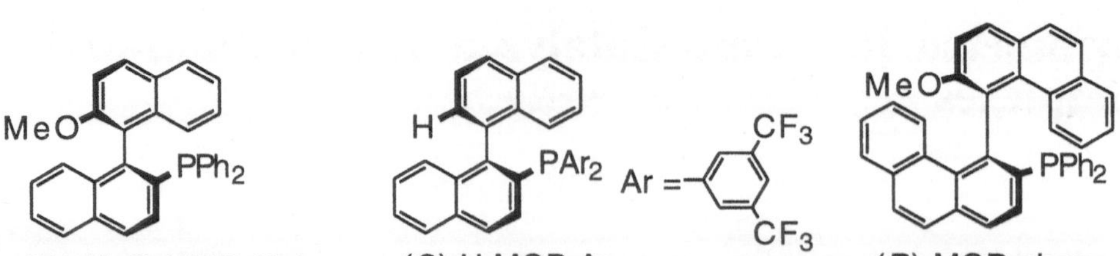

(R)-MeO-MOP (**1a**)   (S)-H-MOP-Ar   Ar =   (R)-MOP-phen

Herein I describe the preparation of the chiral monodentate phosphine ligands and their use in palladium-catalyzed asymmetric reactions, mainly reduction of allylic esters with formic acid.

Scheme 1

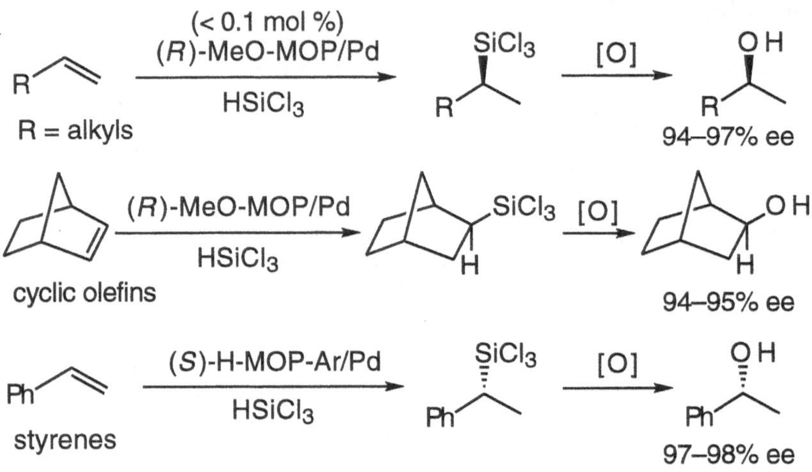

R   (< 0.1 mol %)
(R)-MeO-MOP/Pd
————————————→   SiCl₃   [O]   OH
HSiCl₃   94–97% ee
R = alkyls

cyclic olefins   (R)-MeO-MOP/Pd
————————————→   SiCl₃   [O]   OH
HSiCl₃   94–95% ee

styrenes   (S)-H-MOP-Ar/Pd
————————————→   SiCl₃   [O]   OH
HSiCl₃   97–98% ee
Ph

Scheme 2

R   (R)-MeO-MOP/Pd
————————————→
HB(O)(O)
    <61% ee

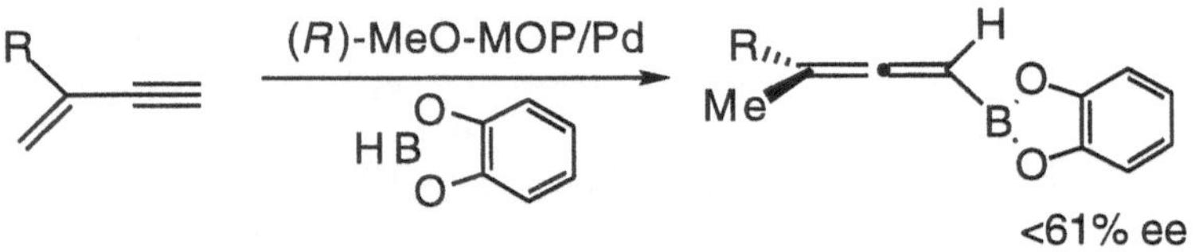

Scheme 3

R²   OCO₂R   (R)-MOP-phen/Pd
————————————→
R¹   HCOOH/NR₃
    <93% ee

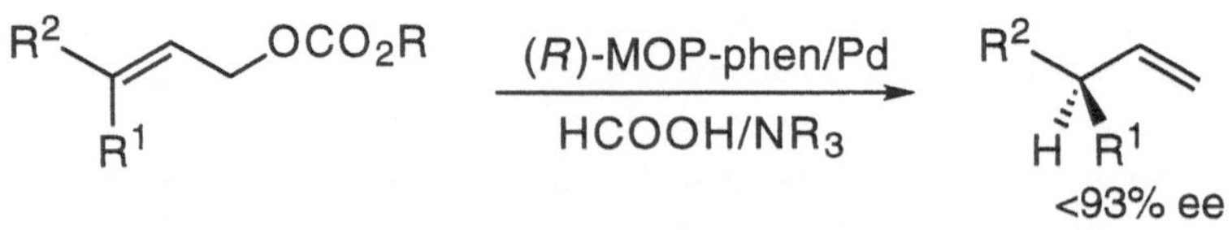

2 Preparation of MOP Ligands

The chiral binaphthyl skeleton was chosen as a basic structure of the monodentate phosphine ligand since in the case of using axially chiral binaphthyl compounds to construct an effective chiral template for asymmetric reactions there are numerous examples documented in the literature[1]. We have also attached importance to the introduction of a functional group onto the chiral ligand thus in doing so the latter would be expected to interact attractively with a functional group in the substrate[7]. Morgans and coworkers[8] have reported the selective monophosphinylation of 2,2'-bis(trifluoromethanesulfonyloxy)-1,1'-binaphthyl (2) with diphenylphosphine oxide in the presence of palladium catalyst giving a high yield of 2-diphenylphosphinyl-2'-trifluoromethanesulfonyloxy-1,1'-binaphthyl (3), which attracted our attention as a versatile starting compound for the preparation of chiral monophosphine ligands. Furthermore the triflate group on 3 was considered to be a convenient functionality for the introduction of various types of functional groups onto the binaphthyl.

The conversion of 3 into 2-(diphenylphosphino)-2'-methoxy-1,1'-binaphthyl (MeO-MOP, 1a) was achieved[9,10] in a high yield by the three step reactions shown in Scheme 4. Thus, triflate (S)-3 was hydrolyzed with aqueous sodium hydroxide to give 99% yield of alcohol, and its phenolic hydroxy group was alkylated by treatment with methyl iodide in the presence of potassium carbonate in acetone to give 99% yield of methyl ether (S)-4a. Reduction of phosphine oxide with trichlorosilane and triethylamine in refluxing xylene led to (S)-MeO-MOP (1a) in 97% yield. The overall yield from 2,2'-dihydroxy-1,1'-binaphthyl was about 90%. Similar phosphines containing benzyl ether and isopropyl ether, (S)-1b and (S)-1c, were also prepared by alkylation of the phenol oxygen with benzyl bromide and isopropyl iodide, respectively, followed by reduction of the phosphine oxide.

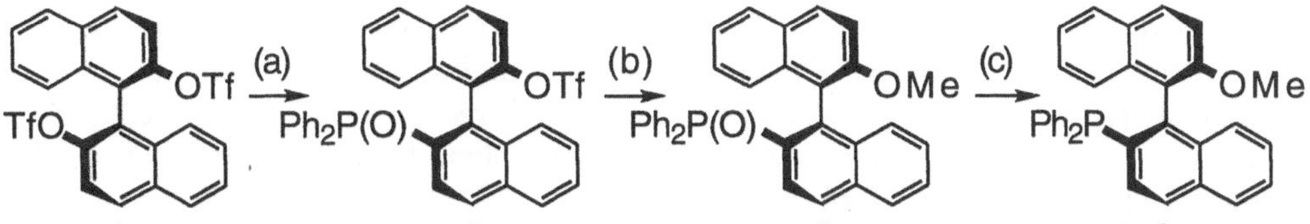

(a) Ph$_2$POH (2 eq), Pd(OAc)$_2$ (5 mol %), dppb (5 mol %), *i*-Pr$_2$NEt (4 eq), DMSO, 100 °C, 12 h (3, 95%). (b) (i) 4N NaOH, 1,4-dioxane, methanol. (ii) MeI (4 eq), K$_2$CO$_3$ (4 eq), acetone, reflux, 3 h (4a, 99%). (c) Et$_3$N (20 eq), HSiCl$_3$ (5 eq), xylene, 120 °C, 5 h, (1a, 97%).

Scheme 4

The trifluoromethanesulfonyloxy group on the 2' position can be substituted with an alkyl group by the nickel-catalyzed cross-coupling with the Grignard reagent. Introduction of ethyl group on 3 with ethylmagnesium bromide followed by the reduction with trichlorosilane gave (S)-1d in 64% overall yield. A cyano group can be also introduced at the 2' position of 3 in a quantitative yield by the nickel-catalyzed cyanation with potassium cyanide to give **1e** after the reduction of the phosphine oxide[11]. The reduction of the cyano group with diborane followed by methylation with formaldehyde/formic acid gave aminophosphine **1f**. The MOP ligands **1g** and **1h** which contain an ester and carboxylic acid, respectively, were prepared through palladium-catalyzed monocarbonylation of bis(triflate) **2** giving 2-carbomethoxy-2'-trifluoromethanesulfonyloxy-1,1'-binaphthyl[11]. The homochiral mono-phosphine containing biphenanthryl skeleton, MOP-phen (**5**), was also prepared by a sequence of the reactions from 3,3'-dihydroxy-4,4'-biphenanthryl which are essentially the same as those for the binaphthyl analog **1a**[12].

1a: X = OMe	**1f**: X = CH$_2$NMe$_2$
1b: X = OCH$_2$Ph	**1g**: X = COOMe
1c: X = OPr-i	**1h**: X = COOH
1d: X = Et	**1i**: X = OH
1e: X = CN	**1j**: X = H

(*S*)-MOP (**1**)

(*R*)-MOP-phen (**5**)

3 Asymmetric Reduction of Allylic Esters with Formic Acid

Palladium-catalyzed reduction of allylic esters with formic acid, a reaction which has been developed by Tsuji and coworkers[13], provides a convenient method for regioselective synthesis of less-substituted olefins. Mechanistic studies[14] on the catalytic reduction have revealed that the olefin is produced by reductive elimination from the key intermediate, Pd(II)(π-allyl)(hydrido)(L), which is generated by the decarboxylation of the palladium formate complex and that the use of monodentate phosphine ligand is essential for the high regioselectivity. We found that the catalytic asymmetric reduction forming optically active olefins is attained by use of the chiral monodentate phosphine ligand, (*R*)-MOP (**1**), and its biphenanthryl analog, (*R*)-MOP-phen (**5**)[12,15].

Reaction of geranyl methyl carbonate ((*E*)-**6**) with formic acid (2.2 equiv) and 1,8-bis(dimethylamino)naphthalene (1.2 equiv) in the presence of 1 mol % of a palladium catalyst, generated in situ by reacting Pd$_2$(dba)$_3$•CHCl$_3$ with (*R*)-MeO-MOP (P/Pd = 2/1), in dioxane at 20 °C for 16 h proceeded regioselectively to give a quantitative yield of (*S*)-3,7-dimethyl-1,6-octadiene (**7**) in 76% ee (Scheme 5). The reduction of *Z* carbonate, neryl methyl carbonate ((*Z*)-**6**), under the same reaction conditions gave the olefin (*R*)-**7** which has essentially the same enantiomeric purity (75% ee) but the opposite absolute configuration. Similarly, the reduction of racemic linalyl carbonate *dl*-**8** gave 82% yield of (*S*)-**7** in 55% ee. In contrast to the high catalytic activity and high regioselectivity observed with the MOP/palladium catalyst, the reduction is very slow and not regioselective with chelating bisphosphine ligands such as (*R*)-BINAP[2].

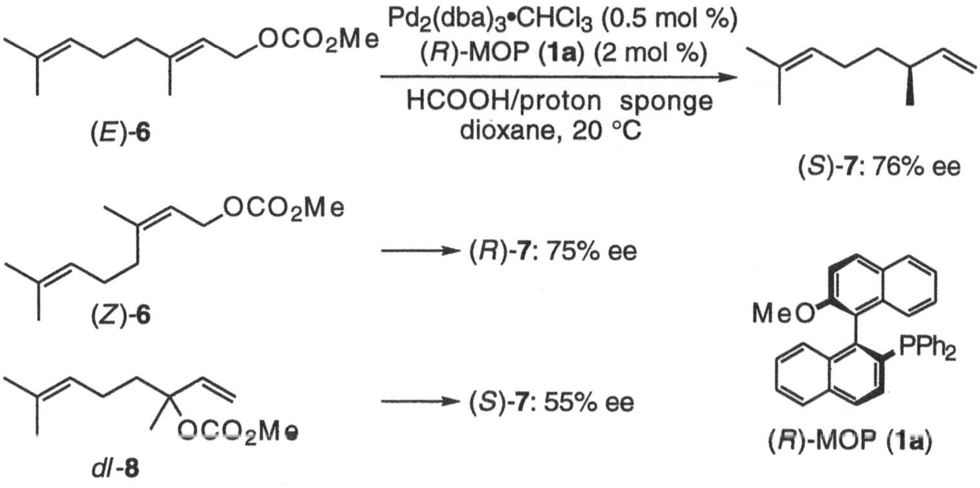

Scheme 5

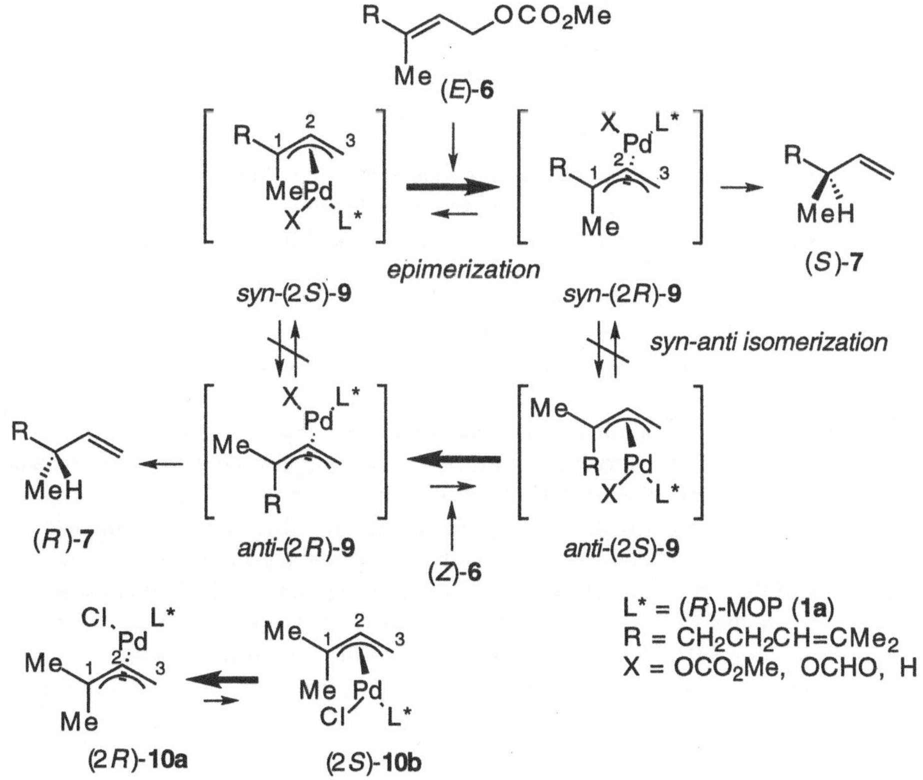

Scheme 6

The reduction of **6** must proceed via a π-{1-(4-methyl-3-pentenyl)-1-methylallyl}palladium(II) intermediate **9** which presumably undergoes *syn-anti* isomerization (*syn*-**9** ↔ *anti*-**9**) and epimerization (((2*R*)-**9** ↔ (2*S*)-**9**) by the σ-π-σ mechanism[16] (Scheme 6). The reversal of configuration of **7** observed for the asymmetric reduction of (*E*)-**6** and (*Z*)-**6** demonstrates that the rate of *syn-anti* isomerization of **9** is much slower than the rate of reduction in forming **7**. As a model for the key intermediate in the asymmetric reduction, the structure of PdCl(η³-1,1-dimethylallyl)((*R*)-MeO-MOP) (**10**), which is readily obtained by the reaction of [PdCl(η³-1,1-di-methylallyl)]₂ with (*R*)-MeO-MOP, was studied in solution and in a crystalline state.

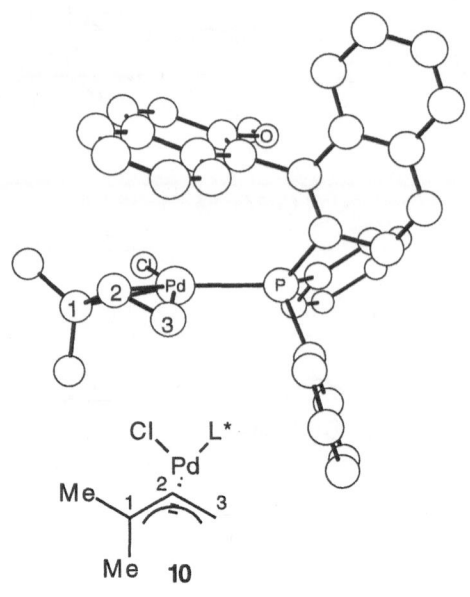

Figure 1. Molecular structure of PdCl(η³-1,1-dimethylallyl)((*R*)-MeO-MOP) (**10**)

¹H and ³¹P NMR studies of **10** in CDCl₃ revealed that the π-allylpalladium **10** exists as a mixture of isomers which are in an equilibrium state between –60 °C and 20 °C, the ratio of major isomer to minor isomer being 4.5 : 1, 5.1 : 1, and 6.5 : 1, at 20 °C, –20 °C, and -60 °C, respectively. Both isomers are determined by a large coupling ($^4J_{\text{H-P}}$ = *ca.* 9 and 5 Hz) between the methyls on C-1 carbon and the phosphorus atom which is consistent with the structures **10a** and **10b** (Scheme 6) whereby the C-1 carbon on the π-allyl is *trans* to phosphine. One of the significant features in the ¹H NMR was that the proton on the C-2 carbon of major diastereoisomer **10a** appeared at an unusually high field (2.67 ppm) compared with that of the minor isomer **10b** (4.55 ppm) and normal PdCl(π-allyl)(PR₃) complexes. The X-ray crystal structure of the palladium complex **10** (Figure 1) obtained by recrystallization from benzene and ether shows that the C-2 proton is in close proximity to the naphthyl ring substituted with methoxy group, which will cause the high field shift of the C-2 proton. Thus, the conformation in the X-ray crystal structure is assumed to be similar to that of the major isomer **10a** in solution. The stereochemical outcome in forming (*S*)-**7** observed in the catalytic asymmetric reduction of (*E*)-**6** is accounted for by the reductive elimination of the hydrido and π-allyl from the intermediate *syn*-(2*R*)-**9** (X = H) after the equilibration between *syn*-(2*R*)-**9** and *syn*-(2*S*)-**9**, the former possessing the same configuration of π-allyl moiety as **10a**. Similarly, (*R*)-**7** is formed from *anti*-(2*R*)-**9** after the equilibration with *anti*-(2*S*)-**9**. An epimerization ensuing from this equilibration was demonstrated by the reduction of racemic linalyl carbonate **8** which gave nonracemic product (55% ee (*S*)).

The monodentate phosphine (R)-MOP-phen (**5**) that contains axially chiral biphenanthryl skeleton was found to be a more enantioselective ligand than the binaphthyl ligand MeO-MOP (**1a**) for the palladium-catalyzed asymmetric reduction of allylic esters. Thus, the use of **5** for the reduction of (E)-**6** and (Z)-**6** increased the enantioselectivity to 85% ee and 82% ee, respectively (Scheme 7). The enantioselectivity in the reduction of (E)-3-cyclohexyl-2-propenyl carbonate (**11a**) and (E)-3-phenyl-2-propenyl carbonate (**11b**) was also improved by use of MOP-phen (**5**).

Asymmetric syntheses of optically active olefins which all bear a deuterium atom at the stereogenic center in the allylic position were also successful by use of formic acid-d_2 (DCOOD). Thus, the reaction of (E)-**6** and (E)-**11a** with formic acid-d_2 in the presence of 1,8-bis(dimethylamino)naphthalene and the palladium-MOP-phen catalyst introduced deuterium selectively at the stereogenic center to give the corresponding deuterated olefins in 84~85% ee, the enantioselectivity being essentially the same as the reduction with HCOOH. No deuterium scrambling was observed in the reduction products.

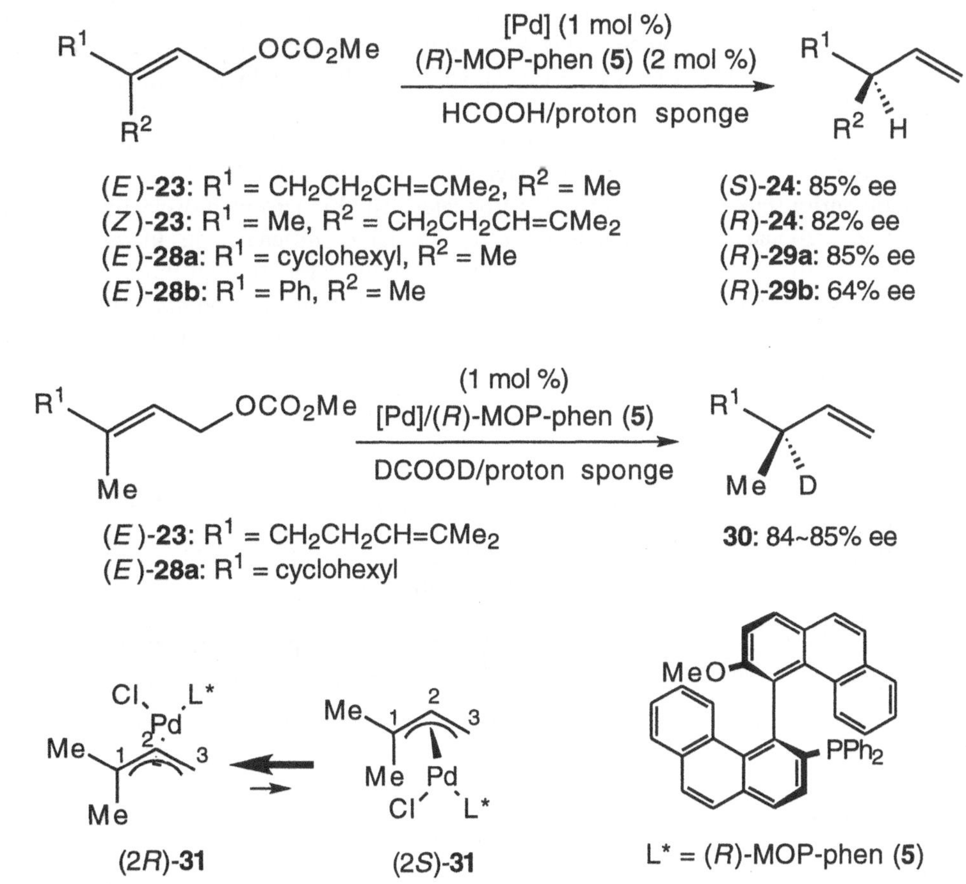

$$R^1 \diagdown \!\!\!\diagup OCO_2Me \xrightarrow[\text{HCOOH/proton sponge}]{\substack{\text{[Pd] (1 mol \%)} \\ (R)\text{-MOP-phen (5) (2 mol \%)}}} R^1 \diagdown \!\!\!\diagup$$

with R² below starting material and R² H below product.

(E)-**23**: R¹ = CH₂CH₂CH=CMe₂, R² = Me   (S)-**24**: 85% ee
(Z)-**23**: R¹ = Me, R² = CH₂CH₂CH=CMe₂   (R)-**24**: 82% ee
(E)-**28a**: R¹ = cyclohexyl, R² = Me   (R)-**29a**: 85% ee
(E)-**28b**: R¹ = Ph, R² = Me   (R)-**29b**: 64% ee

$$R^1 \diagdown \!\!\!\diagup OCO_2Me \xrightarrow[\text{DCOOD/proton sponge}]{\substack{\text{(1 mol \%)} \\ \text{[Pd]/}(R)\text{-MOP-phen (5)}}} R^1 \diagdown \!\!\!\diagup$$

with Me below starting material and Me D below product.

(E)-**23**: R¹ = CH₂CH₂CH=CMe₂
(E)-**28a**: R¹ = cyclohexyl

30: 84~85% ee

(2R)-**31**   (2S)-**31**   L* = (R)-MOP-phen (**5**)

Scheme 7

The manner in which the MOP-phen ligand coordinates to π-allylpalladium was also studied by X-ray crystal structure analysis of the palladium complex PdCl(η^3-1,1-dimethylallyl)((*R*)-MOP-phen)) (**14**) [12]. The basic structure around palladium atom in **14** is similar to that of its binaphthyl analog **10** (see Figure 1). Thus, the η^3-1,1-dimethylallyl group coordinates to palladium with the absolute configuration of *R* at the C-2 position, and the C-1 carbon on the π-allyl, which is substituted with two methyl groups, is *trans* to the phosphorus. ^{1}H and ^{31}P NMR studies of PdCl(η^3-1,1-dimethylallyl)((*R*)-MOP-phen) (**14**) in $CDCl_3$ revealed that the π-allylpalladium **14** exists as a mixture of isomers which are in an equilibrium state between –60 °C and 20 °C, the ratios of main isomer (2*R*)-**14** to minor isomer (2*S*)-**14** are 6 : 1, 10 : 1, and 13 : 1 at 20 °C, –20 °C, and –60 °C, respectively. The ratios are higher than those observed for the palladium-MOP complex **10** (*vide supra*), which is consistent with the higher enantioselectivity of MOP-phen (**5**) than MeO-MOP (**1a**) for the catalytic asymmetric reduction.

The catalytic asymmetric reduction was applied to the synthesis of optically active allylic silanes that have a stereogenic carbon center at the α-position and thus would be considered as useful chiral reagents in organic synthesis[17] (Scheme 8). For example, the reduction of 3-alkyl-3-trialkylsilyl-2-propenyl methyl carbonates **15** with formic acid in the presence of the palladium/MOP-phen catalyst proceeded at 20 °C to give high yields (>90%) of allylsilanes **16** in up to 91% ee. The allylsilanes were treated with trimethylacetaldehyde in the presence of titanium tetrachloride to give optically active homoallyl alcohols by chirality transfer during the S_E' reaction[18]. The asymmetric reduction of the allyl carbonates bearing *E* double bond under the same reaction conditions gave the corresponding allylsilanes with lower enantiomeric purity than those from the *Z* carbonates. The higher enantioselectivity of the reaction of *Z* carbonates was accounted for by the larger difference in thermodynamic stability between the epimeric pairs of *anti*-π-allylpalladium complex than those of the *syn* isomer, which was demonstrated by ^{1}H and ^{31}P NMR studies of *syn* and *anti* isomers of a model π-allylpalladium complex, PdCl(η^3-1-methyl-1-(trimethylsilyl)allyl)((*R*)-MeO-MOP) (**17**).

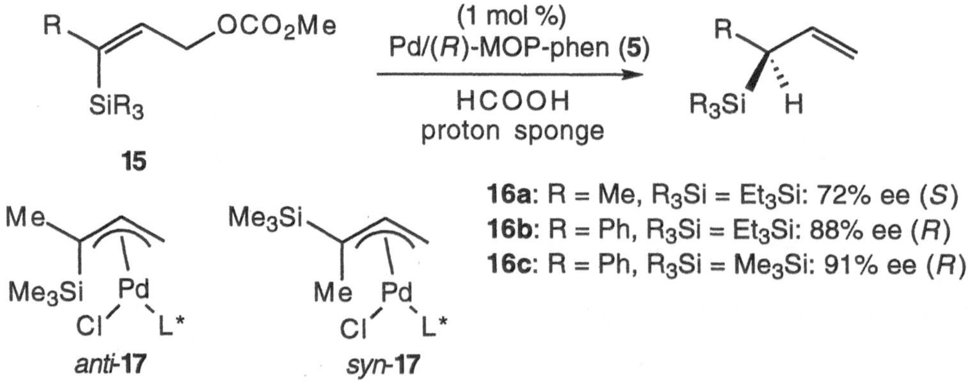
R
OCO2Me
SiR3
15
(1 mol %)
Pd/(R)-MOP-phen (5)
HCOOH
proton sponge
R
R3Si
H
16a: R = Me, R3Si = Et3Si: 72% ee (S)
16b: R = Ph, R3Si = Et3Si: 88% ee (R)
16c: R = Ph, R3Si = Me3Si: 91% ee (R)
Me
Me3Si
Pd
Cl
L*
anti-17
Me3Si
Me
Pd
Cl
L*
syn-17

Scheme 8

The allylic esters used for the asymmetric reduction are limited to those with a geometrically pure *E* or *Z* double bond for the high enantioselectivity, because the *E* and *Z* isomers produce the enantiomeric olefins. The reversal of the configuration is exemplified by the reaction of geranyl and neryl esters (see Scheme 5). However it was found that racemic tertiary allylic esters can be also used for the asymmetric reduction, if one of the alkyl groups at the α position is a sterically bulky group (Scheme 9)[19]. Since the racemic tertiary allylic esters are readily obtained through a ketone and the vinyl Grignard reagent, it provides a practical method for the synthesis of optically active olefins. For example, the racemic ester **18a**, obtained from tetralone, gave reduction product **19a** (87% yield) with 93% enantiomeric purity. Interestingly the asymmetric reduction of *dl*-**18a** is much faster than that of its regioisomeric ester, 3,3-disubstituted-2-propenyl carbonate (*E*)-**20**.

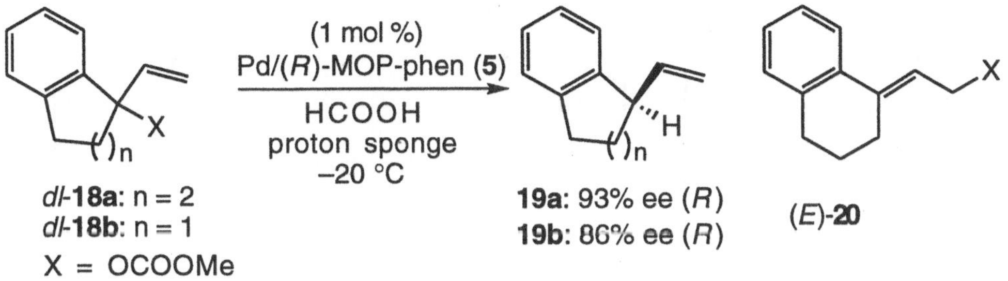

(1 mol %)
Pd/(R)-MOP-phen (5)
HCOOH
proton sponge
–20 °C
X
n
H
n
X
dl-18a: n = 2
dl-18b: n = 1
X = OCOOMe
19a: 93% ee (R)
19b: 86% ee (R)
(E)-20

Scheme 9

The reduction of (*E*)-**20** did not take place at the reaction temperature of 0 °C or lower. At 20 °C it gave (*R*)-**19a** in 83% ee, the stereoselectivity being essentially the same as that for *dl*-**18a** at 20 °C. The lower reactivity of (*E*)-**20** is ascribed to the two alkyl substituents at the 3 position of (*E*)-**20**. The steric hindrance retards the oxidative addition step in the catalytic cycle which takes place in an S_N' manner[20].

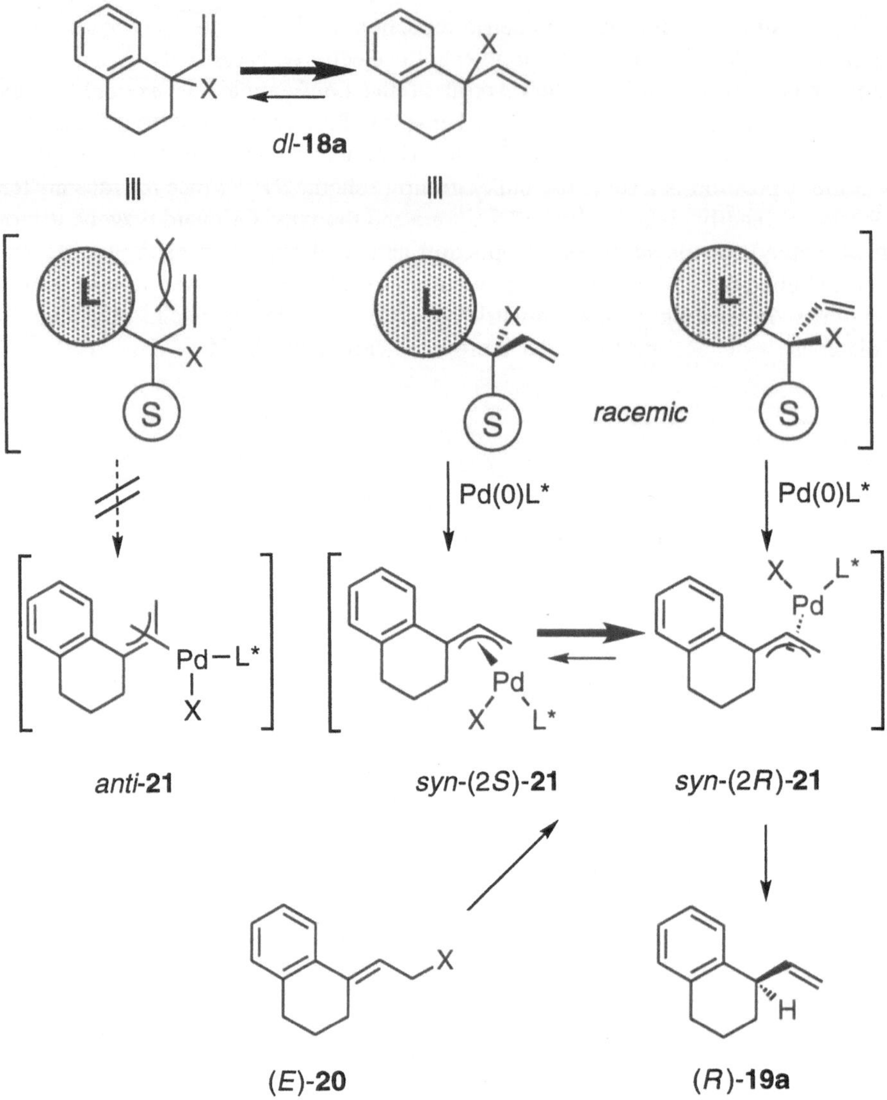

Scheme 10

The stereochemical results in the reduction of *dl*-**18a** and (*E*)-**20** is illustrated in Scheme 10. The π-allylpalladium intermediate resulting from (*E*)-**20** should be *syn*-**21**, which contains the aromatic ring at *syn* position with respect to the hydrogen at 2 position of π-allyl. The same stereochemical outcome in the reaction of (*E*)-**20** and *dl*-**18a** indicates that the π-allylpalladium intermediate formed from *dl*-**18a** is also *syn*-**21**, and the configuration *R* of the product **19a** indicates that the configuration of the predominant π-allylpalladium intermediate is *syn*-(2*R*)-**21** in both cases. In the reaction of racemic

180

1,1-disubstituted-2-propenyl ester *dl*-**18** where one of the substituents on the 1 position is much bigger than the other, the allyl ester undergoes the oxidative addition with the conformation forming a π-allylpalladium intermediate with the bigger alkyl group substituted at the *syn* position. After the epimerization between *syn*-(2R)-**21** and *syn*-(2S)-**21** the product (*R*)-**19a** is formed from thermodynamically more stable *syn*-(2R)-**21**.

The palladium-MOP catalyst is also effective for the asymmetric reduction of racemic 5-carbomethoxy-2-cyclohexenyl methyl carbonates **22** which proceeds through palladium intermediates **24** that have meso-type π-allyl. The reduction of *cis*-**22** and *trans*-**22** gave (*S*)-4-carbomethoxycyclohexene (**23**) (87% ee) and (*R*)-**23** (77% ee), respectively (Scheme 11).

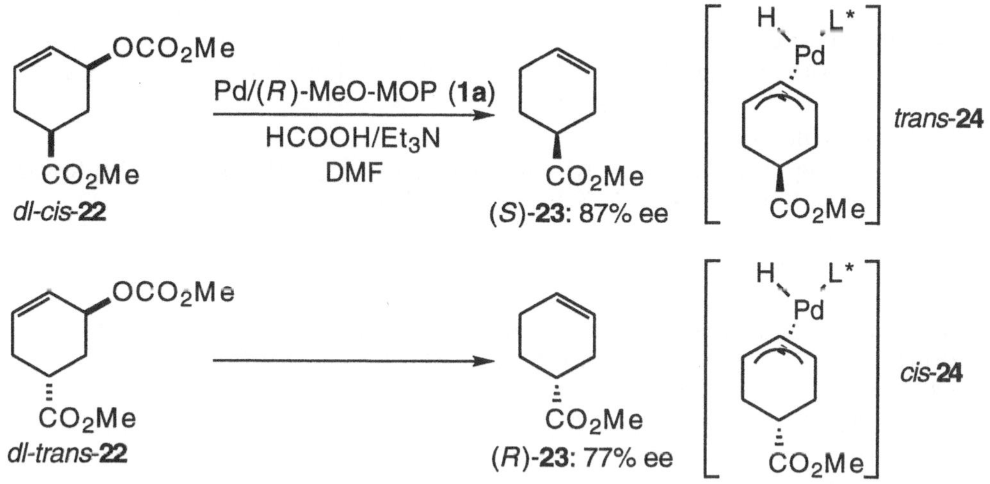

Scheme 11

4 Other Catalytic Asymmetric Reactions

Monodentate optically active phosphine ligand MeO-MOP (**1a**) have also been found to be very useful for palladium-catalyzed asymmetric hydrosilylation of olefins. High stereoselectivity was observed in both enantioface-selective hydrosilylation of alkyl-substituted terminal olefins[21] and enantioposition-selective reaction of bicyclic olefins such as norbornene[22]. The catalytic asymmetric hydrosilylation provides the most efficient and practical route to optically active alcohols (92–97% ee). For the hydrosilylation of styrene derivatives, H-MOP-Ar was much more enantioselective than MeO-MOP to give the corresponding benzylic alcohols of over 97% ee[23]. Conjugated dienes such as cyclopentadiene also underwent the asymmetric hydrosilylation in the presence of MOP-phen–palladium catalyst to give allylic silanes of up to 80% ee[24]. Palladium-catalyzed addition of catecholborane to but-1-en-3-ynes in the presence of MeO-MOP ligand proceeds in a 1,4-fashion to give axially chiral allenylboranes[25].

For the palladium-catalyzed asymmetric reactions shown above, any chelating bisphosphine ligands can not be used. The reduction of allylic esters are very slow with bisphosphines and the regioselectivity in forming terminal olefins is low. Hydrosilylation does not take place even at higher reaction temperature with bisphosphine ligands. The hydroboration of but-1-en-3-ynes proceeds in a 1,2-fashion giving dienylboranes in place of allenylboranes. They must be carried out in the presence of a monodentate phosphine ligand. The MOP ligands will find further application in some new catalytic asymmetric reactions.

References

1. For recent reviews: (a) Brunner, H. *Synthesis* **1988**, 645. (b) Brunner, H. *Top. Stereochem.* **1988**, *18*, 129. (c) Noyori, R.; Kitamura, M. *Modern Synthetic Methods*; Scheffold, R., Ed.; Springer-Verlag: New York, **1989**; Vol. 5, p 115. (d) Ojima, I.; Clos, N.; Bastos, C. *Tetrahedron* **1989**, *45*, 6901. (e) Ojima, I. *Catalytic Asymmetric Synthesis*; VCH: New York, **1993**. (f) Noyori, R. *Asymmetric Catalysis in Organic Synthesis*; Wiley: New York, **1994**.

2. (*R*)-2,2'-Bis(diphenylphosphino)-1,1'-binaphthyl: (a) Takaya, H.; Mashima, K.; Koyano, K.; Yagi, M.; Kumobayashi, H.; Taketomi, T.; Akutagawa, S.; Noyori, R. *J. Org. Chem.* **1986**, *51*, 629. (b) Noyori, R. *Chem. Soc. Rev.* **1989**, *18*, 187 and references cited therein. (c) Noyori, R.; Takaya, H. *Acc. Chem. Res.* **1990**, *23*, 345 and references cited therein.

3. (a) Hayashi, T.; Kishi, K.; Yamamoto, A.; Ito, Y. *Tetrahedron Lett.* **1990**, *31*, 1743. (b) Hayashi, T.; Yamamoto, A.; Ito, Y.; Nishioka, E.; Miura, H.; Yanagi, K. *J. Am. Chem. Soc.* **1989**, *111*, 6301. (c) Yamamoto, A.; Ito, Y.; Hayashi, T. *Tetrahedron Lett.* **1989**, *30*, 375. (d) Hayashi, T.; Kanehira, K.; Hagihara, T.; Kumada, M. *J. Org. Chem.* **1988**, *53*, 113. (e) Hayashi, T.; Yamamoto, A.; Ito, Y. *Tetrahedron Lett.* **1988**, *29*, 99. (f) Hayashi, T.; Yamamoto, A.; Ito, Y. *Chem. Lett.* **1987**, 177. (g) Hayashi, T.; Yamamoto, A.; Hagihara, T.; Ito, Y. *Tetrahedron Lett.* **1986**, *27*, 191.

4. (a) Hayashi, T.; Sawamura, M.; Ito, Y. *Tetrahedron* **1992**, *48*, 1999. (b) Sawamura, M.; Ito, Y.; Hayashi, T. *Tetrahedron Lett.* **1989**, *30*, 2247. (c) Ito, Y.; Sawamura, M.; Hayashi, T. *Tetrahedron Lett.* **1988**, *29*, 239. (d) Ito, Y.; Sawamura, M.; Shirakawa, E.; Hayashizaki, K.; Hayashi, T. *Tetrahedron Lett.* **1988**, *29*, 235. (e) Ito, Y.; Sawamura, M.; Shirakawa, E.; Hayashizaki, K.; Hayashi, T. *Tetrahedron* **1988**, *44*, 5253. (f) Ito, Y.; Sawamura, M.; Hayashi, T. *Tetrahedron Lett.* **1987**, *28*, 6215. (g) Ito, Y.; Sawamura, M.; Hayashi, T. *J. Am. Chem. Soc.* **1986**, *108*, 6405. (h) Hayashi, T.; Uozumi, Y.; Yamazaki, A.; Sawamura, M.; Hamashima, H.; Ito, Y. *Tetrahedron Lett.* **1991**, *32*, 2799.

5. Examples of optically active monophosphine ligands: (a) (*S*)-(*o*-methoxyphenyl)-cyclohexylmethylphosphine ((*S*)-CAMP): Knowles, W. S.; Sabacky, M. J.; Vineyard, B. D. *J. Chem. Soc., Chem. Commun.* **1972**, 10. (b) Neomentyldiphenylphosphine: Morrison, J. D.; Burnett, R. E.; Aguiar, A. M.; Morrow, C. J.; Phillips, C. *J. Am. Chem. Soc.* **1971**, *93*, 1301.

6. Hayashi, T.; Hayashizaki, K.; Kiyoi, T.; Ito, Y. *J. Am. Chem. Soc.* **1988**, *110*, 8153.

7. (a) Hayashi, T. *Pure Appl. Chem.* **1988**, *60*, 7. (b) Sawamura, M.; Ito, Y. *Chem. Rev.* **1992**, *92*, 857.

8. Kurz, L.; Lee, G.; Morgans, D., Jr.; Waldyke, M. J.; Ward, T. *Tetrahedron Lett.* **1990**, *31*, 6321.

9. Uozumi, Y.; Hayashi, T. *J. Am. Chem. Soc.* **1991**, *113*, 9887.

10. Uozumi, Y.; Tanahashi, A.; Lee, S.-Y.; Hayashi, T. *J. Org. Chem.* **1993**, *58*, 1945.

11. Uozumi, Y.; Suzuki, N.; Ogiwara, A.; Hayashi, T. *Tetrahedron* **1994**, *50*, 4293.

12. Hayashi, T.; Iwamura, H.; Uozumi, Y.; Matsumoto, Y.; Ozawa, F. *Synthesis* **1994**, 526.

13. (a) Tsuji, J.; Yamakawa, T. *Tetrahedron Lett.* **1979**, 613. (b) Tsuji, J.; Shimizu, I.; Minami, I. *Chem. Lett.* **1984**, 1017. (c) Tsuji, J.; Minami, I.; Shimizu, I. *Synthesis* **1986**, 623. (d) Mandai, T.; Matsumoto, T.; Kawada, M.; Tsuji, J. *J. Org. Chem.* **1992**, *57*, 1326.

14. Oshima, M.; Shimizu, I.; Yamamoto, A.; Ozawa, F. *Organometallics* **1991**, *10*, 1221.

15. Hayashi, T.; Iwamura, H.; Naito, M.; Matsumoto, Y.; Uozumi, Y.; Miki, M.; Yanagi, K. *J. Am. Chem. Soc.* **1994**, *116*, 775.

16. For reviews: (a) Consiglio, G.; Waymouth, R. M. *Chem. Rev.* **1989**, *89*, 257. (b)Frost, C. G.; Howarth, J.; Williams, J. M. J. *Tetrahedron Asymmetry* **1992**, *3*, 1089.

17. Hayashi, T.; Iwamura, H.; Uozumi, Y. *Tetrahedron Lett.* **1994**, *35*, 4813.

18. (a) Hayashi, T.; Konishi, M.; Ito, H.; Kumada, M. *J. Am. Chem. Soc.* **1982**, *104*, 4962. (b) Hayashi, T.; Konishi, M. Kumada, M. *J. Org. Chem.* **1983**, *48*, 281. (c) Hayashi, T.; Konishi, M.; Okamoto, Y.; Kabeta, K.; Kumada, M. *J. Org. Chem.* **1986**, *51*, 3772.

19. Hayashi, T.; Kawatsura, M.; Iwamura, H.; Yamaura, Y.; Uozumi, Y. *J. Chem. Soc., Chem. Commun.* **1996**, 1767.

20. Hayashi, T.; Yamamoto, A.; Ito, Y.*Tetrahedron Lett.* **1987**, *28*, 4837.

21. (a) Uozumi, Y.; Hayashi, T. *J. Am. Chem. Soc.* **1991**, *113*, 9887. (b) Uozumi, Y.; Kitayama, K.; Hayashi, T., Yanagi, K.; Fukuyo, E. *Bull. Chem. Soc. Jpn.* **1995**, *68*, 713.

22. (a) Uozumi, Y.; Lee, S.-Y.; Hayashi, T. *Tetrahedron Lett.* **1992**, *33*, 7185. (b) Uozumi, Y.; Hayashi, T. *Tetrahedron Lett.* **1993**, *34*, 2335.

23. (a) Uozumi, Y.; Kitayama, K.; Hayashi, T. *Tetrahedron Asymmetry* **1993**, *4*, 2419. (b) Kitayama, K.; Uozumi, Y.; Hayashi, T. *J. Chem. Soc., Chem. Commun.* **1995**, 1533. (c) Kitayama, K.; Hirate, S.; Uozumi, Y.; Hayashi, T. to be submitted.

24. Kitayama, K.; Tsuji, H.; Uozumi, Y.; Hayashi, T. *Tetrahedron Lett.* **1996**, *37*, 4169.

25. Matsumoto, Y.; Naito, M.; Uozumi, Y.; Hayashi, T. *J. Chem. Soc., Chem. Commun.* **1993**, 1468.

New Chiral Ferrocenyl Ligands in Asymmetric Catalysis: Applications and Mechanistic Studies

Antonio Togni

Laboratorium für Anorganische Chemie der ETH-Zürich,
ETH-Zentrum, CH-8092 Zürich, Switzerland

1 Introduction

Chiral ferrocenyl ligands have been known for more than two decades and constitute nowaday one of the most important classes of auxiliaries used in asymmetric catalysis[1], as they find application in a variety of transition-metal assisted reactions. From, e.g., Rh-catalyzed hydrogenations, to Ni- and Pd-catalyzed Grignard cross-coupling reactions, to the remarkable Au(I)-catalyzed aldol-type formation of oxazolines from aldehydes and isocyanoacetates, these ligands often afford remarkably high stereoselectivities.

From a structural point of view, one can identify three major types of ferrocenyl ligands, as illustrated by the specific examples **1**, **2**, and **3-4** below.

Derivative **1** is a representative of the first generation, mainly developed by Hayashi and co-workers[2]. It contains the 1,1'-bis(diphenylphosphino)ferrocene unit and bears a functionalized side-chain attached to the pseudo benzylic stereogenic center at position 2 of the ferrocene core. The latter substituent is the most prominent feature, because it may be readily varied, in order to fulfil the purpose of a secondary interaction with the substrate in the course of the catalytic reaction[3]. Less important are the related compounds lacking the 1'-phosphino group. It is to be noted that virtually all known derivatives of type **1** have in common the *diphenyl*phosphino group(s). The C_2-symmetric biferrocene **2** and its derivative containing alkyl phosphine substituents have been developed in recent years by Ito and co-workers[4]. These ligands have the peculiarity to span a *trans* coordination geometry.

The discussion in the present article will be focussed on compounds of type **3** and **4**, constituting the new generation[5]. These bidentate, asymmetric ligands are characterized by the presence of two ligand atoms (P,P or P,N) imbedded in different and easily modifiable steric and electronic environments.

From a synthetic point of view all compounds of type **1-4** have in common the chiral, enantiopure ferrocenyl amine **5** as starting material. As illustrated in Eq. 1, ortho lithiation of **5** occurs diastereoselectively[6], leading, after reaction with a suited electrophile, to a *planar-chiral* derivative[7], that can usually be isolated as single stereoisomer after recrystallization. The second important synthetic step in this chemistry is the S_N1-type substitution reaction of the dimethylamino group, affected with a variety of heteroatom nucleophiles in polar solvents. It is important to note that this nucleophilic substitution occurs with retention of configuration, by virtue of the participation of the ferrocene moiety in stabilizing the intermediate carbocationic species[8].

$$\text{5} \qquad \text{6} \qquad \text{7} \tag{1}$$

2 Synthesis, Structure, and Coordination Behavior of New - Generation Ferrocenyl Ligands

The preparation of ligands of type **3** and **4** follows the general scheme of Eq. 1. Thereby is to note that the reaction of the intermediate amine **6** with a) a secondary phosphine, and b) a pyrazole occurs cleanly in acetic acid at temperatures between 50° and 90°, typically. Only the use of this solvent allows to circumvent the conversion of amine **6** to the corresponding 1-ferrocenylethyl acetate which is known to undergo substitution reactions.

A series of diphosphines have been prepared in high yields, as shown in Scheme 1. These compounds are usually crystalline, air-stable materials[9]. The preparation of *Phobyphos* is worth noting, since this derivative is an example of a ligand containing the rare fragment [3.3.1]-9-phosphabicyclonon-9-yl (see Scheme 2). The reaction of an excess of „Phobane" with the amine **6** leads to a kinetic separation of the two „Phobane" isomers[10], otherwise not directly available in pure form.

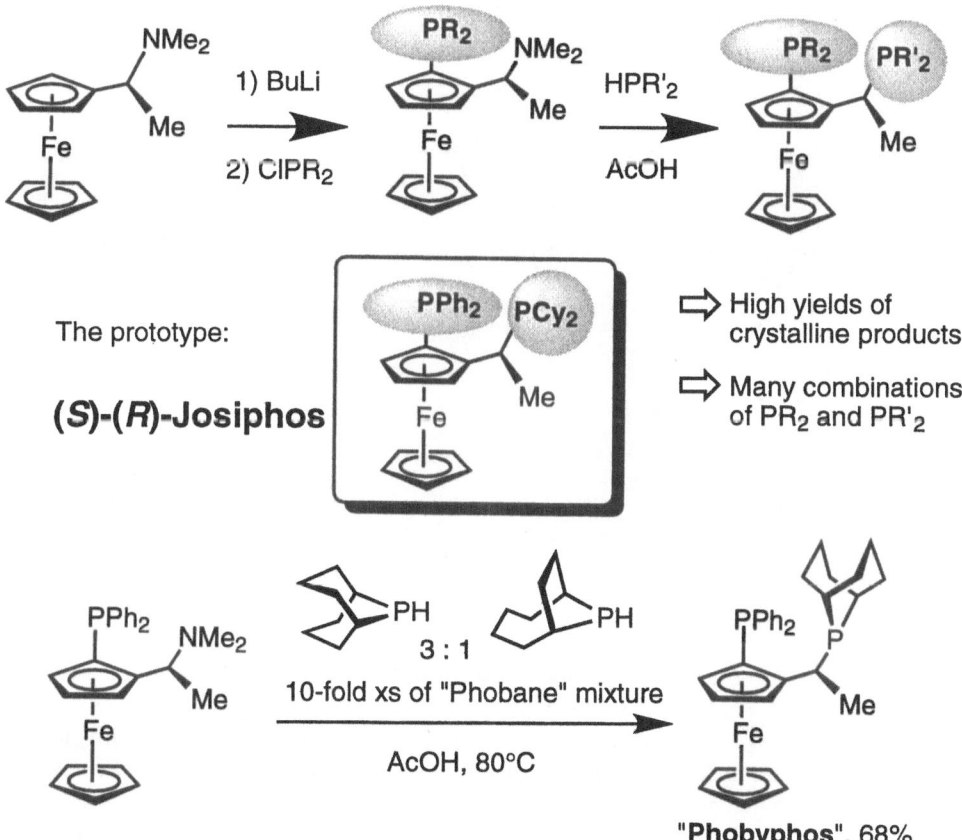

Scheme 1. Preparation of ferrocenyl diphosphines

Several transition-metal complexes containing these chelating diphosphines have been characterized by X-ray crystallography[11]. The generally observed P-M-P bite-angle of the ligands is typically in the range between 93° and 97°, although values slightly below 90° have already been detected[12]. An important feature of these compounds is their ability to maintain a well-defined conformation, no matter which metal they are coordinated to and despite the varying nature of the ancillary ligands. In the case of Josiphos, the schematic superposition of several structures depicted in Figure 1, illustrates this behavior. Thus, one recognizes the constant orientation of the substituents on the P-atoms. The diphenyl-phosphino fragments displays an *endo-axial-* and an *exo-equatorial* orientation of the phenyl groups (*endo* and *exo* refer to the position relative to the ferrocene core), whereas the PCy_2 fragment always adopts an axial orientation, thus minimizing steric interactions with the ferrocene. These conformational characteristics lead also to a maximization of the metal-ferrocene distance.

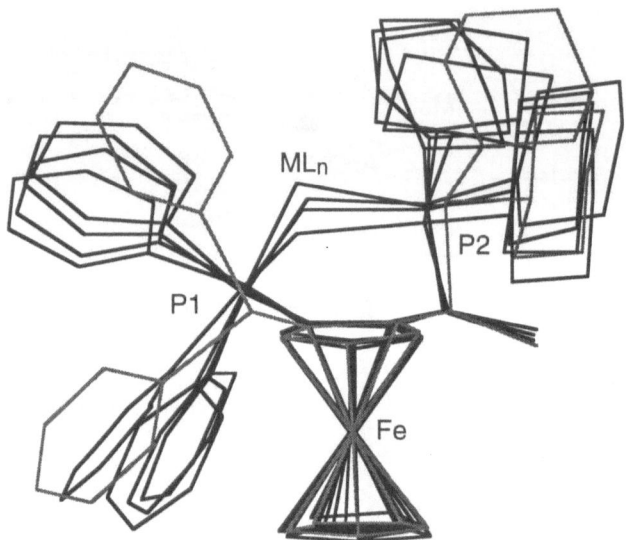

Figure 1. Superposition of the X-ray structures of Josiphos (dashed lines) and four of its complexes (M = Ru, Rh, Pd, and Pt)

Ligands containing a pyrazole instead of a phosphino group attached to the stereogenic center, are prepared in a much similar way, as compared to their diphosphine congeners[13]. Pyrazoles have been chosen because of the opportunity to easily modulate their steric and electronic properties, by variation of the substituents, in particular at the positions 3 and 5, adjacent to the nitrogen atoms (see Scheme 2).

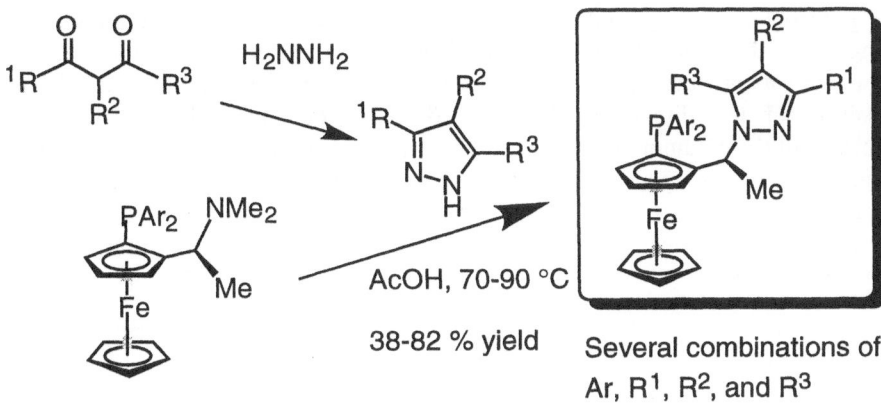

Scheme 2. Synthesis of pyrazole-containing ferrocenyl ligands

The reaction of the pyrazoles with ferrocenylamine **6** shows a clean regioselectivity, in case of different substituents R^1 and R^3. The sterically more bulky and/or more electron-withdrawing substituent always takes the position of R^1, i.e. remote from the ferrocene. An interpretation of this observation, as well as the discussion of few exceptions to this rule, have been reported previously[13]. Pyrazoles are prepared by reacting hydrazine with either a 1,3-diketone or, in the case of 3(5)-monosubstituted derivatives, a ketoenamine[14], as illustrated by the particular example of Scheme 3[15].

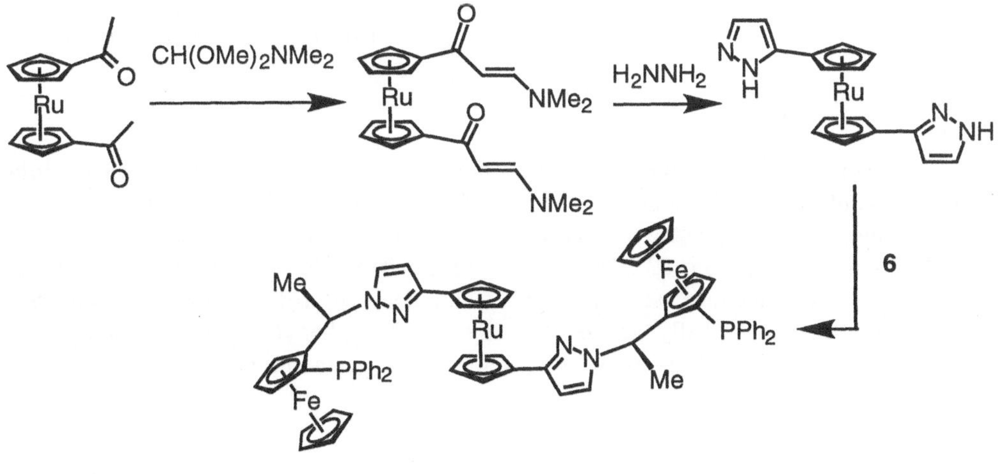

Scheme 3. Synthesis of a bis(ferrocenyphosphine/pyrazole) ligand

The pyrazole-containing ligands are structurally well behaved and show a typical conformation in the solid state in which the heterocycle and one phosphine phenyl group occupy pseudo axial positions with respect to the ferrocene moiety. Figure 2 shows a superposition of ten different derivatives. From it one recognizes that the overall conformation of the common backbone is essentially not influenced by substituents on the pyrazole, the phenyl groups, or the „lower" Cp ring. This is an important structural aspect, since it allows to use the substituents 1) for controlling the electronic properties of the ligand, and 2) for introducing specific steric interactions within the corresponding metal complexes (see below).

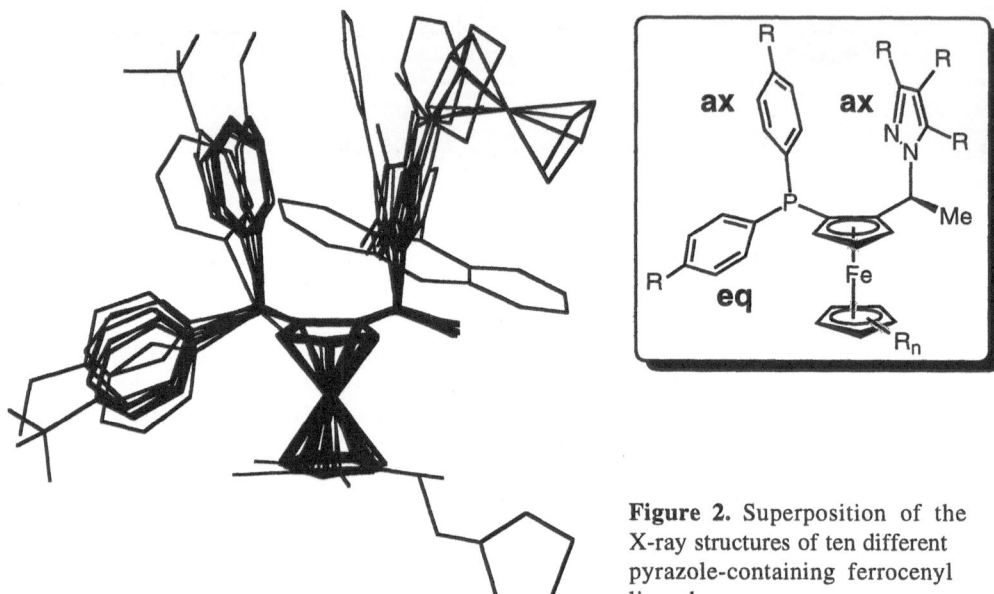

Figure 2. Superposition of the X-ray structures of ten different pyrazole-containing ferrocenyl ligands

The pyrazole ligands coordinate in a chelating fashion to transition-metals, maintaining the general conformational features observed in their free state. This is particularly the case for Rh(I) complexes. Despite the relatively large seven-membered chelate ring, the bite angle P-Rh-N is in the range 85°-87°. In the case of Pd(II)-π-allyl complexes, however, the corresponding bite angle increases to 92°-95°. A consequence of this is the less pronounced differentiation between axially and equatorially oriented phenyl groups on phosphorus. Examples of structures of Rh(I) and Pd(II) complexes are given in Figures 3-4, respectively.

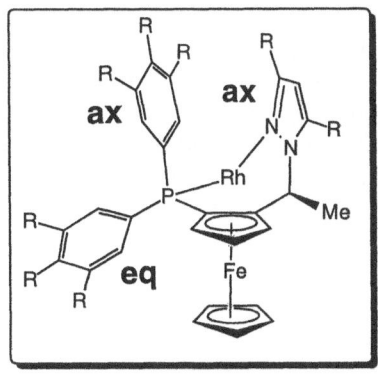

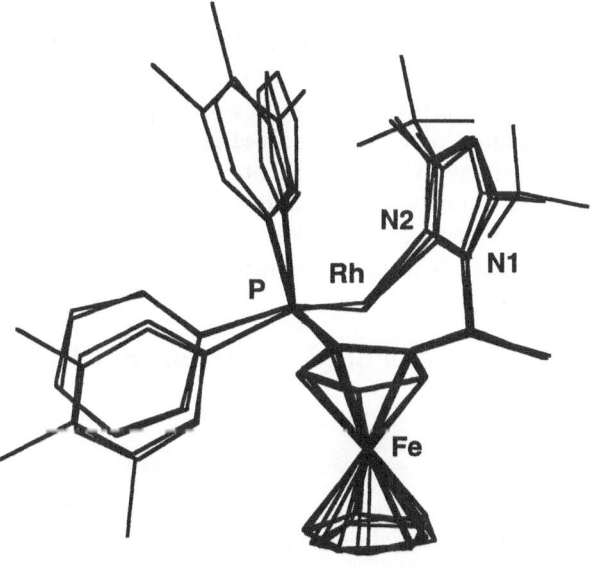

Figure 3. Superposition of the X-ray structures of four different Rh(I) complexes containing pyrazole ferrocenyl ligands. Ancillary ligands are omitted for clarity

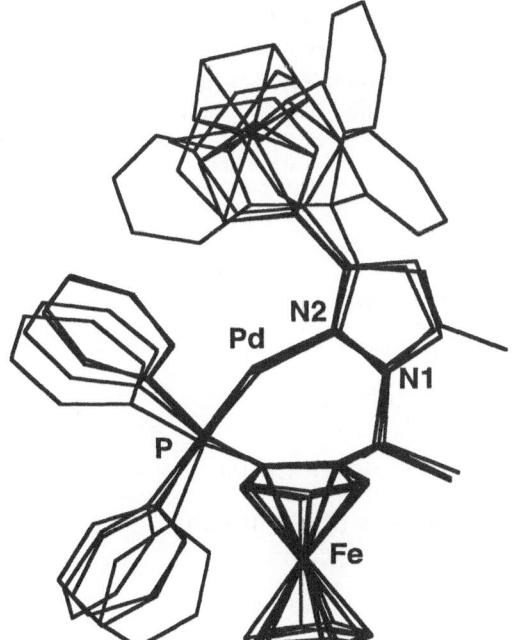

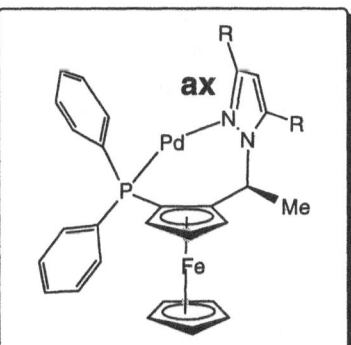

Figure 4. Superposition of the X-ray structures of four different Pd(II)-π-allyl complexes containing pyrazole ferrocenyl ligands. Allyl fragments are omitted for clarity

It is interesting to note that the main coordination plane of Rh in the complexes of Figure 2 forms a tilting angle of 51°-56° with the „upper" Cp ring and is *endo*-oriented with respect to the ferrocene, i.e., the metal center is located slightly below the Cp ring. On the other hand, for the Pd derivatives of Figure 3 the corresponding angle is 34°-37°, and the orientation is rather to be defined as *exo* with the Pd center located above the „upper" Cp ring), a situation similar to that observed for Josiphos complexes (vide supra). Finally, a common feature to all ligands discussed so far is the „lifting" of the PPh_2 phosphorus atom out of coplanarity with the Cp ring in the complexes that have been structurally characterized. This distortion is typically in the range of 0.1 Å to 0.4 Å.

3 Applications of New Ferrocenyl Ligands in Asymmetric Catalysis

3.1 Industrial Applications in Hydrogenation Reactions

Two ferrocenyl diphosphines of the new generation have already found an application in production scale hydrogenation processes. Chronologically the first one relates to a new industrial synthesis of the vitamine biotin, developed at LONZA Ltd.[16]. The catalytic process and the specific ligand, however, have been developed first at CIBA, in a nice example of industrial collaboration. As illustrated in Scheme 4 a Josiphos analogue containing a bulky (*t*-butyl)$_2$P group is used in the Rh-catalyzed, highly diastereoselective hydrogenation of a fully substituted olefin. This reaction introduces the two new stereogenic centers of the bicyclic system of biotin, and constitutes a rare example of the successful reduction of this kind of substrates[17].

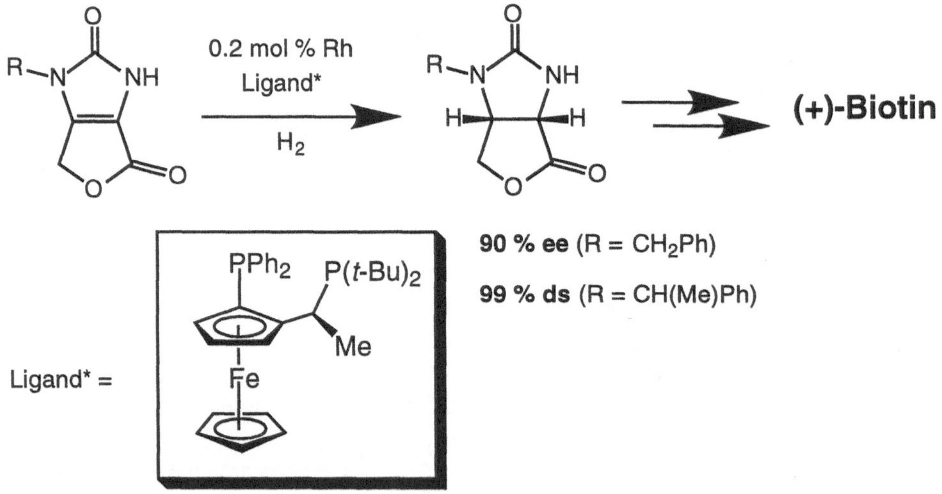

90 % ee (R = CH_2Ph)

99 % ds (R = CH(Me)Ph)

Scheme 4. Industrial scale Rh-catalyzed hydrogenation in a new synthesis of (+)-biotin

A second, and probably more impressive application is shown in Scheme 5. The Ir-catalyzed ketimine hydrogenation depicted constitutes the so far largest asymmetric catalytic process generating more than 10,000 tons of the herbicide (*S*)-Metolachlor per year[18]. Although the ee's obtained do not go beyond ca. 80%, the efficiency of the catalyst plays a much more decisive role. With substrate to catalyst ratios of up to one million and full conversions within hours, this is also one of the fastest catalysts known. The development of this process also illustrates the eminent role of empiricism in asymmetric catalysis. Indeed, the Ir-systems needs co-catalytic amounts of iodide and acid, in order to attain full efficiency. However, the exact function of these components is not fully understood[19].

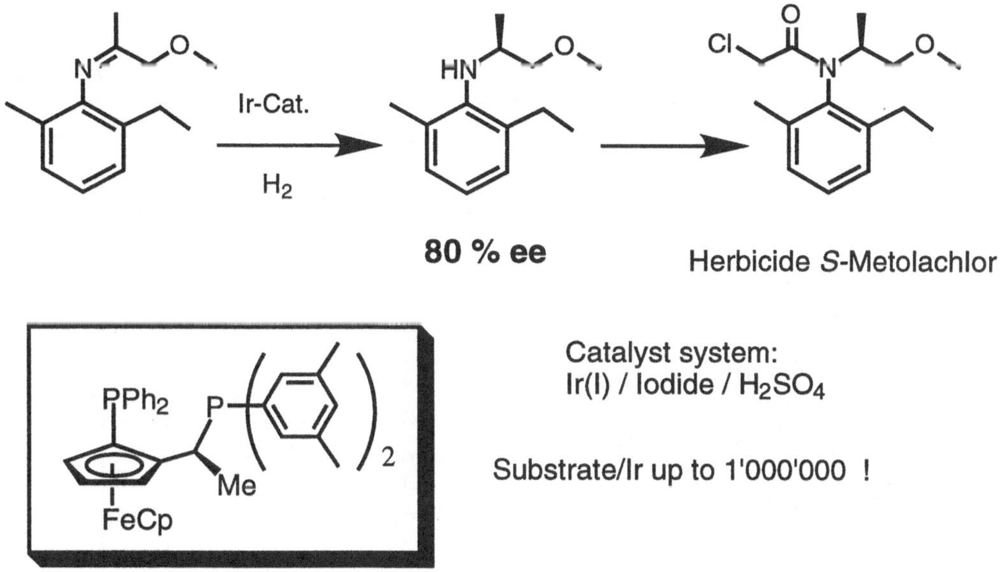

80 % ee   Herbicide *S*-Metolachlor

Catalyst system:
Ir(I) / Iodide / H_2SO_4

Substrate/Ir up to 1'000'000 !

Scheme 5. Iridium-catalyzed imine hydrogenation in the production of the herbicide (*S*)-Metolachlor

3.2 Rhodium-Catalyzed Hydroboration of Styrenes

The diphosphine Josiphos has been found to furnish relatively high ee's in the hydroboration of styrene with catecholborane[9]. However, enantioselectivities exceeding 90% could only be obtained when the reaction was conducted at low temperature. Much better results are given by the pyrazole-containing ligands. Indeed, our system affords the highest known enantioselectivity in this particular reaction at room temperature[20]. An important drawback, however, is the observed relatively low regioselectivity of the reaction, usually generating up to 40% of the undesired primary achiral product.

Ligand electronic effects on enantioselectivity are very important in this reaction. Indeed, it has been found that the full range of ee's from almost 0% to 98.5% is covered by ligand derivatives differing from one another essentially in their electronic properties by virtue of different peripheral substituents, while maintaining practically identical steric characteristics. It is important to note that no ligand was found so far able to switch enantioselectivity solely based on an alteration of its electronics, i.e., the sense of asymmetric induction is still determined by the absolute configuration of the ferrocenyl ligand backbone. Representative examples are depicted in Scheme 6.

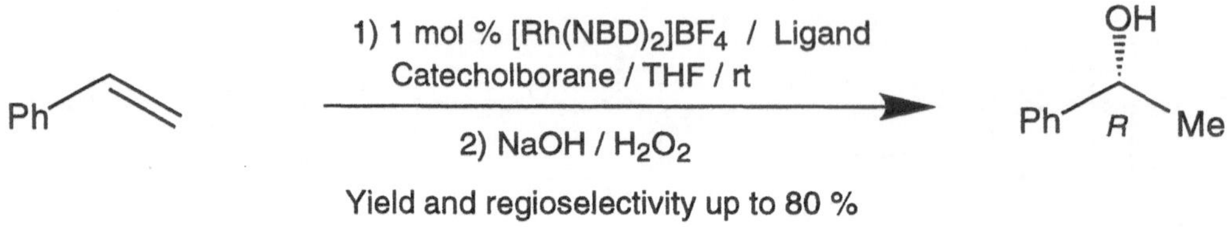

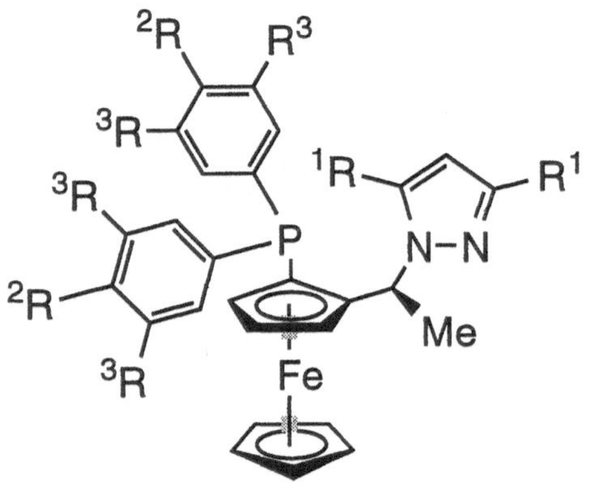

1) 1 mol % [Rh(NBD)$_2$]BF$_4$ / Ligand
Catecholborane / THF / rt

2) NaOH / H$_2$O$_2$

Yield and regioselectivity up to 80 %

Ligand			% ee
R^1	R^2	R^3	
Me	H	CF$_3$	98.5
Me	CF$_3$	H	98
Me	H	H	95
i-Pr	H	H	92
Me	OMe	H	90
Me	NMe$_2$	H	72
CF$_3$	H	H	33
CF$_3$	OMe	H	5

Scheme 6. Influence of ligand electronic properties on the enantioselectivity of the Rh-catalyzed hydroboration of styrene

In order to obtain optimum enantioselectivities, it is clear that the pyrazole nitrogen atom coordinating to the metal has to be a good σ-donor, and significant π-accepting properties at phosphorus are also required. This is easily obtained using electron-donating and -withdrawing substituent, respectively. The inverse combination is very much deleterious and leads to very low selectivities. A more extended range of substituents on the pyrazole seems not to be applicable, since stereoselectivity is very sensitive to the size of the group at position 3 of the heterocycle. In fact, already on going beyond the bulk of an ethyl or isopropyl group the ee's drop significantly.

The origin of such large electronic effects are not fully understood. In view of the poor ligand quality of the nitrogen atom in those systems possessing trifluoromethyl groups on the pyrazole (ligands leading to low ee's), one must ask the question whether or not a possible dissociation of the pyrazole from the first coordination sphere of Rh could be responsible for the bad stereochemical outcome. In other words, if such a dissociation were to take place, the monodentate behavior of the ferrocenyl ligand would possibly explain the poor inductions observed. We addressed this question by examining the relative thermodynamic stability of the model cationic complexes [Rh(P,N-ligand)(COD)]$^+$ [20]. Indeed, the more electron rich ligands are able to displace the electron poor ones from Rh in fast reactions in simple NMR-tube experiments, such that a qualitative correlation exists between complex stability and the enantioselectivity afforded by the corresponding ligand. However, we could not identify any species containing any of the P,N-ligands coordinated in a monodentate fashion. Good monodentate ligands such as PEt$_3$ or PPh$_3$ were found to liberate the P,N-ligand when added to the Rh complex, even in understoichiometric amounts. On the other hand, excess styrene did not undergo any observable reaction with the model complexes. We therefore conclude that, if any dissociation of the electron-poor ligands is relevant to catalysis, then a highly active species not containing the chiral ligand at all must be formed, however, in quantities not detectable by common spectroscopic techniques. Although improbable, this event cannot be excluded.

Assuming that the observed electronic effects are indeed due to an alteration of the electronic properties of the catalytically active species containing the P,N-ligand, we prepared another series of neutral Rh(I) model complexes containing a carbonyl ligand in *trans* position to the pyrazole moiety, from the reaction of [Rh$_2$(CO)$_4$Cl$_2$] and the P,N-system[21]. The CO-ligand may thus be used as diagnostic tool for examining the electronic environment induced at the metal center by the chiral ligand, since the CO-stretching frequency is sensitive to the donor/acceptor properties of the latter. Of course the observed ν-(CO) reflects the combined influence of both the pyrazole and the phosphine, thus care must be exercized while interpreting the results and only trend indications may be derived. A representative selection of the data so far collected is given below.

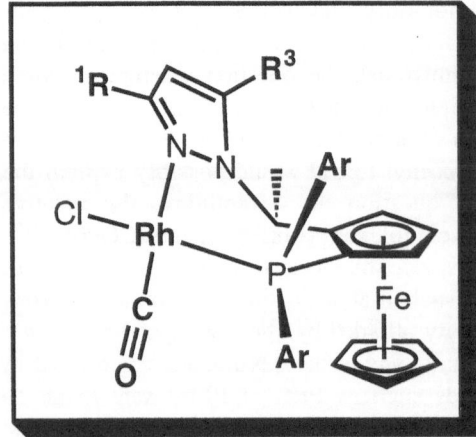

R^1	R^3	Ar	v(CO) (cm^{-1})	%ee
CH$_3$	CH$_3$	Ph	1989	95
CF$_3$	CH$_3$	Ph	1994	44
CF$_3$	CF$_3$	Ph	2000	33
CH$_3$	CH$_3$	p-MeO-Ph	1987	87
CH$_3$	CH$_3$	p-Me-Ph	1991	94
CH$_3$	CH$_3$	3,5-(CF$_3$)$_2$-Ph	2003	98.5

Figure 5. IR-data for a selection of [RhCl(CO)(P,N)] complexes and the enantioselectivity obtained in the hydroboration of styrene utilizing the corresponding ligand

From the data of Figure 5 it clearly results that the alteration of the donor capacity of the pyrazole nitrogen, while maintaining constant the nature of the phosphine, has a significant effect on v-(CO). Thus, upon decreasing the electron density at nitrogen and hence its donor ability, an increase of the CO bond order is observed, indicating a less important π-back bonding contribution from Rh, as one would expect. A similar trend, albeit not surprisingly slightly more pronounced, is observed when gradually enhancing the π-acceptor qualities of the phosphine in 3,5-dimethyl pyrazole derivatives.

In conclusion, and although the strong electronic effects observed in the hydroboration reaction with these ligands are poorly understood on a mechanistic level, it is clear that one is dealing with subtle changes of the electronic properties of the metal center induced by the substituents attached to the periphery of the chiral ligands. Our observations indicate more generally that the empyrical tuning of the ligands in asymmetric catalysis should always include the possibility of influencing stereoselectivity by variations of the donor properties of the inducing agent.

3.3 Palladium-Catalyzed Allylic Amination

Whereas asymmetric Pd-catalyzed allylic substitution reactions involving soft carbon nucleophiles have been the subject of extensive studies in recent years[22], the corresponding transformation utilizing amine derivatives (C-N bond forming reaction) has received much less attention[23]. We recently found that pyrazole-containing ligands afford very high enantioselectivities in the amination of 1,3-diphenylallyl acetate (or the corresponding ethyl carbonate) with benzylamine, as illustrated in Scheme 7[24].

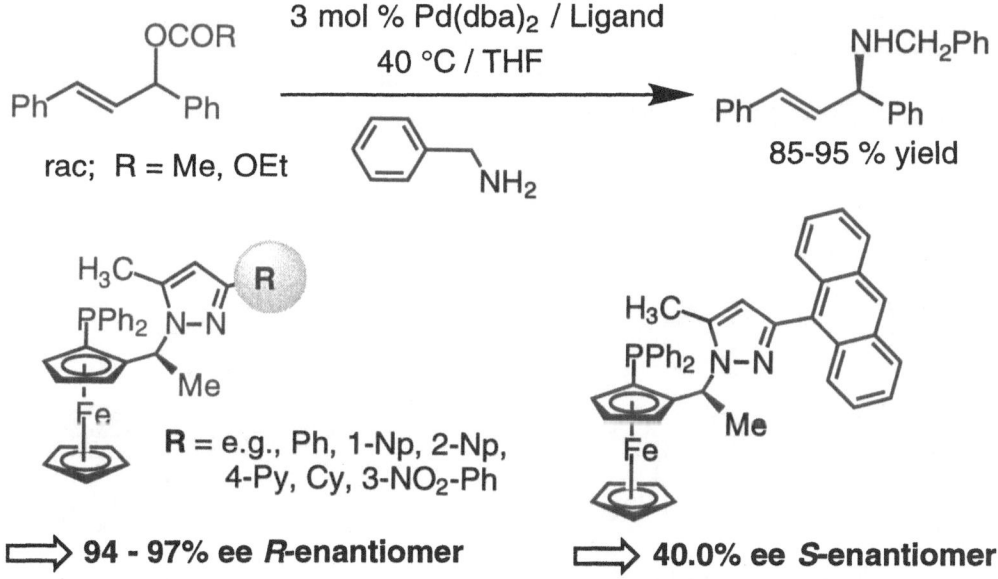

Scheme 7. Asymmetric Pd-catalyzed allylic amination with pyrazole-containing ferrocenyl ligands

Thus, the allylamine product is obtained in 94-97% optical purity utilizing a variety of ligands, containing a fairly large substituent at position 3 of the pyrazole (the methyl group in position 5 is irrelevant). Up to the bulk of a 1- or 2-naphthyl group the enantioselectivity is not influenced to any very significant extent. The *R*-enantiomer of the allylamine is formed preferentially when ferrocenyl ligands possessing the (*S*)-(*R*) absolute configuration are used. However, the introduction of a 9-anthryl group leads to an unexpected reversal of enantioselectivity. The *S*-form is now produced with an ee of 40%. In order to understand this drastic and, as we will see, purely steric effect on the stereoselectivity, we performed a 2D NMR study of the intermediate Pd-π-allyl complexes formed during the catalytic reaction[24, 25]. Thus, it was found that the anthryl-containing complex exists as mixture of two diastereoisomeric compounds, differing in the absolute configuration at the allyl terminus *trans* to phosphorus, as shown in Scheme 8. The configuration of these two stereoisomers, occuring in relative amounts of 29% and 71%, may be defined as *exo-syn-syn* and *exo-syn-anti*, respectively[24].

Assuming that nucleophilic attack takes place at the allylic carbon *trans* to P exclusively, then the ratio of these two diastereoisomers reflects the enantiomer ratio of the product within 1-2%. This would indicate that the reaction of benzylamine with these two isomers takes place with equal rates.

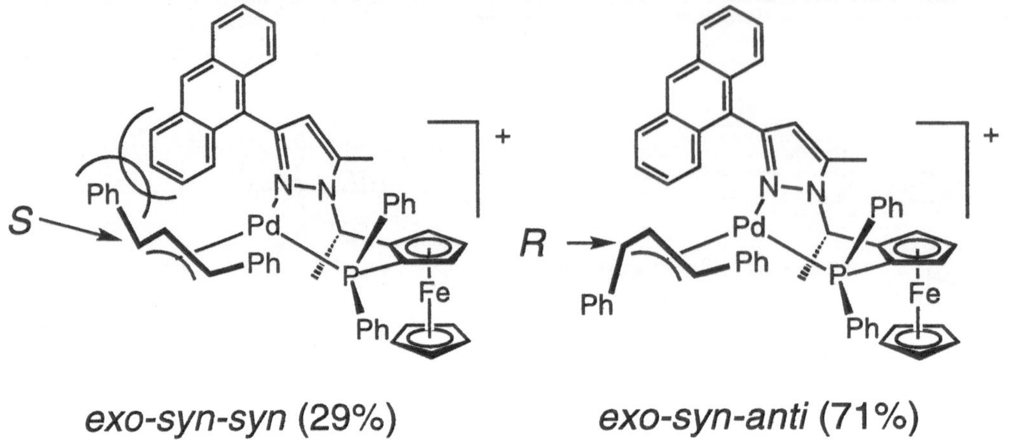

exo-syn-syn (29%)   *exo-syn-anti* (71%)

Scheme 8. Configuration of Pd-π-allyl complexes containing a 9-anthryl substituted pyrazole, as found in solution by 2D NMR methods

The configuration inversion leading to the major *exo-syn-anti* isomer can only be explained in terms of steric repulsion between the 9-anthryl fragment and the phenyl attached to the allyl carbon *trans* to P. Configurational analysis of several other Pd-π-allyl complexes in solution utilizing 2D NMR methods revealed the *exo-syn-syn* form to be present in at least 80% for those ligands affording high ee's[25]. In these cases, however, no clear relationship between isomers- and product enantiomers ratio is apparent.

A 3-(1-adamantyl) substituted pyrazole ligand was found to give the highest ee's for this particular reaction (> 99%). For the corresponding π-allyl complex only the *exo-syn-syn* form could be identified, indicating that the nucleophilic attack indeed takes place at the position *trans* to phosphorus exclusively (see Scheme 9).

The hexafluorophosphate salts of several cationic Pd-π-allyl complexes were characterized by X-ray crystallography. The configuration found in the solid state corresponds to that of the major isomer in solution. For all complexes analyzed[24, 25], a deviation of the π-allyl geometry from its expected orientation was observed. The allyl ligand turns out to be „rotated" in such a way that two carbon atoms are almost coplanar with the Pd, P, and N atoms, whereas the position of the allyl terminus *trans* to phosphorus is significantly out of plane (average distance 0.53 Å for four different complexes). This distortion appears to be dictated by the steric influence of both the diphenylphosphino group and the substituent in position 3 of the pyrazole fragment. The effect of the latter seems to be more subtle (and probably more important) than the one due to the phosphine. Indeed, as described above, the size of this substituent is very crucial, as it may induce a configurational change. On the other hand, we think that the major role of the PPh$_2$ group mainly consists in stabilizing the *exo* form, by virtue of an optimum „inter-phenyl" steric interaction.

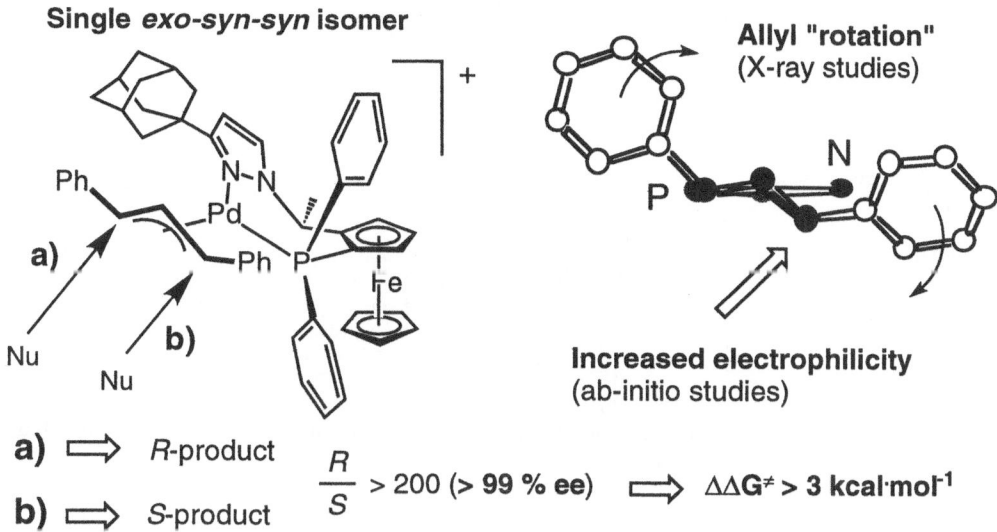

Scheme 9. Configuration and distortion of the Pd-π-allyl complex containing the 1-adamantyl substituted ferrocenyl ligand

Searching for the ultimate reason leading to the pronounced site-selectivity of the nucleophilic attack, we performed first-principles calculation on a model, non-distorted Pd(P,N) system[26]. The results are summarized in Scheme 10.

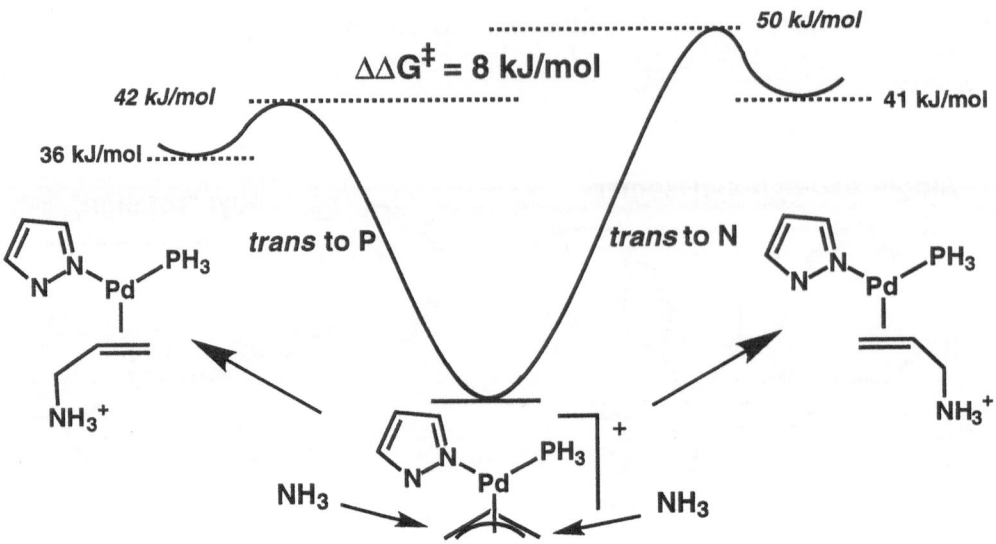

Scheme 10. Calculated energetics for the nucleophilic attack of ammonia onto a non-distorted Pd-π-allyl complex in the gas phase

It clearly turns out that the nucleophilic attack *trans* to phosphorus is preferred. However, the theoretical difference in activation energy between the two possible pathways - *trans* to P and *trans* to N - is not sufficiently high to account for the observed enantioselectivities. The calculated 8 kJ/mol in $\Delta\Delta G^*$ would lead to an ee of less than 95%, whereas the best experimental values go beyond 99%. We therefore included the experimentally observed distortion (allyl rotation) into our calculations and found that the preference for the nucleophilic attack at the center out of plane is very pronounced, whether or not this center is *trans* to phosphorus. In other words, according to theory, the main factor governing site-selectivity is indeed the distortion of the allyl ligand, with a drastically enhanced electrophilicity for the carbon atom out of plane, as schematically shown in Figure 6.

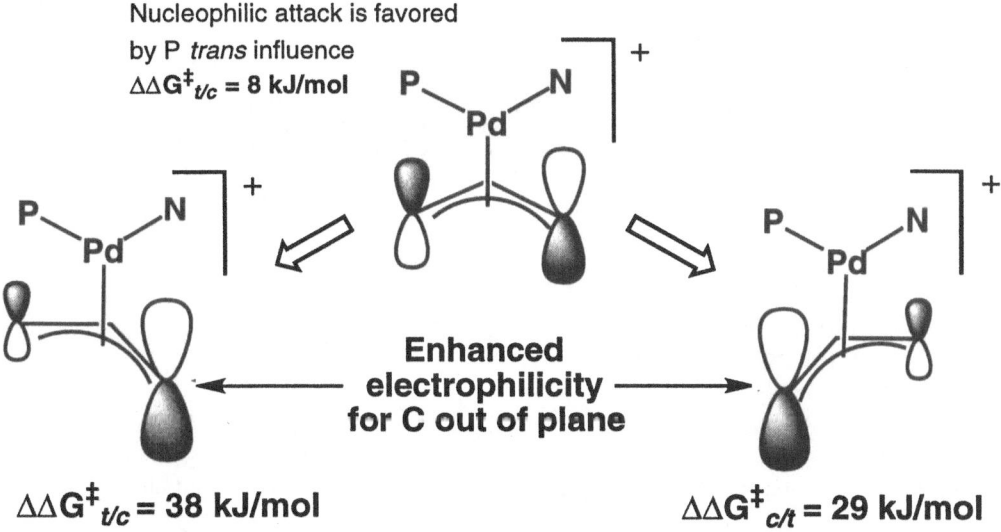

Nucleophilic attack is favored
by P *trans* influence
$\Delta\Delta G^{\ddagger}_{t/c} = 8$ kJ/mol

Enhanced
electrophilicity
for C out of plane

$\Delta\Delta G^{\ddagger}_{t/c} = 38$ kJ/mol

$\Delta\Delta G^{\ddagger}_{c/t} = 29$ kJ/mol

Figure 6. Schematic representation of the effect of allyl rotation on allyl LUMO and hence on the preferred site of nucleophilic attack

Theory thus predicts that it should be possible for the nucleophile to attack at the position *trans* to nitrogen in such systems, provided that the necessary distortion is present. However, no ligand could be found so far for which this could be verified.

The experimentally observed allyl rotation of ca. 15°-20° around the Pd-allyl axis takes the two carbon atoms forming the olefinic double bond in the final product into an orientation much similar to that they are likely to assume in the Pd(0) complex generated by the nucleophilic attack, for wich one would expect a planar geometry. Thus, allyl distortion anticipates an important structural feature of the product.

As discussed above, the reversal of enantioselectivity observed with the 9-anthryl substituted ligand is due to a sterically induced inversion of configuration at the allyl center *trans* to phosphorus. However, the reversal of selectivity is not complete, since a significant amount of the *exo-syn-syn* diastereoisomer is still present. With the aim of completely suppressing the formation of this configurational isomer, in favor of the *exo-syn-anti* form, we prepared a ligand containing a triptycyl substituent at position 3 of the pyrazole[24]. In fact, it turns out that the corresponding Pd-(1,3-diphenylallyl) complex displays the expected *exo-syn-anti* configuration, both in solution and in the solid state (see Figure 7).

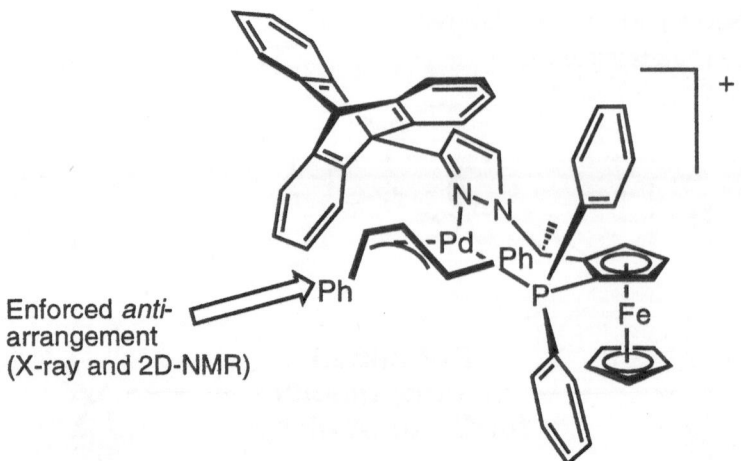

Enforced *anti-*
arrangement
(X-ray and 2D-NMR)

Figure 7. Schematic representation of the structure of a triptycyl substituted Pd-allyl complex

With this information in hand, we were expecting the triptycyl ligand to induce the formation of the opposite enantiomer of the allylamine in high ee's (*S*-enantiomer with ligand absolute configuration (*S*)-(*R*)). However, the use of catalysts containing this ligand did not afford any product at all, even when the reaction was conducted at reflux temperature in THF for prolongued periods of time. An explanation of this, at first sight disappointing result, is given in the cartoon of Scheme 11.

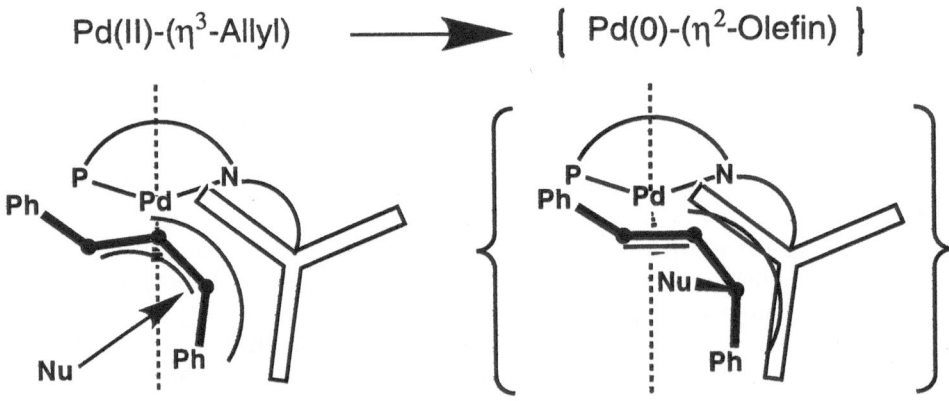

Scheme 11. Steric constraints connected with the nucleophilic attack onto the Pd-allyl complex containing the triptycyl substituted pyrazole ligand

Upon nucleophilic attack at the carbon atom *trans* to phosphorus, and in order to form a trigonal planar Pd(0) olefin complex, the allyl fragment will have to perform a side-on movement, towards the bulky triptycyl fragment. However, the steric fitting of the *anti*-configurated half of the allyl ligand between two blades of the propeller-shaped triptycyl is almost ideal, such that the necessary sliding of the allyl becomes impossible. Furthermore, attack at the opposite allyl terminus is hampered because of the allyl rotation, enormously decreasing the electrophilicity at that particular center.

The detailed structural analyses of cationic Pd-π-allyl complexes described above have been performed utilizing the corresponding PF$_6^-$ salts. Since it is generally accepted that such cationic species form during the catalytic reaction, one would expect that their use as catalyst precursors should give identical results, as compared to reactions carried out with the catalysts formed *in situ* from the common Pd(0) source [Pd$_2$(dba)$_3$·CHCl$_3$] and the chiral ligand. However, it turns out that, when the well-characterized hexafluorophosphate salts of the intermediate Pd-allyl complexes were used, very low enantioselectivities were obtained. This surprising observation could be traced back to a counterion effect[27, 28]. In particular, we found that several anions added to the catalyst system, typically in form of tetraalkyl-ammonium salts, were able to influence stereoselectivity. Our findings are summarized in Scheme 12, by the crucial experiments carried out utilizing the particular ligand containing a 3-ferrocenylpyrazole ligand (similar trends are observed for other ligands). Extremely negative effects are exerted by relatively large, virtually non-coordinating anions such as PF$_6^-$, whereas a positive influence may be obtained when small, hard and potentially coor-dinating ions such as fluoride or BII$_4^-$ are added, other anions displaying intermediate effects.

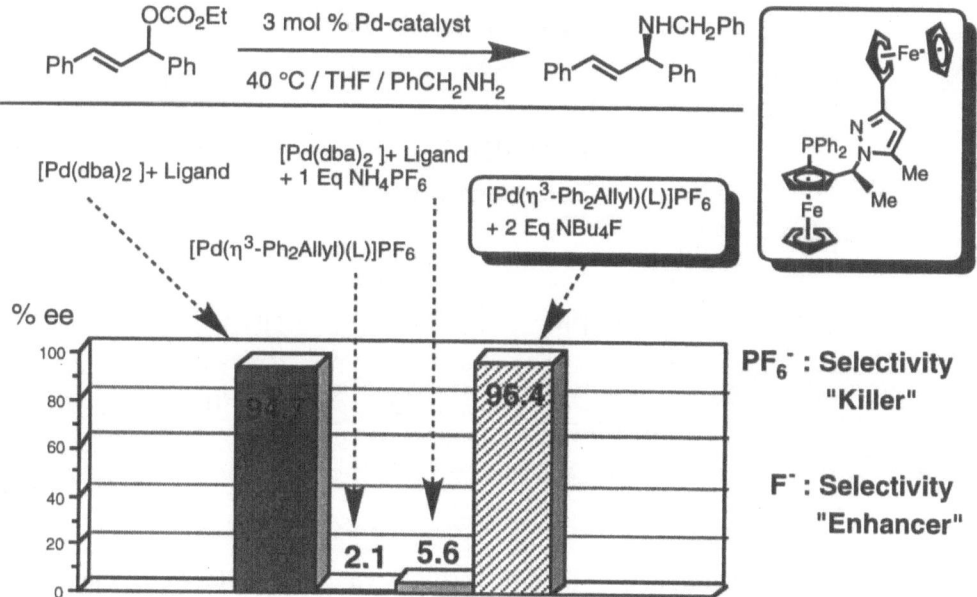

Scheme 12. Anion effects on enantioselectivity in the Pd-catalyzed allylic amination

203

A tentative and simple interpretation of the anion effect, taking into particular consideration the differences between fluoride and hexafluorophosphate, is given in Scheme 13. Having verified that no kinetic resolution of the racemic substrate takes place, it must be assumed that upon oxidative addition of the allylic carbonate diastereoisomeric π-allyl complexes are formed. As illustrated, these could be represented by the observed *exo-syn-syn* form and, e.g., the corresponding *endo-syn-syn* isomer, although the presence of other configurational isomer cannot be excluded. Upon reaction with the amine at the allylic carbon *trans* to phosphorus, the different diastereoisomers would lead to opposite product enantiomers, unless fast interconversion is operating. Exactly in determining this equilibration, we see the crucial role of the anion. Hexafluorophosphate does not accelerate equilibration, as it is confirmed by most of the NMR studies performed on PF_6^- salts[24, 25]. Fluoride, however, probably by virtue of revesible coordination to Pd, is enhancing the dynamic behavior of the system, thus allowing the rapid interconversion of the diastereoisomers to give the most stable *exo-syn-syn* form. This means that fluoride ensures that the system is operating under Curtin-Hammett regime[29].

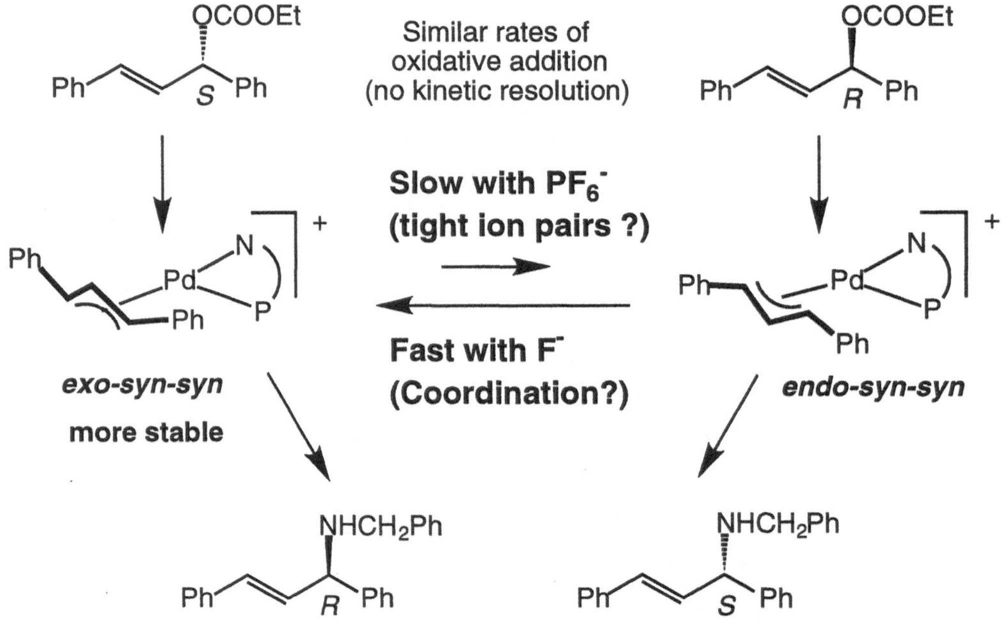

Scheme 13. A tentative explanation of the anion effect in Pd-catalyzed allylic amination

In summary, the studies described above have shown how subtle steric interactions between peripheral parts of the chiral ligand and the π-allyl fragment within the cationic Pd-allyl intermediate may induce 1) a drastic change of the electronic properties of the allyl, leading to a pronounced site-selectivity for nucleophilic attack, and 2) a change in configuration at one allyl carbon inducing a reversal of enantioselectivity of the catalytic reaction. Our system is thus a prominent example where the interplay between steric and electronic effects on stereoselectivity of a catalytic reaction can be unraveled.

4 Additional Ligands

In addition to the two types of ligands described so far, we also prepared the triphosphine „Pigiphos"[30], as well as a series of pentamethylated derivatives containing either iron[31] or ruthenium[32], as shown below.

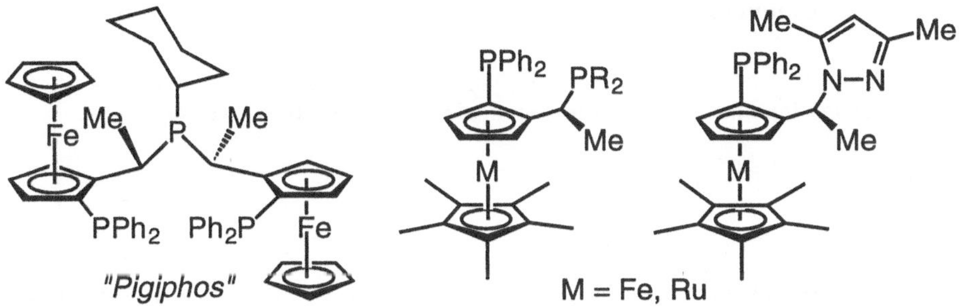

Pigiphos forms Ni(II) and Pd(II) dicationic monosolvento complexes displaying Lewis-catalytic activity in hetero Diels-Alder reactions, however affording poor stereoselectivities. The Fe-Cp* derivatives were not found to offer any particualr advantages and behaved at best as their non-methylated congeners in selected catalytic reactions. Furthermore, the corresponding ruthenocenyl ligands appear not to adopt the same conformational constraints as the Fe derivatives, and are in general much poorer chiral auxiliaries. Figure 8 shows the very different conformational features of Pd-allyl complexes containing Me_5Josiphos and its Ru congeners, respectively. For the latter an *endo-equatorial* orientation of both phenyl groups at phosphorus becomes possible.

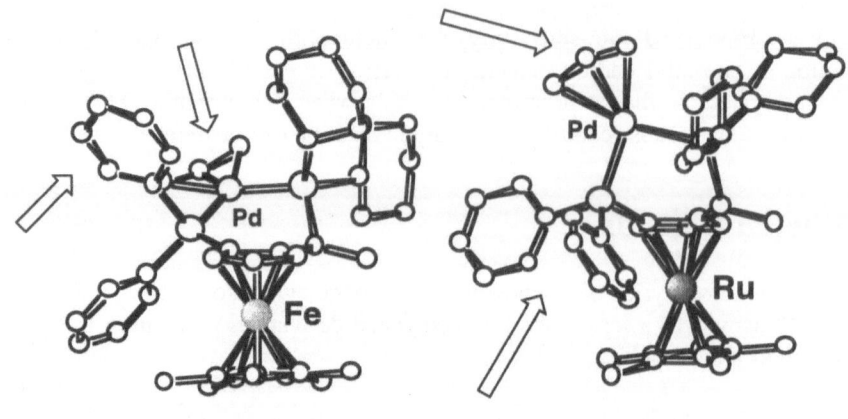

Common Conformation

Figure 8. The structure of cationic Pd-allyl complexes containing Me₅Josiphos and its Ru congener, showing the uncommon conformation adopted by the Ru derivative

References

1. For a recent reviews , see: a) A. Togni, *Chimia* **1996**, *50*, 86. b) A. Togni, *Angew. Chem.* **1996**, *108*, 1581 (*Angew. Chem. Int. Ed. Engl.* **1996**, *35*, 1475).
2. For a review, see: a) T. Hayashi In *FERROCENES: Homogeneous Catalysis, Organic Synthesis, Materials Science*, A. Togni, T. Hayashi, Eds., VCH, Weinheim (New York), **1995**, 105-142, and references cited therein. First report: b) T. Hayashi, K. Yamamoto, K. Kumada, *Tetrahedron Lett.* **1974**, 4405.
3. M. Sawamura, Y. Ito, *Chem. Rev.* **1992**, *92*, 857-871.
4. a) M. Sawamura, H. Hamashima, Y. Ito, *Tetrahedron: Asymmetry* **1991**, *2*, 593. b) M. Sawamura, H. Hamashima, Y. Ito, *J. Am. Chem. Soc.* **1992**, *114*, 8295. c) M. Sawamura, H. Hamashima, Y. Ito, *Tetrahedron* **1994**, *50*, 4439. d) M. Sawamura, R. Kuwano, Y. Ito, *Angew. Chem.* **1994**, *106*, 92 (*Angew. Chem. Int. Ed. Engl.* **1994**, *33*, 111). e) M. Sawamura, H. Hamashima, M. Sugawara, R. Kuwano, Y. Ito, *Organometallics* **1995**, *14*, 4549.
5. A personal account about the development of these ligands is given in ref. 1a).
6. D. Marquarding, H. Klusacek, G. Gokel, P. Hoffmann, I. Ugi, *J. Am. Chem. Soc.* **1970**, *92*, 5389.
7. For a brief review on planar chiral ferrocene derivatives, see ref. 1b). For a discussion of ferrocene chemistry from a stereochemical point of view, see the classical: K. Schlögl, *Top. Stereochem.* **1967**, *1*, 39.
8. G. Gokel, D. Marquarding, I. Ugi, *J. Org. Chem.* **1972**, *37*, 3052.
9. A. Togni, C. Breutel, A. Schnyder, F. Spindler, H. Landert, A. Tijani, *J. Am. Chem. Soc.* **1994**, *116*, 4062.

10. H. C. L. Abbenhuis, U. Burckhardt, V. Gramlich, C. Köllner, R. Salzmann, P. S. Pregosin, A. Togni, *Organometallics* **1995**, *14*, 759.

11. a) A. Togni, F. Spindler, G. Rihs, N. Zanetti, M. C. Soares, T. Gerfin, V. Gramlich, *Inorg. Chim. Acta* **1994**, *222*, 213. b) N. C. Zanetti, F. Spindler J. Spencer, A. Togni, G. Rihs, *Organometallics*, **1996**, *15*, 860.

12. M.C. Soares, A. Togni, unpublished results.

13. U. Burckhardt, L. Hintermann, A. Schnyder, A. Togni, *Organometallics* **1995**, *14*, 5415.

14. H. Brunner, T. Scheck, *Chem. Ber.* **1992**, *125*, 701.

15. U. Burckhardt, A. Togni, to be submitted.

16. J. McGarrity, F. Spindler, R. Fuchs, M. Eyer, Eur. Pat. Appl. EP 624 587 A2, (*LONZA AG*), *Chem. Abstr.* **1995**, *122*, P81369q.

17. For other approaches to the hydrogenation of fully substituted olefins utilizing A) ferrocenyl ligands, see, e.g.: a) T. Hayashi, N. Kawamura, Y. Ito, *J. Am. Chem. Soc.* **1987**, *109*, 7876. b) T. Hayashi, N. Kawamura, Y. Ito, *Tetrahedron Lett.* **1988**, *29*, 5969. c) M. Sawamura, R. Kuwano, Y. Ito, *J. Am. Chem. Soc.* **1995**, *117*, 9602. B) other ligands: d) M.J. Burk, M.F. Gross, J.P. Martinez, *J. Am. Chem. Soc.* **1995**, *117*, 9375.

18. For comments about recent developments in the chemistry of chiral ferrocenes and about the successful application of ferrocenyl ligands in industry, see: *Chemical & Engineering News*, July 22 **1996**, pp. 38-40.

19. For an account about the successful development of this process, see: F. Spindler, B. Pugin, H.-P. Jalett, H.-P. Buser, U. Pittelkow, H.-U. Blaser, In *Catalysis of Organic Reactions*, R.E. Malz, Jr., Ed. (Chem. Ind. Vol. *68*), Dekker, New York, **1996**, pp. 153-166.

20. A. Schnyder, L. Hintermann, A. Togni, *Angew. Chem.* **1995**, *107*, 996 (*Angew. Chem. Int. Ed. Engl.* **1995**, *34*, 931).

21. A. Schnyder, A. Togni, U. Wiesli, *Organometallics*, in print.

22. For a recent review, see: B.M. Trost, D.L. Van Vranken, *Chem. Rev.* **1996**, *96*, 395, and references cited therein.

23. For successful approaches utilizing 1,3-disubstituted acyclic allylic substrates, see: a) T. Hayashi, A. Yamamoto, Y. Ito, E. Nishioka, H. Miura, K. Yanagi, *J. Am. Chem. Soc.* **1989**, *111*, 6301. b) T. Hayashi, K. Kishi, A. Yamamoto, Y. Ito, *Tetrahedron Lett.* **1990**, *31*, 1743. c) T. Hayashi, A. Yamamoto, Y. Ito, *Tetrahedron Lett.* **1988**, *29*, 99. d) P. von Matt, O. Loiseleur, G. Koch, A. Pfaltz, C. Lefeber, T. Feucht, G. Helmchen, *Tetrahedron: Asymmetry* **1994**, *5*, 573. Furthermore, Trost and co-workers developed the amination of cyclic allylic substrates, based on the enantiotopic group differentiation. See: e) B.M. Trost, D.L. Van Vranken, C. Bingel, *Angew. Chem. Int. Ed. Engl.* **1992**, *31*, 228 (*Angew. Chem.* **1992**, *104*, 194). f) B.M. Trost, D.L. Van Vranken, C. Bingel, *J. Am. Chem. Soc.* **1992**, *114*, 9327. g) B.M. Trost, D.L. Van Vranken, *J. Am. Chem. Soc.* **1993**, *115*, 444.

24. A. Togni, U. Burckhardt, V. Gramlich, P. S. Pregosin, R. Salzmann, *J. Am. Chem. Soc.* **1996**, *118*, 1031.

25. U. Burckhardt, V. Gramlich, P. Hofmann, R. Nesper, P. S. Pregosin, R. Salzmann, A. Togni, *Organometallics* **1996**, *15*, 3496.

26. P. E. Blöchl, A. Togni, *Organometallics* **1996**, *15,* 4125. For a more classical extended-Hückel approach, see: T.R. Ward, *Organometallics* **1996**, *15*, 2836.

27. U. Burckhardt, M. Baumann, A. Togni, submitted.

28. Counterion effect in Pd-allyl chemistry heve been observed previously, albeit less pronounced than in our case. See, e.g.: (a) T. Hayashi, A. Yamamoto, Y. Ito, E. Nishioka, H. Miura, K. Yanagi, *J. Am. Chem. Soc.* **1989**, *111*, 6301. (b) M. Bovens, A. Togni, L.M. Venanzi, *J. Organomet. Chem.* **1993**, *451*, C28. (c) B.M. Trost, M.G. Organ, G.A. O'Doherty, *J. Am. Chem. Soc.* **1995**, *117*, 9662. (d) B.M. Trost, R.C. Bunt, *J. Am. Chem. Soc.* **1996**, *118*, 235. For the effect of Cl⁻ on π-allyl rotation in stoichiometric systems, see: (e) S. Hansson, P.-O. Norrby, M.P.T. Sjögren, B. Åkermark, M.E. Cucciolito, F. Giordano, A. Vitagliano, *Organometallics* **1993**, *12*, 4940. (f) A. Gogoll, J. Örnebro, H. Grennberg, J.E. Bäckvall, *J. Am. Chem. Soc.* **1994**, *116*, 3631. (g) P.G. Andersson, A. Harden, D. Tanner, P.-O. Norrby, *Chem. Eur. J.* **1995**, *1*, 12. For a general reference, see also: A. Loupy, B. Tchoubar, *Salt Effects in Organic and Organometallic Chemistry*, VCH: Weinheim (Germany), **1992**.

29. See, e.g.: (a) P.R. Auburn, P.B. Mackenzie, B. Bosnich, *J. Am. Chem. Soc.* **1985**, *107*, 2033. (b) P.B. Mackenzie, J. Whelan, B. Bosnich, *J. Am. Chem. Soc.* **1985**, *107*, 2046. For a general treatment of the Curtin-Hammett principle, see: (c) J.I. Seeman, *Chem. Rev.* **1983**, *83*, 83.

30. P. Barbaro, A. Togni, *Organometallics* **1995**, *14*, 3570.

31. H. C. L. Abbenhuis, A. Togni, B. Müller, A. Albinati, U. Burckhardt, V. Gramlich, *Organometallics* **1994**, *13*, 4481.

32. H. C. L. Abbenhuis, U. Burckhardt, V. Gramlich, A. Martelletti, J. Spencer, I. Steiner , A. Togni, *Organometallics*, **1996**, *15*, 1614.

Mechanisms of the Catalyzed Mukaiyama Cross-Aldol Reaction

William W. Ellis and B. Bosnich

Department of Chemistry, The University of Chicago, 5735 South Ellis Avenue, Chicago, Illinois 60637

Abstract

The mechanism of the catalyzed Mukaiyama reaction has been investigated using a variety of Lewis acid catalysts. The transition metal Lewis acid, $[Ti(Cp)_2(OTf)_2]$, induces rapid catalysis but it is shown that the complex is not the real catalyst. It serves to produce Me_3SiOTf which is the real catalyst. The Me_3SiOTf catalyst is formed by hydrolysis of the complex when small amounts of water are present. Triflic acid is produced which generates Me_3SiOTf by reaction with the silyl enol ether. When the hydrolysis is suppressed, Me_3SiOTf is formed as a consequence of the mechanism. Similarly, it is shown that Ph_3COTf is not a catalyst, it serves to produce Me_3SiOTf by hydrolysis. These observations indicate that mild Lewis acids should obviate the formation of Me_3SiOTf. A mechanistic study of mild Lewis acid lanthanide complexes has shown that the Mukaiyama reaction can be induced without release of "Me_3Si^+". These reactions proceed by a quasi-concerted silyl transfer mechanism and are found to form [2 + 2] adducts initially. These [2 + 2] adducts are in equilibrium with the substrates and are converted to the Mukaiyama product. The factors which control the enantioselective Mukaiyama reaction are discussed.

1 Introduction

A variety of Lewis acids catalyze the Mukaiyama cross-aldol reaction (eq. 1). Among these are the catalysts Me_3SiOTf[1], Me_3SiI[2], $Me_3SiCl/SnCl_2$[3], $Ph_3CCl/SnCl_2$[4], Ph_3CClO_4[5], Ph_3COTf[6], a variety of rhodium complexes[7], lanthanide triflates[8], iron complexes[9] and complexes of titanium (IV)[10] and many others. In principle, the Mukaiyama reaction is a powerful carbon-carbon bond forming reaction if the diastereoselection and enantioselection of the catalytic reactions could be controlled. Although a number of very successful catalytic enantioselective cross-coupling reactions have been reported, the origins of the enantioselection have not been defined at a mechanistic level. Any rational development of the selectivity of the reaction requires a detailed understanding of the mechanisms which control the rates of formation and disappearance of the putative intermediates of reaction.

$$X = R, OR, SR \quad (1)$$

A related carbon-carbon bond forming reaction was discovered by Hosomi and Sakurai[11] where carbonyl compounds, ketals, and acetals couple with an allylic silane (eq. 2) in the presence of a Lewis acid. Generally, Lewis acids which catalyze the Mukaiyama reaction also catalyze the Sakurai reaction[12].

Some time ago we embarked on a study of Lewis acids derived from transition metal complexes[13]. They were chosen for a variety of reasons. Unlike conventional Lewis acids, transition metal Lewis acids can be constructed to be insensitive to water, can be made to have known stable structures and can be tuned both structurally and electronically by ligand variation. These characteristics allow for easy manipulation and provide an unambiguous basis for defining the origins of stereoselection.

The catalysts $[Ti(Cp)_2(OTf)_2]$[14] ,**1**, $[Zr(Cp)_2(OTf)_2THF]$[14] ,**2**, and $[Ru(salen)(NO)H_2O]SbF_6$[15] ,**3**, were used to catalyze the Diels-Alder, Mukaiyama and Sakurai reactions. All three of these catalysts are stable to water and catalysis proceeds in the presence of water. The titanium and zirconium catalysts have weakly coordinating triflate

210

ligands which are readily displaced by carbonyl groups[13,14] and the oxophilic nature of the metals leads to activation of the carbonyl bearing substrate to reaction. The ruthenium catalyst **3** contains ruthenium (II) which is generally electron rich and therefore would not be expected to be a Lewis acid. The catalyst **3** is an excellent Lewis acid because of the presence of a positive charge, the electron withdrawing properties of the NO+ ligand and the presence of "hard" donor atoms, namely oxygen and nitrogen. All of these conspire to make the ruthenium (II) atom in **3** electron deficient and therefore a Lewis acid. Incorporating "soft" ligands such as sulfur, phosphorus or arsenic tends to quench the Lewis acidity. For example, the positively charged complex, [Ru(diphos)$_2$Cl]+, diphos = Ph$_2$PCH$_2$CH$_2$PPh$_2$, does not catalyze the Diels-Alder nor the Mukaiyama reactions. A further important characteristic of the catalyst **3** is the presence of the NO+ ligand trans to the aquo ligand which engenders lability to the aquo ligand. Consequently, although the aquo group is stable, it is labile and it has been shown that the rate of aquo exchange is very fast[15]. Rapid exchange also occurs when the aquo ligand is replaced by aldehydes or ketones and, as a result, the catalytic turnover frequency is not impeded by slow exchange of substrates or products.

Having developed these transition metal based Lewis acids and having demonstrated that the titanium **1** and zirconium **2** catalysts promoted the Mukaiyama and Sakurai reactions very rapidly[13,14] at catalyst loading of 0.5 mol% in CH$_2$Cl$_2$ solutions at 25°C, we proceeded to prepare chiral analogues of these catalysts. Three of these chiral catalysts are illustrated, **4**[16], **5**[17], **6**[18].

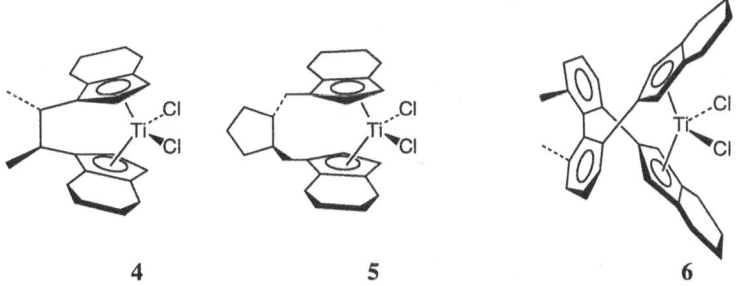

It will be noted that all of the ligands in these complexes are chiral and that, for systems **4** and **5**, diastereomeric complexes are formed because two centers of chirality are present, one at the ligand and the other about the metal. As a result, one chirality of the ligand tends to give a preferred chirality about the metal so that resolution of the complex is obviated. For the complex **6**, a further refinement is introduced where the chiral ligand can only give a single isomer of the complex and the complex has C$_2$ symmetry. All three of these complexes act as efficient catalysts for the Mukaiyama and Diels-Alder reactions. Ketones and aldehydes couple with silyl enol ethers and ketene acetals in the Mukaiyama reaction and α,β-unsaturated aldehydes and ketones undergo the Diels-Alder reaction with dienes. For the Diels-Alder reaction, high ee's were obtained with the catalyst 6[19]. None of the chiral catalysts gave any significant ee for the Mukaiyama reaction when a variety of substrate combinations were used. This was a surprising observation and the fact that appreciable ee was observed for the Diels-Alder reaction with 6 suggested that the mechanism of the Mukaiyama reaction was not that commonly supposed. It is usually supposed that, after carbon-carbon bond formation between the carbonyl group and the silyl enol ether, the silicon atom transfers from one oxygen atom to the other (eq. 1). It was expected that such a mechanism should give at least modest ee's with these catalysts. The absence of enantioselection led us to investigate the mechanism of these reactions[20].

2 Mechanism

The two possible mechanisms for the Mukaiyama coupling are illustrated in Scheme 1 where a Lewis acid triflate, MOTf, is employed as the catalyst.

Scheme 1

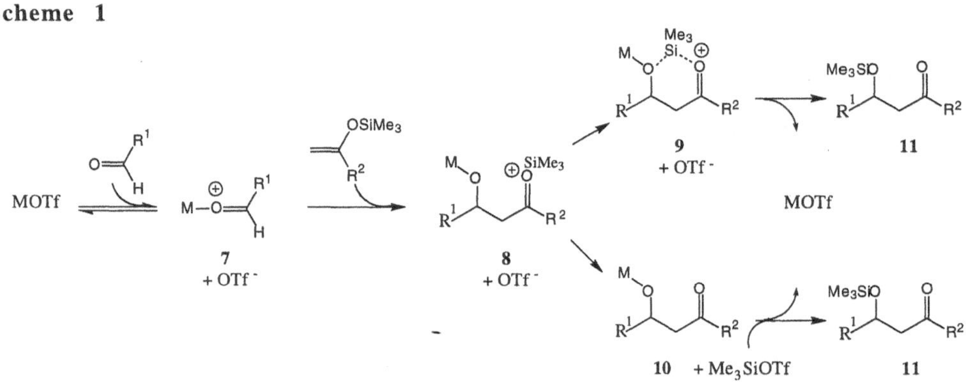

Consistent with our observations with the titanium and zirconium catalysts 1 and 2, the aldehyde is assumed to reversibly bind to the metal by displacing the triflate ligand to form the adduct 7. Coupling of the adduct with a silyl enol ether leads to the formation of the silylated aldolate intermediate 8. This aldolate 8 can then transform to the product 11 by intramolecular silicon transfer, 9. This is the generally assumed path. Alternatively, the trimethylsilyl group can be captured by OTf⁻ to give the bound neutral aldolate 10 and trimethylsilyl triflate. If these events were to occur, two possible reactions could ensue. The Me₃SiOTf could displace the aldolate to give the product and the regenerated catalyst or the Me₃SiOTf could act as an achiral catalyst[1] for the Mukaiyama reaction. The extent to which either of the paths obtain will depend on the relative rates of reaction. Thus, if the rate of reaction of the bound aldolate **10** with Me$_3$SiOTf were very much faster than the Me$_3$SiOTf catalyzed path, the enantioselection would be the same as for the intramolecular path because the chirality of the product is irreversibly fixed at the carbon-carbon bond forming step. If, however, Me$_3$SiOTf catalysis is very much faster than the reaction between the aldolate and the Me$_3$SiOTf, the ee will be diminished by the extent to which the reaction followed the achiral Me$_3$SiOTf path. It should be noted the Me$_3$SiOTf is a very powerful catalyst for the Mukaiyama reaction[1] and hence the intermolecular removal of the aldolate from the catalyst requires to be very fast to compete with Me$_3$SiOTf catalysis. This mechanistic background provides the basis for further discussion of the mechanism.

3 [Ti(Cp)$_2$(OTf)$_2$] Catalysis

If the ratio of substrates to catalyst is about 3:1, the reaction can be followed by ^{1}H NMR spectroscopy at 25°C in CD$_2$Cl$_2$ solution. It was found that during catalysis using [Ti(Cp)$_2$(OTf)$_2$], benzaldehyde and the silyl enol ether **12**, Me$_3$SiOTf was formed. We have identified two mechanisms of reaction for the production of Me$_3$SiOTf, one originating from hydrolysis of the catalyst and the other arising from the mechanism.

$$Ph \diagdown \diagup OSiMe_3$$

12

The hydrolysis mechanism was demonstrated by identifying the products formed from mixing [Ti(Cp)$_2$(OTf)$_2$] and the silyl enol ether **12** in CD$_2$Cl$_2$ solution. The CD$_2$Cl$_2$ solvent was dried by distillation over CaH$_2$. Aside from the catalyst and the silyl enol ether, the solution was found to contain Me$_3$SiOTf, (Me$_3$Si)$_2$O, acetophenone and an oligomeric titanium complex of the formula, [Ti(Cp)$_2$O][21]. Despite the normal drying of the solvent, small amounts of water enter the system. The amount of [Ti(Cp)$_2$(OTf)$_2$] that remains after **12** is added depends on the amount of water present. The reactions which are believed to occur when water is present are illustrated in eqs. 3-7.

$$[Ti(Cp)_2(OTf)_2] + H_2O \;\rightleftharpoons\; [Ti(Cp)_2(OTf)OH_2]OTf \qquad (3)$$

$$[Ti(Cp)_2(OTf)OH_2]OTf \;\rightleftharpoons\; [Ti(Cp)_2(OTf)OH] + HOTf \qquad (4)$$

$$[Ti(Cp)_2(OTf)OH] \;\rightleftharpoons\; [Ti(Cp)_2O] + HOTf \qquad (5)$$

$$HOTf + \underset{Ph}{\overset{OSiMe_3}{\diagup}} \;\longrightarrow\; O{=}\underset{Ph}{\diagup} + Me_3SiOTf \qquad (6)$$

$$Me_3SiOTf + 1/2\,H_2O \;\longrightarrow\; 1/2\,(Me_3Si)_2O + HOTf \qquad (7)$$

The reactions (3), (4) and (5) generate two equivalents of triflic acid which, in turn, react with the silyl enol ether to give the aldehyde (6) and then Me_3SiOTf rapidly reacts with water (7). It should be noted that eqs. 6 and 7 represent a dehydration process so that a completely dry solution containing neither Me_3SiOTf nor HOTf would result at the expense of the silyl enol ether. Because the $[Ti(Cp)_2O]$ complex is found to remain in these solutions, the dehydrated solution should contain $[Ti(Cp)_2O]$ and Me_3SiOTf in the ratio of 1:2. This is observed. Since Me_3SiOTf is a powerful catalyst, only small amounts of water need to be present in the solvent in order to cause the intervention of the Me_3SiOTf catalyst. Given this, one could either attempt to exhaustively dry the system or one could use a hindered base to quench the triflic acid. The latter is a far more practical proposition and we have used the hindered base **13**. Thus, when excess of **13** is added to a solution of $[Ti(Cp)_2(OTf)_2]$ followed by the silyl enol ether **12**, the solution was found to contain protonated base but no Me_3SiOTf was present. The base, therefore, effectively quenches the reactivity of triflic acid.

13

Addition of benzaldehyde to the CD_2Cl_2 solution containing [Ti(Cp)$_2$(OTf)$_2$], the hindered base and the silyl enol ether leads to the production of the Mukaiyama product. As the reaction proceeds, Me$_3$SiOTf is formed in equal proportion to the titanium aldolate **14**.

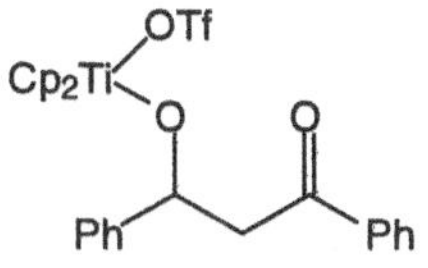

14

After the reaction is complete, the aldolate and Me$_3$SiOTf remain and do so for many hours at 25°C. It is concluded that the aldolate and Me$_3$SiOTf are formed by triflate capture of Me$_3$Si$^+$ from the intermediate **15**.

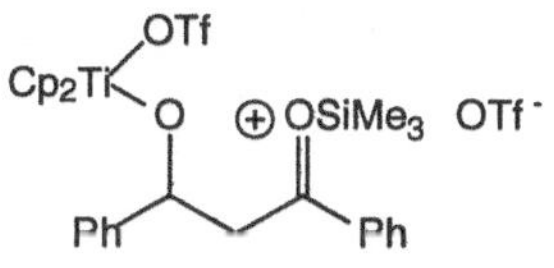

15

Further, Me$_3$SiOTf does not remove the aldolate from the titanium atom. Since the titanium aldolate **14** is not formed from the Mukaiyama product, it is concluded that **14** and Me$_3$SiOTf are formed as a consequence of the mechanism. Moreover, the rate of catalysis of the silyl enol ether with benzaldehyde by Me$_3$SiOTf was found to be the same as the rate of catalysis using the putative [Ti(Cp)$_2$(OTf)$_2$] catalyst. The conclusion seems inescapable; [Ti(Cp)$_2$(OTf)$_2$] is not the catalyst and serves only as an initiator for the production of Me$_3$SiOTf which is the sole catalyst of reaction. It is therefore clear why no ee was observed when chiral titanium complexes were used to initiate the reaction.

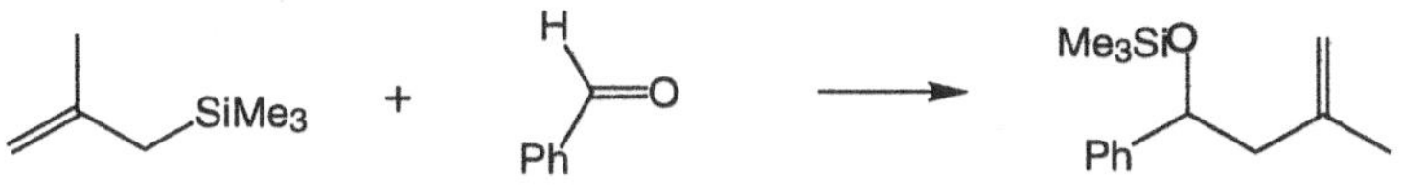

(8)

215

A wholly analogous situation obtains for the Sakurai reaction (eq. 8) in the presence of $[Ti(Cp)_2(OTf)_2]$. Mixing $[Ti(Cp)_2(OTf)_2]$ with the allylic silane in CD_2Cl_2 leads to the formation of $[Ti(Cp)_2O]$, $(Me_3Si)_2O$, Me_3SiOTf and the corresponding olefin which is produced by the reaction of the allylic silane with triflic acid (eq. 9).

$$\text{SiMe}_3 \quad + \text{ HOTf} \quad \longrightarrow \quad \quad + \text{ Me}_3\text{SiOTf} \quad (9)$$

The sequence of reactions illustrated in eq. 3-7, where eq. 6 is replaced by eq. 9 is believed to account for these observations. When the hindered base is added to $[Ti(Cp)_2(OTf)_2]$, benzaldehyde and the allylic silane, no olefin is produced. The Sakurai reaction proceeds and produces the intermediate **16** and an equivalent of Me_3SiOTf.

16

The intermediate **16** remains for many hours in the presence of Me_3SiOTf, it is not formed from the Sakurai product, and the rate of catalysis with $[Ti(Cp)_2(OTf)_2]$ can be totally accounted for by the rate of Me_3SiOTf catalysis. Hence, it is concluded that $[Ti(Cp)_2(OTf)_2]$ is not a catalyst but serves as an initiator for the true catalyst, Me_3SiOTf.

Given these unexpected results, we began to question the validity of a number of reported catalysts. We describe here mechanistic studies on two reported systems.

4 Trityl Triflate Catalyst

Mukaiyama reported[6] that Ph_3COTf catalyzed the reaction shown in eq. 10 where the erythro:threo ratio was found to be 69:31.

$$\text{(10)}$$

$$E:T = 69:31$$

Perhaps significantly, it was later reported that this same coupling was catalyzed by Me_3SiOTf under similar conditions and gave a similar E:T ratio[1b]. Although as we shall see presently the situation is more complicated than it appears, the fact that the E:T ratio is very similar in the two cases raises the suspicion that Me_3SiOTf is the catalyst for both reported couplings.

The Ph_3COTf reaction was followed by ^1H NMR spectroscopy in CD_2Cl_2 solution at - 80°C. It was found that the solution contained Ph_3COH, $(Me_3Si)_2O$, Me_3SiOTf and cyclohexanone. The product was found to have an E:T ratio of 60:40. The concentrations of the Ph_3COH, $(Me_3Si)_2O$, Me_3SiOTf and cyclohexanone increased with increasing water content of the solutions. The amount of Ph_3COH was equal to that of Me_3SiOTf. Addition of the silyl enol ether of cyclohexanone to CD_2Cl_2 solutions of Ph_3COTf at -80°C gave Ph_3COH, $(Me_3Si)_2O$, Me_3SiOTf and cyclohexanone upon mixing. This result is wholly analogous to that observed with the $[Ti(Cp)_2(OTf)_2]$ complex and the results are interpreted in a similar way as illustrated in eq. 11-13.

$$Ph_3COTf + H_2O \rightleftharpoons Ph_3COH + HOTf \qquad \text{(11)}$$

$$\text{(12)}$$

$$Me_3SiOTf + 1/2\ H_2O \longrightarrow 1/2\ (Me_3Si)_2O + HOTf \qquad \text{(13)}$$

As before, it was found that in the absence of acid, Me_3SiOTf does not remove the hydroxyl group of Ph_3COH and, as a result, equal concentrations of Me_3SiOTf and Ph_3COH remain in solution.

We have demonstrated that Ph$_3$COTf is neither a catalyst nor an initiator of the Mukaiyama reaction by two methods. It was found that the rate of catalysis is linearly dependent on the concentration of Me$_3$SiOTf but showed no dependence on the concentration of Ph$_3$COTf. When no Me$_3$SiOTf is present, no catalysis occurs in the presence of Ph$_3$COTf. This kinetic conclusion was confirmed when the reaction was attempted in the presence of hindered base. No reaction occurred because the hindered base quenched the triflic acid and prevented the formation of Me$_3$SiOTf (eq. 12).

Although Me$_3$SiOTf and the Ph$_3$COTf initiated reactions give similar E:T ratios, it turns out that the coincidence is accidental. Using pure Me$_3$SiOTf, we found that the rate of the reaction (eq. 10) was very much faster than the reaction observed with Ph$_3$COTf as the initiator. Surprisingly, the E:T ratio was found to be 30:70, the reverse of that found by Mukaiyama. When the OTf$^-$ ion concentration is increased, the Me$_3$SiOTf catalysis slows and the E:T ratio changes. When the concentration of OTf$^-$ ions reaches that present in the Ph$_3$COTf promoted reaction, the Me$_3$SiOTf catalysis proceeds at a comparable rate and the E:T ratio is similar to that reported by Mukaiyama. The rate dependence of the Me$_3$SiOTf catalysis on OTf$^-$ concentration can be understood in terms of the common ion effect (eq. 14) which leads to a lower concentration of the aldehyde adduct (Me$_3$SiOCHPh$^+$) as the concentration of OTf$^-$ increases.

$$Me_3SiOTf + PhCHO \rightleftharpoons Me_3SiOCHPh^+ + OTf^- \quad (14)$$

The dependence of the E:T ratio on the OTf$^-$ ion concentration is more subtle and may be related to the formation of ion pairs (Me$_3$SiOCHPh·OTf) or perhaps ion clusters. It could be that the ion pair leads to one E:T ratio while the cationic Me$_3$SiOCHPh$^+$ adduct leads to the opposite ratio so that the outcome is dependent on the relative rates of reaction of the two species. If this is true, the question as to why the two species should give different stereoselection remains to be answered.

5 Ph$_3$CClO$_4$ Catalyst

It was reported[12c] that Ph$_3$CClO$_4$ catalyzed the reaction shown in eq. 15. Following this reaction by ^{1}H NMR spectroscopy revealed that the reaction is complex, giving in addition to the product, a variety of other molecules, some of which are derived in parallel to the Sakurai coupling.

$$\text{SiMe}_3 + \text{PhCH(OMe)}_2 \xrightarrow[\text{CH}_2\text{Cl}_2, -25\ ^\circ\text{C}]{5\ \text{mol \% Ph}_3\text{CClO}_4} \underset{\text{Ph}}{\overset{\text{OMe}}{\diagup}} + \text{Me}_3\text{SiOMe} \quad (15)$$

When the reaction is carried out in the presence of the hindered based, the rate is orders of magnitude slower than when the base is absent. No Me_3SiClO_4 could be detected by ^1H NMR spectroscopy but it was found that Me_3SiClO_4 is an exceedingly powerful catalyst and that the rate of reaction could be accounted for by the presence of 7 x $10^{-5}M$ Me_3SiClO_4. In the presence of base, the $HClO_4$ that is formed from hydrolysis of Ph_3CClO_4 would be quenched with respect to reaction with the allylsilane and hence no Me_3SiClO_4 would form by this route. Hence, Ph_3CClO_4 could be a catalyst unless another method of generating Me_3SiClO_4 exists. This is the case, for we found that the relatively slow reaction shown in eq. 16 occurs.

The rate of this reaction was found to be sufficient to account for all of the catalysis observed when base is present. It would appear, therefore, that Ph_3CClO_4 is not the real catalyst, rather it serves to generate the true catalyst, Me_3SiClO_4, by the allylation reaction (eq. 16).

$$\text{\raisebox{0pt}{\includegraphics}}\;SiMe_3 \;+\; Ph_3CClO_4 \;\longrightarrow\; Ph_3C\text{\raisebox{0pt}{}} \;+\; Me_3SiClO_4 \qquad (16)$$

6 Enantioselection

The work described here presents a somewhat disconcerting picture for developing enantioselective Mukaiyama reactions. Production of the "Me_3Si^+" can occur by a variety of means. Very small amounts of water can generate acid by hydrolysis of the Lewis acid aquo complex, eq. 17.

$$MX + H_2O \;\rightleftharpoons\; X^- + MOH_2^+ \;\rightleftharpoons\; MOH + HX \qquad (17)$$

The acid then reacts exceedingly rapidly with the silyl enol ether or the allylic silane to produce the "Me_3Si^+" catalyst. In principle, if the "Me_3Si^+" is capable of removing the hydroxyl group from MOH, complete dehydration will occur and no acid nor "Me_3Si^+" will remain (eq. 18-25). This process depends on the ability of "Me_3Si^+" to scavenge the water that is bound to the Lewis acid. This is not the case for the Lewis acids, $[Ti(Cp)_2(OTf)_2]$ and Ph_3COTf because the former forms $[Ti(Cp)_2O]$ and the latter gives Ph_3COH, neither of which reacts with "Me_3Si^+". Consequently, Me_3Si^+ remains in solution to promote catalysis. Even when the acid produced from hydrolysis of the Lewis acid is quenched by addition of base, "Me_3Si^+" can form either as part of the mechanism as in the case of the titanium catalyst (Scheme 1) or as in the case of the Sakurai coupling the Lewis acid may react with the allylic silane (eq. 16).

$$M^+ + H_2O \rightleftharpoons MOH_2^+ \tag{18}$$

$$MOH_2^+ \rightleftharpoons MOH + H^+ \tag{19}$$

$$H^+ + \underset{R}{\overset{OSiMe_3}{\diagup\diagdown}} \longrightarrow \underset{R}{\overset{O}{\diagup\diagdown}} + Me_3Si^+ \tag{20}$$

$$Me_3Si^+ + MOH \longrightarrow M^+ + Me_3SiOH \tag{21}$$

$$Me_3SiOH \rightleftharpoons Me_3SiO^- + H^+ \tag{22}$$

$$H^+ + \underset{R}{\overset{OSiMe_3}{\diagup\diagdown}} \longrightarrow \underset{R}{\overset{O}{\diagup\diagdown}} + Me_3Si^+ \tag{23}$$

$$Me_3SiO^- + Me_3Si^+ \longrightarrow (Me_3Si)_2O \tag{24}$$

Net: $\quad H_2O + 2 \underset{R}{\overset{OSiMe_3}{\diagup\diagdown}} \longrightarrow 2 \underset{R}{\overset{O}{\diagup\diagdown}} + (Me_3Si)_2O \tag{25}$

One conclusion that might be drawn from these observations is that, for the Mukaiyama catalysis, the Lewis acid-oxygen bond should be kinetically labile at least with respect to capture by Me_3Si^+. Clearly, the titanium catalyst does not belong to this category because the aldolate intermediates **14** and **16** are stable in the presence of Me_3SiOTf. It would appear, therefore, that in order to avoid the production of "Me_3Si^+" Lewis acids which have kinetically weak oxygen bonds are required. Such weak bonds will allow for rapid transfer of Me_3Si^+ to the oxygen atom of the aldolate by an intramolecular path or by an intermolecular reaction. In addition, it is probable that neutral Lewis acids will tend to promote capture of the aldolate oxygen atom by Me_3Si^+ because there is no counter ion to capture Me_3Si^+ and the Lewis acid-aldolate bond is expected to be weaker in most cases. Of course, even with neutral Lewis acids, the Me_3Si^+ can be captured by the carbonyl bearing substrate and induce catalysis by "Me_3Si^+". Similarly, polar solvents, particularly those possessing oxygen atoms, could capture Me_3Si^+. Thus the mechanism of the Mukaiyama reaction as depicted in Scheme 1 implies that a delicate balance exists between a pathway which involves the Lewis acid catalyst and the "Me_3Si^+" path.

Despite this complexity, a number of very successful enantioselective catalysts have been devised for the Mukaiyama reaction[22]. All of these catalysts appear to be neutral species although in no case was the actual catalyst identified. Many require low temperatures (-78°C) and high catalyst loadings (20 mol%) for success. Some of these catalysts require that the substrates be added very slowly in order to achieve high ee's. It is probable that the slow addition is required because "Me_3Si^+" is released and reacts slowly with the Lewis acid-aldolate. Thus, the Me_3Si^+ removal of the aldolate reaction requires to be completed before more substrate is added. Otherwise the catalysis will proceed via the achiral "Me$_3$Si$^+$" path. Other catalysts proceed at higher temperatures (-10 to 0°C) and at moderate catalyst loadings (2-5 mol%) without slow addition of the substrates. These are neutral Lewis acids of titanium (IV) and are probably mild Lewis acids. The mechanisms of these reactions are not known but it may be that the intramolecular Me$_3$Si$^+$ transfer path occurs. There is, however, another possible mechanistic path for the Mukaiyama reaction which we have recently discovered and which may operate in some of the successful enantioselective reactions.

7 The [2 + 2] Mechanism

On the assumption that mild, neutral Lewis acids were likely to lead to the desired Me$_3$Si$^+$ transfer path, we investigated a variety of complexes which we supposed possessed the necessary characteristics. A number of neutral lanthanide, zinc and magnesium complexes were investigated and all showed similar behavior for both aromatic and aliphatic aldehydes coupling with ketene acetals. No reaction was observed with ketones nor were silyl enol ethers reactive. We report the results found using the europium catalyst **17** as representative.

17

When benzaldehyde and the ketene acetal are added to the catalyst, a slow transformation occurs initially (eq. 26). The initial products are the cyclic orthoester isomers which are kinetically formed in equal amounts.

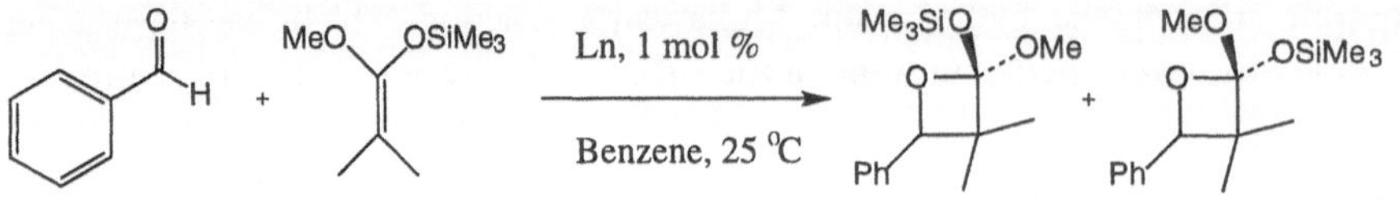

$$\text{(26)}$$

1 M   1 M

Over two hours, the ratio of these isomers changes to the equilibrium ratio of 60:40. At equilibrium, only 60% of the substrates are converted to these products. If after equilibration the solution is diluted ten-fold at constant Ln, the same 60:40 ratio of isomers is maintained but the amount of conversion diminishes to 20%. This result suggests that the catalyzed equilibrium (eq. 27) is established with K=3.

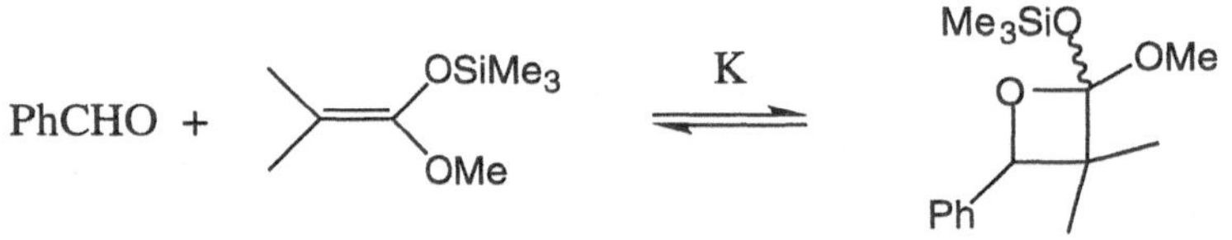

$$\text{(27)}$$

Further, since the reaction is formally a reversible [2 + 2] addition, the isomer equilibration must occur by fragmentation and recombination of the substrates rather than by ring opening and closing of the putative intermediate 18.

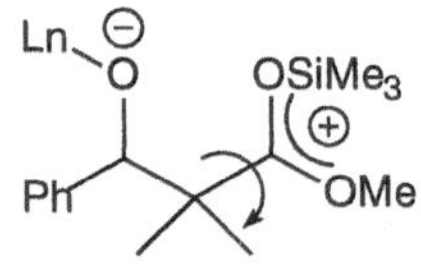

18

222

That retro-[2 + 2] fragmentation occurs is supported by a number of observations including the results of the equilibration of the products of the catalysis shown in eq. 28. The isomers are initially formed in a ratio of 30:70 and after equilibration, the isomer ratio is reversed to 90:10.

$$PhCHO \; + \qquad \underset{\text{Ln}}{\overset{\text{Ln}}{\rightleftharpoons}} \qquad (28)$$

Isomeration of the product shown in eq. 28 can only occur by breaking of the requisite carbon-carbon bond, implying that equilibration occurs by complete fragmentation to the substrates.

In all cases, the catalysis proceeds irreversibly to the Mukaiyama product over 20 hours at 25°C. No substrates nor cyclic orthoesters remain. These results indicate that the mechanism outlined in Scheme 2 operates. It is proposed that the first step involves the formation of the zwitterionic intermediate **19** which can collapse to the cyclic orthoester. Alternatively, **19** may convert to the silicon bridging intermediate **20** which transfers the silyl group to give the Mukaiyama product irreversibly. For the case of the Ln catalyst, the reversible formation of the cyclic orthoester is a much faster process than the silyl transfer step. We have found that for other catalysts, the rate of orthoester formation and the rate of formation of the Mukaiyama product are comparable. Thus, for the bis-hexafluoro-acetylacetonate zinc (II) catalyst, equilibration of the cyclic orthoester is not achieved because the formation of the Mukaiyama product occurs at a rate faster than orthoester equilibration. It is conceivable therefore that, depending on the catalyst, the catalysis could occur without detectable formation of the cyclic orthoesters.

Scheme 2

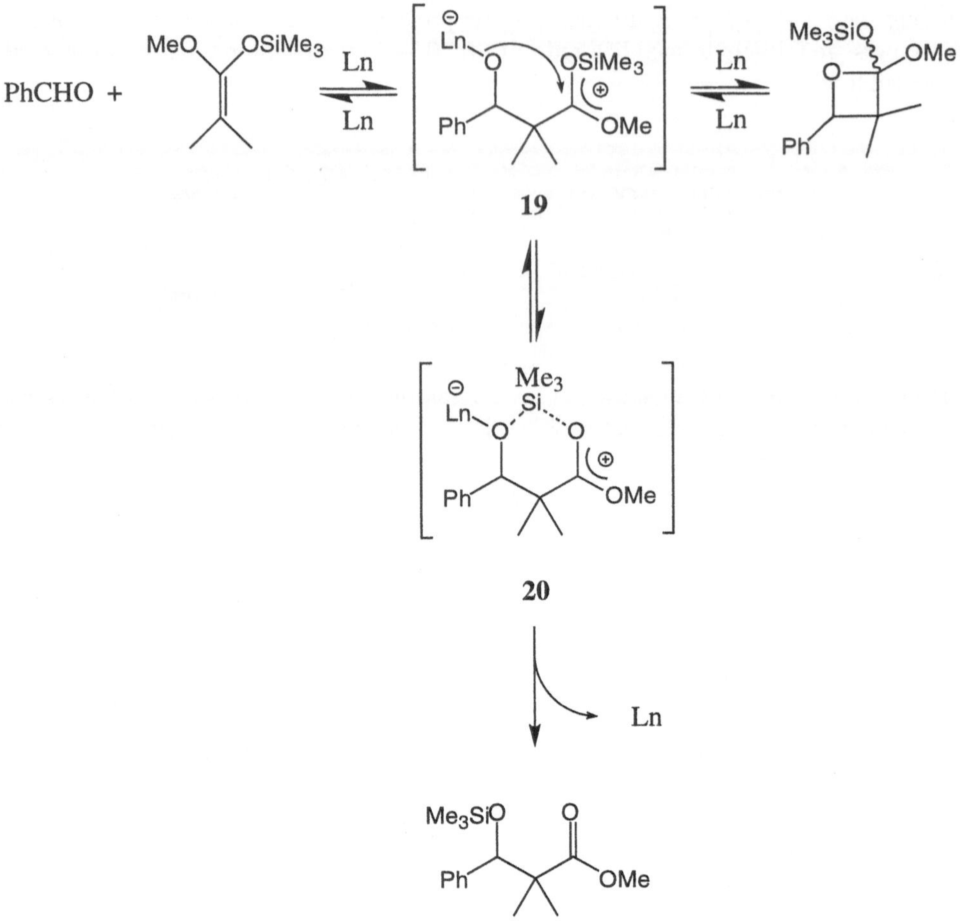

For the reaction shown in Scheme 2, we find that the Ln catalyst gives 15% ee. A 5% ee is obtained from the ortho esters provided the reaction is quenched before equilibration of the isomers occurs. Of course, 0% ee is obtained from the ortho esters after isomeric equilibrium is achieved. In general, the enantioselectivity from the mechanism outlined in Scheme 2 must depend on a complex mix of rate constants associated with the various species in solution. In order to put this complexity in perspective, we outline the steps involved in enantioselection in Scheme 3. Enantioselection depends on the relative rates of production of the R and S isomers. The ee will depend on the relative rates of the R and S silyl transfer steps to give the R and S products and on the rate at which the two silyl transfer intermediates are formed from the substrates and from the corresponding cyclic orthoesters.

224

Scheme 3

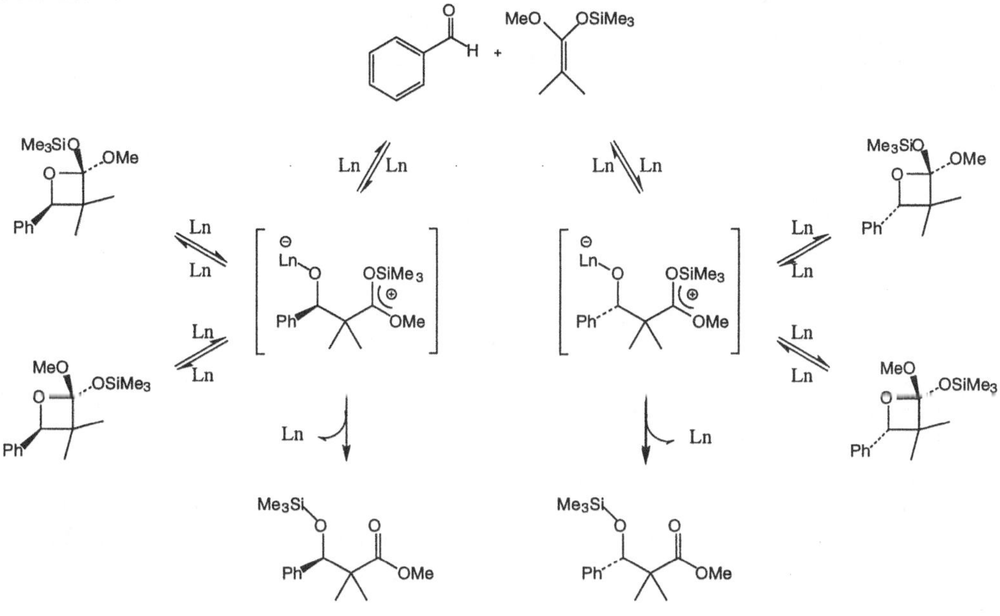

Therefore the interpretation of the origins of the enantioselection will depend on the relative rates at which a particular chiral catalyst interconverts the various species among each other and on the rate of product formation.

8 Discussion

The work described here illustrates the potential complexity of the catalyzed Mukaiyama reaction. If strong Lewis acids are used in the sense that they form kinetically robust Lewis acid-aldolate bonds or become acidic upon hydrolysis, it is likely that catalysis will proceed by the "Me₃Si⁺" path, as has been shown here. Thus the Lewis acid requires to be calibrated carefully. It appears that neutral lanthanide and zinc complexes as well as complexes of elements with similar characteristics are likely to catalyze the Mukaiyama reaction without "Me₃Si⁺" intervention. The formation of the [2 + 2] adducts is likely with these types of Lewis acids. The [2 + 2] adducts are ideal intermediates since the silyl group is stable in these compounds. It is perhaps surprising that these [2 + 2] adducts have not been detected before because [2 + 2] adducts formed from aldehydes and dialkyl ketene acetals have been known for some time[23]. The present work may provide a guide for the design of new stereoselective catalysts for the Mukaiyama reaction.

Acknowledgement

This work was supported by grants from the National Institutes of Health.

References

1. (a) Murata, S.; Suzuki, M.; Noyori, R. *J. Am. Chem. Soc.* **1980**, *102*, 3248.
   (b) Mukai, C.; Hashizume, S.; Nagami, K.; Hanaoka, M. *Chem. Pharm. Bull.* **1990**, *38*, 1509.
2. Sakurai, H.; Sasaki, K.; Hosomi, A. *Bull. Chem. Soc. Jpn.* **1983**, *56*, 3195.
3. Iwasawa, N.; Mukaiyama, T. *Chem. Lett.* **1987**, 463.
4. Mukaiyama, T.; Kobayashi, S.; Tamura, M.; Sagawa, Y. *Chem. Lett.* **1987**, 491.
5. (a) Mukaiyama, T.; Kobayashi, S.; Murakami, M. *Chem. Lett.* **1985**, 447.
   (b) Mukaiyama, T.; Kobayashi, S.; Murakami, M. *Chem. Lett.* **1984**, 1759.
6. Kobayashi, S.; Murakami, M.; Mukaiyama, T. *Chem. Lett.* **1985**, 1535.
7. Reetz, M. T.; Vougioukas, A. E. *Tetrahedron Lett.* **1987**, *28*, 793. Sato, S.; Matsuda, I.; Izumi, Y. *Tetrahedron Lett.* **1987**, *28*, 6657. Sato, S.; Matsuda, I.; Izumi, Y. *Tetrahedron Lett.* **1986**, *27*, 5517. Mukaiyama, T.; Soga, T.; Takenoshita, H. *Chem. Lett.* **1989**, 1273.
8. Kobayashi, S.; Hachiya, I.; Takahori, T. *Synthesis* **1993**, 371. Kobayashi, S.; Hachiya, I.*Tetrahedron Lett.* **1992**, *33*, 1625. Kobayashi, S. *Chem. Lett.* **1991**, 2187. Kobayashi, S.; Hachiya, I.; Ishitani, H.; Araki, M. *Synlett.* **1993**, 472.
9. Bach, T.; Fox, D. N. A.; Reetz, M. T. *J. Chem. Soc.; Chem. Commun.* **1992**, 1634.
10. Hara, R.; Mukaiyama, T. *Chem. Lett.* **1989**, 1909. Mukaiyama, T.; Hara, R. *Chem.Lett.* **1989**, 1171.
11. Hosomi, A.; Sakurai, H. *Tetrahedron Lett.* **1976**, *16*, 1295. Hosomi, A.; Endo, M.; Sakurai, H. *Chem. Lett.* **1976**, 941.
12. (a) Tsunoda, T.; Suzuki, M.; Noyori, R. *Tetrahedron Lett.* **1980**, *21*, 71. (b) Sakurai, H.; Sasaki, K.; Hosomi, A. *Tetrahedron Lett.* **1981**, *22*, 745. 0(c) Mukaiyama, T.; Nagaoka, H.; Murakami, M.; Ohshima, M. *Chem. Lett.* **1985**, 977. Ohshima, M. *Chem. Lett.* **1985**, 977.
13. Hollis, T. K.; Robinson, N. P.; Whelan, J.; Bosnich, B. *Tetrahedron Lett.* **1993**, *34*, 4309.
14. Hollis, T. K.; Robinson, N. P.; Bosnich, B. *Tetrahedron Lett.* **1992**, *33*, 6423.
15. (a) Odenkirk, W.; Whelan, J.; Bosnich, B. *Tetrahedron Lett.* **1992**, *33*, 5729. (b) Odenkirk, W.; Rheingold, A. L.; Bosnich, B. *J. Am. Chem. Soc.* **1992**, *114*, 6392.
16. Rheingold, A. L.; Robinson, N. P.; Whelan, J.; Bosnich, B. *Organometallics* **1992**, *11*, 1869.
17. Hollis, T. K.; Rheingold, A. L.; Robinson, N. P.; Whelan. J.; Bosnich, B. *Organometallics* **1992**, *11*, 2812.

18. Ellis, W. W.; Hollis, T. K.; Odenkirk, W.; Whelan, J.; Ostrander, R.; Rheingold, A. L.; Bosnich, B. *Organometallics* **1993**, *12*, 4391.

19. Odenkirk, W.; Bosnich, B. *Chem. Commun.* **1995**, 1181.

20. Hollis, T. K.; Bosnich, B. *J. Am. Chem. Soc.* **1995**, *117*, 4570.

21. Bottomley, F. Polyhedron **1992**, *11*, 1707.

22. (a) Kobayashi, S.; Fujishita, Y.; Mukaiyama, T. Chem. *Lett.* **1990**, 1455. (b) Kobayashi, S.; Furuya, M.; Ohtsubo, A.; Mukaiyama, T. *Tetrahedron:Asymmetry* **1991**, 2, 635. (c) Furuta, K.; Maruyama, T.; Yamamoto, H. *J. Am. Chem. Soc.* **1991**, *113*, 1041. (d) Furuta, K.; Maruyama, T.; Yamamoto, H. *Synlett.* **1991**, 439. (e) Corey, E. J.; Cywin, C. L., Roper, T. D. *Tetrahedron Lett.* **1992**, *33*, 6907. (f) Parmee, E. R.; Tempkin, O.; Masamune, S.; Abiko, A. *J. Am. Chem. Soc.* **1991**, *113*, 9366. (g) Parmee, E. R.; Hong, Y.; Tempkin, O.; Masamune, S. *Tetrahedron Lett.* **1992**, *33*, 1729. (h) Kiyooka, S.; Kaneko, Y.; Kume, K. *Tetrahedron Lett.* **1992**, *33*, 4927. (i) Carreira, E. M.; Singer, R. A., Lee, W. *J. Am. Chem. Soc.* **1994**, *116*, 8837.

23. (a) Scheeren, H. W.; Aben, R. W. M.; Ooms, P. H. J.; Nivard, R. J. F. *J. Org. Chem.* **1977**, *42*, 3128. (b) Aben, R. W.; Scheeren, J. W.; *Synthesis*, **1978**, 400. (c) Bakker, C. G.; Scheeren, J. W.; Nivard, R. J. F. *Recl. Trav. Chim. Pays-Bas.* **1981**, *100*, 13. (d) Scheeren, J. W. *Recl. Trav. Chim. Pays-Bas.* **1986**, *105*, 71.

New Rhodium Hydroformylation Catalysts

Piet W.N.M. van Leeuwen and Paul C.J. Kamer

Van 't Hoff Research Institute (pwnm@anorg.chem.uva.nl), Department of Inorganic Chemistry and Homogeneous Catalysis, Nieuwe Achtergracht 166, 1018 WV Amsterdam, the Netherlands

1 Introduction

Before we turn to the recent developments in hydroformylation in our laboratory we will sketch how we got involved in the ligands that we are using today. In the mid-seventies, at Shell in Amsterdam, one of us was involved in a screening programme of phosphorus ligands in the rhodium catalysed hydroformylation. The goals were diverse:

1. Hydroformylation of internal alkenes to terminal products. The best results reported involve the use of tris(2,2,2-trifluoroethyl)phosphite with rhodium yielding up to 70% of linear product with 3-heptene.[1]

2. Hydroformylation of terpenic alkenes and the like as intermediates for aroma chemicals. Bulky phosphites were found to be excellent ligands for this reaction, see below.[2]

3. Selective hydroformylation of petrochemicals. An example is the hydroformylation of butadiene using rhodium complexes of dppe (1,2-bis(diphenylphosphino)ethane) which gave selectively (>95%) linear pent-3-enal, while all other diphosphines tested gave selectivitities for any product not surmounting 25%.[3]

internal alkenes, 62% linear aldehyde
terminal alkenes, 96% linear aldehyde

bulky monophosphites
very fast catalysts

butadiene to linear
monoaldehydes

A decade before, Pruett and Smith at UCC had studied the influence of a wide range of phosphites on the rhodium catalysed hydroformylation of 1-alkenes.[4] From their results one can deduce a tendency to higher linearities when stronger π-acceptor phosphite ligands are used. We found that the highest linearity (96%) for 1-alkenes was reached using tris(2,2,2-trifluoroethyl)phosphite. Phosphite ligands were also extensively studied by Treciak and Ziólkowski.[5] Pruett also looked at ortho-substituted aryl phosphites, but the peculiar activity escaped their attention as they were focusing on high linearities, we presume. Phosphites such as tri-o-tBu-phenyl phosphite were widely available as anti-oxidants used in polypropylene. Our publication[2] on the extremely high catalytic activity triggered their renewed interest in this type of catalysts[6]. UCC introduced bulky bisphenols in the monophosphites rendering a higher hydrolytic stability to the catalyst system[7]. It was mentioned that hydroformylation of 2-butene might be of commercial interest with the bulky phosphite catalyst. Subsequently, a patent appeared from Kuraray[8] also describing the use of bulky phosphites using as substrates vinyl acetate, octenal and 3-methylbut-3-en-1-ol. Since 1987 they use this catalyst for the preparation of 3-methyl-1,5-pentanediol from 3-methylbut-3-en-1-ol followed by a hydrogenation step on a 3000 ton/annum scale. A variety of other substrates has been hydroformylated using bulky phosphite ligands.[9]

An important breakthrough was reported by Bryant and coworkers when they had discovered that bulky *di*phosphites may give very high selectivity to the linear product, albeit at lower rates than bulky monophosphites.[10] Selectivities and rates appeared to be much better than those achieved so far with triphenyl phosphine catalysts. We had also looked at di- and triphosphites (diffids and triffids) at the end of the seventies and we had noted a higher selectivity for linear products ("The diffids are coming" was the title of one of the internal presentations),[11] but we had never found such enormous effects. In the meantime we had left the rhodium catalysis and started investigations into platinum and palladium catalysis.[12] Later, activities at the Universities of Groningen, Nijmegen and Amsterdam were initiated focusing on polymer supported bulky phosphites, [13] rhodium-phosphite complexes as hydrogenation catalysts, [14] and mechanistic studies on bulky phosphite hydroformylation catalysts.[15] We will describe our recent results with mono and diphosphite modified rhodium catalysts, including asymmetric hydroformylation of styrene. An important feature of bidentate ligands is their "bite angle", not only in phosphites but also in diphosphines as we will see.

2 Mono Bulky Phosphites

One of the intermediates in the rhodium catalysed hydroformylation cycle is the five co-ordinate rhodium hydrido complex 1 containing four two-electron donor-ligands which can be carbon moxide or phosphorus ligands. The well known cycle is shown in Figure 1 with the 18-electron hydride species at the top left. Depending on their steric requirements and the vapour pressure of the competing carbon monoxide ligand the number of coordinating phosphites in the complex varies between zero and four. Co-ordination of two or three ligands is most common. In most hydroformylation catalyst systems the formation of species 3, the alkene co-ordination complex, is the rate-determining step. This is a simplification, naturally, which depends on the detailed conditions. One would expect that this step should be further slowed down when the size of the phosphorus ligands is further increased, since co-ordination of an alkene is mainly determined by steric factors.

Surprisingly, when tris-2-*t*butyl-phenyl phosphite was used as the ligand, a highly active catalyst was obtained[2, 15] that gave reaction rates for disubstituted alkenes as high as those for monosubstituted alkenes using triphenylphosphine or phosphite catalysts. Typical reaction conditions are: 60-100°C, 5-30 bar, 1 M initial alkene concentration, 10^{-3-4} M Rh, 50-fold excess of phosphite. Examples of "otherwise unreactive" alkenes are shown in Fig. 2. The reactions using bulky phosphites are several times to two orders of magnitude faster (several thousands $mol \cdot mol^{-1} \cdot h^{-1}$) than those employing triphenylphosphine (<100 $mol \cdot mol^{-1} \cdot h^{-1}$). [16] Unsaturated fatty esters can also be hydroformylated, although a diene content of 14% has a retarding effect.[17] For the synthesis of aldehydes many more applications can be envisaged in additions to the ones mentioned so far.

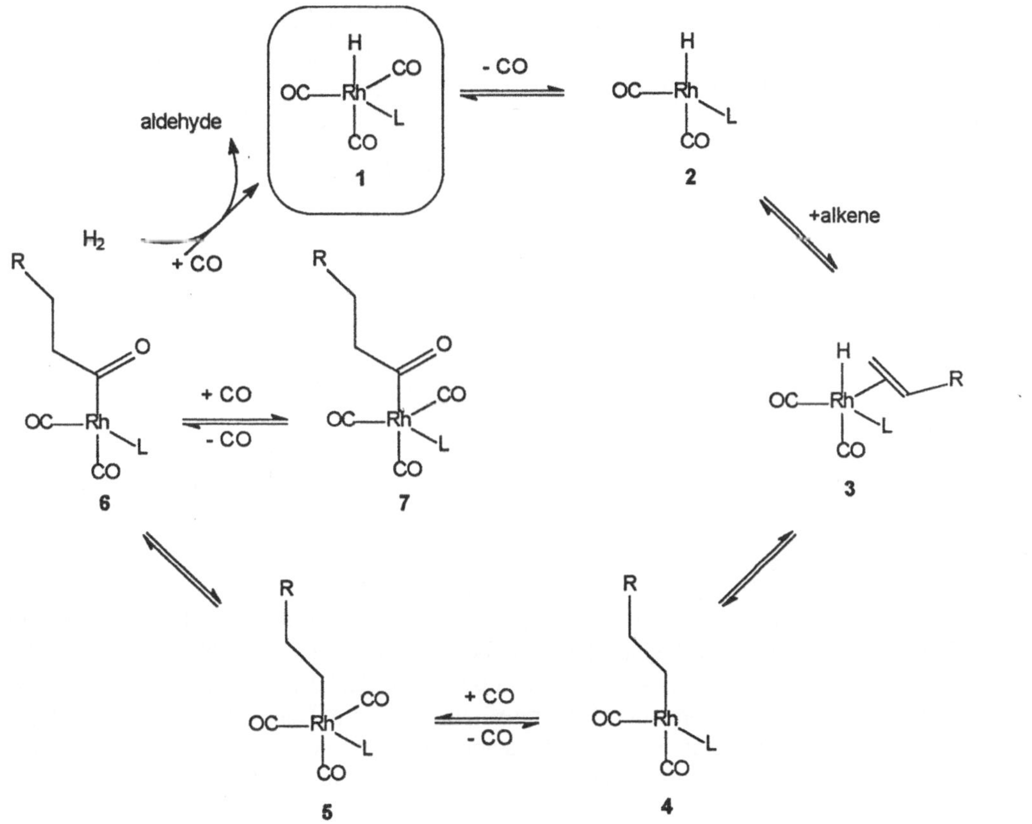

Figure 1. Mechanism of cyclohexene hydroformylation with bulky phosphite rhodium catalyst

The explanation for the high rates is simple. In situ IR and NMR (^{1}H, ^{31}P, ^{103}Rh) spectroscopy measurements have shown that the species present in solution is a monophopshite tricarbonyl rhodium hydride, no matter how large an excess of phosphite was added.[15b] One of the axial position of the trigonal bipyramidal complex is occupied by the hydrido ligand. Thus, there are no two trans positions available for the very bulky ligands with their cone angles aproaching 180°. While the ligand is larger, the resulting complex is less sterically hindered than a complex containing two smaller phosphites.

The predominant species observed is complex 1 shown in the box in Figure 1. Furthermore, CO dissociation should be easier from a mono-phosphite complex for electronic reasons. Strongly electron withdrawing ligands have the same effect.[2] The kinetics show that for hindered alkenes such as cyclohexene the reaction is first order in rhodium and alkene concentration. The reaction is accelerated by applying lower CO pressures.[18] All these data are in accord with the proposal that the conversion of 1 into 3 is rate-determining, but further details about the possible intermediacy of 2 are lacking. As 16-electron species should be accessible for rhodium this route seems reasonable and intramolecular reactions indicate that associatative carbonyl for alkene exchange involving 20-electron species is not very likely in view of the low rates of intramolecular hydroformylation.[19]

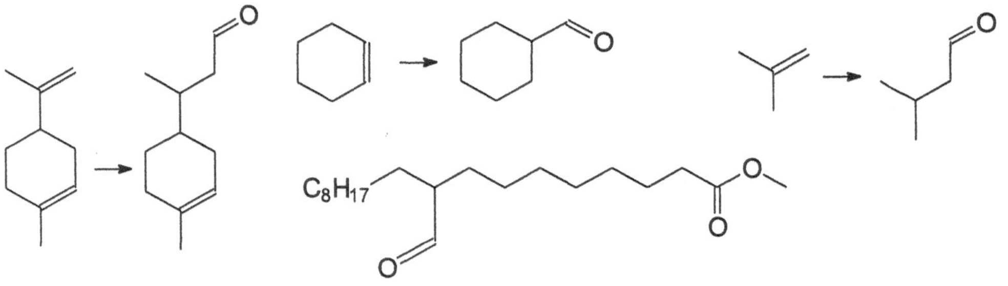

Figure 2. Hydroformylation products of unreactive alkenes with bulky phosphite catalysts

Terminal alkenes (e.g. 1-octene) undergo extremely fast hydroformylation using bulky phosphite modified rhodium catalysts. The ratio between linear and branched product aldehyde is only 2. Turnover frequencies as high as 160,000 mol(aldehyde)·mol^{-1}(rhodium)·h^{-1} have been obtained. A simple comparison of the rates with other catalysts is not possible, because a peculiar kinetic expression was found. Under standard conditions and at alkene concentrations > 0.2 M the rate turned out be independent of the alkene concentration. The rate was approximately first order in dihydrogen pressure and rhodium concentration and it showed an approximate inverse order in carbon monoxide concentration.[18] The predominant species under reaction conditions is not hydride 1, but an acyl species, presumably 7. Hence, the unhindered 1-alkenes undergo a fast reaction to the acyl species, which then must lose CO and react with dihydrogen to form the final product. The observed kinetics are an exception (equation 2); as a rule the kinetics obey equation 1.

rate equation for disubstituted alkenes: $v = k \, [\text{Rh}] \, [\text{cyclohexene}] \, [\text{CO}]^{-0.7}$ (1)

rate equation for 1-alkenes: $v = k' [\text{Rh}] \, [\text{H}_2] \, [\text{CO}]^{-1}$ (2)

As a result of the negative order in CO concentration the reaction accelerates under mass transfer limiting conditions; the reaction is very fast, diffusion of depleting CO is too slow, owing to the kinetics the reaction becomes even faster and thus a runaway occurs consuming all carbon monoxide. A very fast isomerisation occurs when all CO has been consumed. This explains why this reaction may have escaped the attention during previous work involving this catalyst.[4]

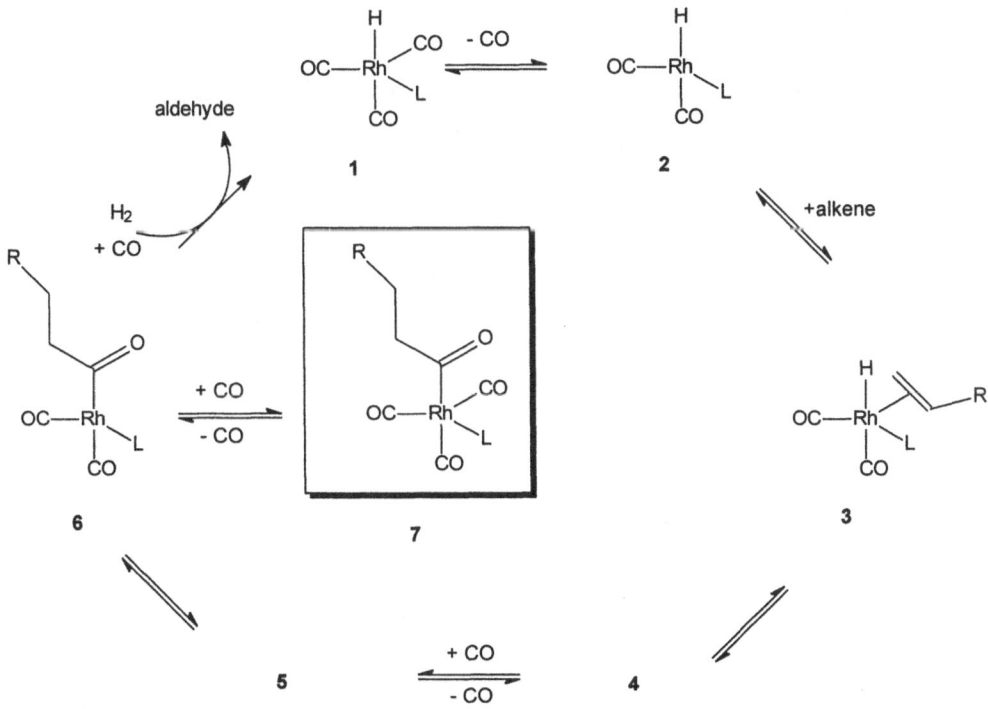

Figure 3. Hydroformylation of 1-alkenes using L = tris(2-t-Bu-phenyl)phosphite

3 Diphosphites

Diphosphites were not considered to be useful ligands until the publication of the first patent in this field by Bryant and co-workers in 1987.[20] It was followed by many other patents from UCC and many other companies initiated activities in the area of diphosphite modified catalysts, not only for hydroformylation. In 1989 we started our activities at the University of Amsterdam aiming at obtaining mechanistic insight and asymmetric hydroformylation of vinyl aromatics. The work was supported by molecular mechanics studies, which not only gave more insight into the structures but also led to the discovery of new diphosphine based catalysts, see below. The first results were presented in 1992.[21] Typical phosphites used in this field are shown in Figure 4.

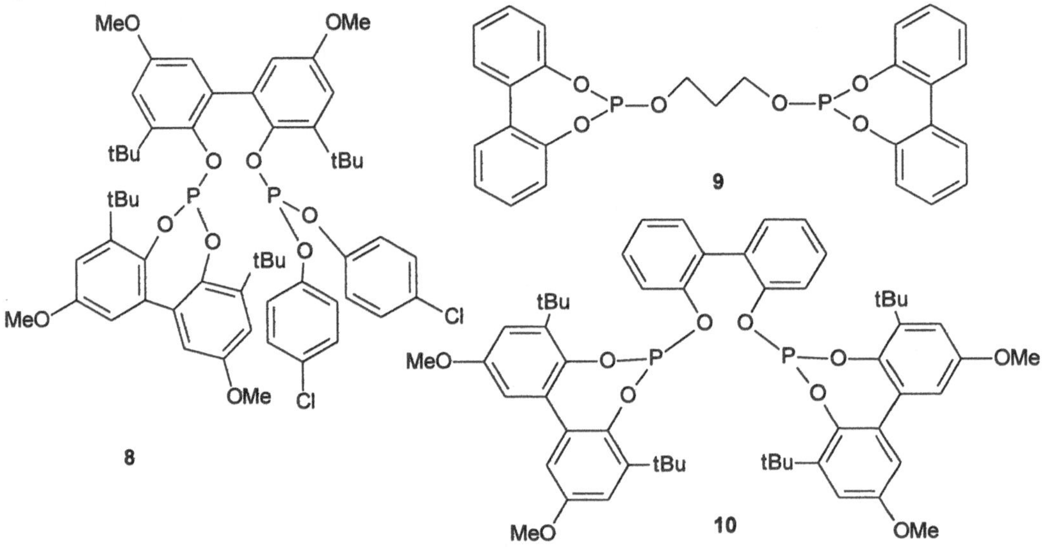

Figure 4. Typical examples of diphosphites used in hydroformylation

As might have been expected, the reaction is much slower when diphosphites are used instead of bulky monophosphites. This is true for both disubstituted and monosubstituted alkenes. Thus, the hydroformylation of 1-octene proceeds according to rate equation 1; the reaction is first order in the alk-1-ene concentration, it has an approximate -0.65 order in CO pressure, and is virtually independent of the dihydrogen pressure. The linearity decreases with incresing CO partial pressure. This is consistent with a kinetic scheme in which the alkene addition to rhodium is rate-determining.[22] UCC reported very high selectivities for linear products.[20] We found that this strongly depended on the precise structure of the ligand and the complexes formed. For example, ligand **8**, reported by UCC, gives very high linearities under the conditions we applied (only linear product was observed together with 18% of 2-octene at 80°C, 20 bar, ligand/Rh=20, H_2/CO = 1), while ligand **9** and **10** behave as common monophosphites, albeit at much lower rates.

The coproduction of internal alkenes when using higher alkenes is not very attractive, because this will eventually lead to mainly branched aldehydes. For lower alkenes such as butenes (including 2-butenes) UCC has achieved high contents of linear products. For propene isomerisation cannot occur. The advantages of the diphosphite system for propene hydroformylation as compared to the commercial triphenylphosphine system are that less ligand is required and that higher rates can be obtained. A problem to be solved is the (autocatalytic) hydrolysis of phosphites in continuous processes; in the laboratory batch experiments ligand decomposition does not play a role.

We and others have studied in detail the solution structures of the diphosphite rhodium hydride complexes by NMR spectroscopy.[22, 23]

11   **12**

Basically, bidentate ligands can give rise to two types of bipyramidal complexes, **11** and **12**, in which the bidentate ligand co-ordinates in a bis-equatorial (e-e) fashion and an equatorial-axial (e-a) fashion. The majority of diphosphines known in literature contain a bridge between the two phosphorus atoms consisting of 2, 3 or 4 carbon atoms. The preferred valence angle of these ligands is around 90° and as a result an equatorial-axial co-ordination mode predominates in these complexes. We studied the influence of the bridge and bulkiness of the ligands on the structure assumed in solution. The phosphorus chemical shifts are often almost the same in these complexes, even when the symmetry of the ligands would not predict this. In addition, both structure **11** and **12** are highly fluxional, which makes the interpretation slightly more difficult and necessitates measuring the spectra at low temperatures. Simulation of the spectra is recommended. General observations are: e-e co-ordination gives a large P-P coupling constant (200-300 Hz), a small hydride phosphorus coupling constant; e-a co-ordination gives one large phosphorus-hdyrogen coupling constant (160 Hz) and one small coupling constant (< 10 Hz), a small phosphorus-to-phosphorus coupling constant (0-70 Hz). The exchange between the two phosphorus atoms is extremely fast, especially in type **12** complexes, and at room temperature averaged spectra are observed (i.e. a triplet in the hydride region with a coupling constant ~ 70 Hz). The variation in these numbers is large, presumably because the structures are not ideally trigonal bipyramidal.

Phosphites containing a 1,2-diol bridge give complexes co-ordinating in an e-a fashion. The regioselectivity achieved using these complexes is comparable to that achieved using monodentates, linear/branched = ~ 2. Ligands such as **9** lacking the steric congestion in the ligand often give rise to mixtures of complexes and the selectivities obtained in the catalysis are

are also low. Diphosphites containing a diphenol bridge and sufficient steric bulk such as **8** give stable complexes and usually only one complex is observed attaining an e-e mode of co-ordination. As mentioned above, these compexes give the highest linear to branched ratios. There are, however, exceptions to this rule. Ligand **10** forms selectively e-e type complexes **11**, but nevertheless the regioselectivity for linear product is only 55%. In conclusion, high linearities are obtained when the ligands co-ordinate in an e-e fashion, but the reverse is not true.

Bisequatorial co-ordination requires that the ligand can accomodate a valence angle around 120°, i.e. it should have a natural bite angle around this value. At this point it is of interest to refer to the work on BISBI, Figure 5. BISBI is one of the few diphosphine ligands giving selectively linear aldehyde products as was discovered by Eastman Kodak.[24] Casey et al. have studied this ligand in depth;[25] they found that the bite angle of this ligand is indeed ~ 120° and that the preferential mode of co-ordination is bisequatorial. Figure 5 shows the structural relation between BISBI and the diphosphites introduced by UCC. Later we will come back to diphosphines having bite angles suited for bisequatorial co-ordination.

Figure 5. Structural relationship between two groups of ligands inducing high selectivity

We have obtained[22,26] two crystal structures of rhodium complexes containing **8**, one is a square planar complex of rhodium acetylacetonate RhAcac(**8**), the other is the rhodium hydrido dicarbonyl, RhH(CO)$_2$(**8**). The observed P-Rh-P angle in **13** is 97°. The compound is highly strained as can be seen from the large P-O-C angels in the phosphite, which are as large as 140°. The bite angle in **14** is 116° and the complex is indeed a trigonal bipyramide containing the diphosphite in the equatorial plane.

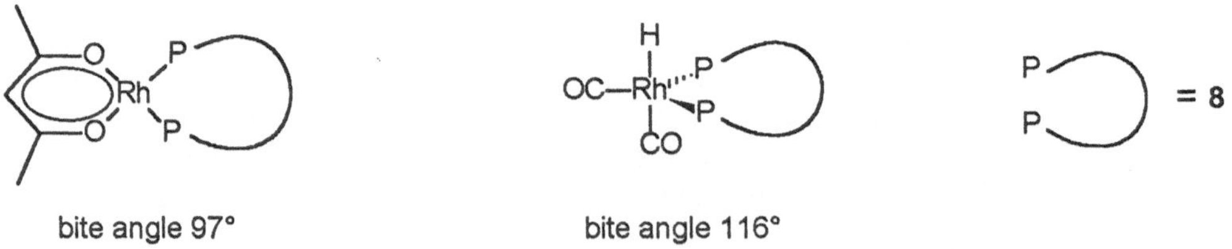

bite angle 97°   bite angle 116°

13   **14**

The structure is somewhat distorted in that the larger ligands bend toward the small hydride. This explains why the coupling constants for the coupling between phopshorus and hydrogen are sometimes larger (up to 70 Hz) than might be expected in a complex showing a pure cis-relationship between hydride and phosphorus ligand (< 10 Hz). This crystal structure confirms the NMR analysis, but so far it is the only X-ray example of a diphosphite rhodium catalyst.

4 Asymmetric Hydroformylation

The application of chiral phosphites in hydroformylation catalysts has been recently exploited as an efficient route to enantiomerically pure aldehydes.[27] The products can be used as precursors for the synthesis of high value-added compounds such as pharmaceuticals, agrochemicals, flavors and fragrances. Wink and co-workers were the first to report on the rhodium catalysed hydroformylation of styrene with chiral diphosphites as ligands. Achiral propanediol bridges, substituted with chiral (bis)dioxaphospholanes as chiral auxiliaries were used but the reaction lacked enantioselectivity.[28] In contrast, Takaya et al. reported on the asymmetric hydroformylation of vinyl acetate with enantioselectivities up to 49% with chiral bis(triarylphosphite)-rhodium(I) complexes.[29]

We started our investigations in this area in 1989 aiming at chiral diphosphites that could be easily made from natural sources and bulky bisphenoxyphosphorus chlorides and that would span a bite angle of ~ 120°.[21] Our first attempts[31,32] concerned backbones such as mannitol (15), tartrates (16), and TADDOL[30] (17) type molecules. Neither the two-carbon nor the three-carbon bridged compounds gave substantial enantioselectivities. The former leads to complexes having the phosphorus ligand in e-a positions, and the negative result may not be surprising. Ligands based on 2,3-butanediol also gave e-a coordination and low e.e.'s.[32]

Ligands such as **17** including 2,5-hexanediol, however, assume an e-e configuration but nevertheless the enantioselectivities were low. When we exchange the tartrate bridge and "bulky" ears for a bulky bridge and tartrate "ears" (**18**) the results are even more disappointing and the e.e.'s drop to a few percent. Phosphites based on sugar backbones such as **19** and **20** gave e.e.'s up to 50% provided that the bisphenols were substituted with t-butyl groups; if these were absent the enantioselectivity was very low. The rhodium hydrido complexes formed are stable and have an e-e configuration. A trans-1,3 relationship between two phosphite groups in a sugar backbone leads to a ligand that cannot function as a bidentate; a mixture of unstable complexes is formed under syn gas pressure and no enantioselectivities are obtained.[33]

237

groups in a sugar backbone leads to a ligand that cannot function as a bidentate; a mixture of unstable complexes is formed under syn gas pressure and no enantioselectivities are obtained.[33]

18   19   20

An important breakthrough was reported by Babin and Whiteker at Union Carbide.[34] They used bulky diphosphites derived from homochiral 2,4-pentanediol and obtained high enantioselectivities in the asymmetric hydroformylation of styrene of around 90%. Apparently a bridge of three carbon atoms in the diol is most effective for obtaining enantioselectivity. We were interested in the solution structures of these highly selective catalysts and prepared a few of the ligands reported by Babin and Whitaker and similar ones all containing the homochiral 2,4-pentanediol (for example 21 and 22).

R = H, t-Bu
R ' = H, t-Bu, OMe   21   22

The active complexes can be conveniently formed from Rh(acac)(CO)$_2$ and the free ligand under hydroformylation conditions (60-80 °C, 10-20 bar CO/H$_2$). The reactions were monitored by IR and NMR spectroscopy and we noted that sometimes the formation of the hydride required several hours (8 h!). This explains the long incubation times of the hydroformylation reactions and perhaps also the variation in the e.e.'s reported by different authors. Hydrido complexes of ligands 21 and 22 containing the less bulky groups were not stable under the reaction conditions. At room temperature the complexes showed a doublet in the phosphorus NMR spectrum having a rhodium to phosphorus coupling constant in the range of 231-237 Hz. This is typical of an equatorial co-ordination of the diphosphite. The averaged coupling constant of a ligand that co-ordinates equatorially-axially is smaller, at ~210 Hz. For both

Complexes having bis-equatorial co-ordination show a small phosphorus to hydride coupling constant (cis relationship). Typically they show IR absorptions at 1990, 2015, and 2075 (+/- 5) cm^{-1}, while axial-equatorial complexes have absorption at 1988 and 2029 cm^{-1}. The latter show the characteristic large coupling constant of phosphorus to hydrogen which averages at high temperature to a triplet of 70-100 Hz. At lower temperatures we are able to freeze the exchange on the NMR time scale and an AB spectrum can be observed in the ^{31}P NMR spectrum. For the axial-equatorial complexes the phosphorus-to-phosphorus coupling constant ranges from 0-70 Hz but for the bis-equatorial complexes formed with ligands 21-22 the coupling constant is typically 240-235 Hz. This is perhaps larger than one would have expected and many workers might have thought they were dealing with a trans-relationship. Interestingly, the exchange process for an e-a bidentate system is an order of magnitude faster than the exchange of an e-e system. As expected, at low temperature the very large (140-200 Hz) coupling constant of the hydrido to the axial phosphorus atom is observed, while that to the cis phosphorus atom is very small. We were able to obtain a detailed explanation of the NMR spectra. These spectra are recorded at high concentrations. Using IR spectroscopy at both high and low ("catalytic") concentrations we were able to show that in the catalytic runs the species discussed above are present indeed.

We noted above the differences in stability of the complexes depending on the length of the bridge of the diphosphite, and the steric bulk of a diphosphite. Together they determine the preferred bite angle of the bidentate. As an example of a 2,3-butanediol derivative that forms less stable hydrido complexes we show here ligand 23 that readily forms complexes of structural type 24 (asignment based on FAB ms, IR 1818 cm^{-1}, $J_{P-Rh} - 330$ Hz), reminiscent of the orange dimers reported by Wilkinson in the triphenyl phosphine system. Obviously, ligands having larger bite angles resist the formation of complexes in which they should co-ordinate as cis-ligands. This is a clear example of ligands stabilising or destabilising the "catalytic" species. The most stable hydrides were found for ligands 21 and 22 containing ortho-t-butyl groups in the bisphenol "ears".

R = t-Bu
R ' = OMe

23

24

We started our investigations with the proposal that the highest enantioselectivities should be obtained for catalysts containing e-e co-ordinating C$_2$ asymmetric ligands. This turned out to be correct, but the reverse is not true. There are many rhodium hydrides having the desired structure (e.g. those based on substituted 1,4-butanediols) that give low e.e.'s. All diphosphites co-ordinating in an e-a fashion have shown, so far, very low e.e.'s. The reason for the latter is explained in Figure 6. In structure *a* only one site is available at which the substrate can co-ordinate, which will be directed by the C$_2$ symmetric ligand (note that the complex has an overall C_1 symmetry). This will lead to enantioselectivity if the steric bulk is sufficient. In structures *b* and *b'* two competing sites are available for the co-ordination of the alkene substrate. Formally these sites are no mirror images of one another, but in practice this may be nearly the case. Hence, these sites will produce the opposite enantiomers. This seems to be the case for diphosphite ligands, but certainly it is not a universal rule.

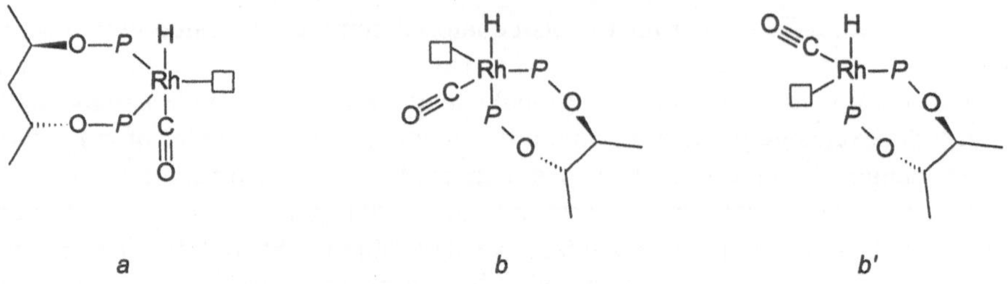

Figure 6. Enantioselective catalyst *a* and non-selective catalyst *b*

Equatorially-axially co-ordinating ligands forming eight-membered hydrido-rhodium phosphine-phosphite catalysts as reported by Takaya and co-workers[35] give, by contrast, high enantioselectivities in the hydroformylation of various substrates. Molecular models of this catalyst seem to indicate that one of the sites in *b* is blocked by the ligand and now high enantioselectivities are obtained. This approach offers new possiblities for ligand design.

5 Matched Diastereoisomeric Diphosphite Ligands

Since the biaryl moieties of the diphosphites discussed in this study contain bulky substituents, hindered rotation around the biaryl axis can be expected. The additional chirality originating from the atropisomeric biaryl substituents results in several possible diastereoisomers. Since low rotational barriers have been reported in compounds containing dibenzo[*d*,*f*][1,3,2]-dioxaphosphepins,[36, 37] it is reasonable to assume that the actual rhodium diphosphite catalyst can consist of several diastereoisomers. This can lead to the formation of matched and mismatched diastereoisomers.[38] A study of Burgess *et al.* on asymmetric hydrogenation with DIOP-DIPAMP hybrid ligands showed moderate differences between matched and mismatched diastereoisomeric ligands.[39]

We have reported on the asymmetric hydroformylation of styrene using homochiral *ortho* substituted bisnaphthol containing diphosphites.[40] To study the effect of double stereodifferentiation we synthesized a series of diphosphites based on homochiral pentane-2,4-diol ligand backbones substituted with bulky 2,2'-bisphenols and bulky 2,2'-bisnaphthols. Since interconversion around the binaphthyl bond is energetically highly unfavorable, stable diastereoisomeric diphosphites could be obtained in optically pure form and the results of the "frozen" bisnaphtol derivatives can be compared with those of the flexible bisphenol derivatives. The absolute configuration of the synthesized diastereoisomers is given in a shorthand notation as: *(S,2R,4R,S), (S,2S,4S,S), (R,2R,4R,R),* and *(S,2R,4R,R).* The indicators *S* and *R* refer to the absolute configurations around the chiral axis while the indicators *2R,4R* and *2S,4S* refer to the

240

absolute configurations of carbon atoms C-2 and C-4 in the pentane-2,4-diol backbone. The structures having the conformation *(S,2R,4R,R)* and *(R,2R,4R,S)* are equivalent as a consequence of the C$_2$ symmetry. Furthermore, in this study we used having increasing steric bulk in the ortho positions (trimethyl-, triethyl- and *tert*-butyldimethylsilyl). The diphopshites **26** are mixtures of diastereomers.

5.1 Solution Structures of Hydroformylation Catalysts

Hydridorhodium diphosphite dicarbonyl complexes [HRhL∩L(CO)$_2$, L∩L = **25a-c** and **27-30**] have been prepared and analyzed *in situ* under standard hydroformylation reaction conditions. At 333 K, the ^{31}P resonance for the complex HRh(**25a**)(CO)$_2$ appeared as a sharp doublet (Δ ϖ$_{1/2}$ = 11 Hz). The ^{31}P NMR signals of HRh(**27**)(CO)$_2$ (absolute configuration *S,2R,4R,S*), appeared already as a sharp doublet (J$_{Rh-P}$= 232 Hz) at room temperature. The enantiomers **28** and **29**, with absolute configurations *(S,2S,4S,S)* and *(R,2R,4R,R)* respectively, showed a rather different behavior upon co-ordination to rhodium.

25a	R=Me	26a	R=Me
25b	R=Et	26b	R=Et
25c	R$_3$=Me$_2$,t-Bu	26c	R$_3$=Me$_2$,t-Bu
25d	R=Ph	26d	R=Ph

27	S,2R,4R,S
28	S,2S,4S,S
29	R,2R,4R,R
30	R,2R,4R,S

Attempts to make HRhL∩L(CO)$_2$ complexes for these ligands resulted in the formation of a complex mixture of rhodium-diphosphite species. HRhL∩L(CO)$_3$ complexes in which the diphosphites act as monodentates are probably formed as side products. Furthermore, considerable ligand decomposition in complexes with **28** and **29** was found. As is the case for **27**, HRh(**30**)(CO)$_2$ is formed quantitatively under standard reaction conditions. For diphosphite L∩L = **26a** (**26a** ≡ **5** : **7** : **8** = 1 : 1 : 2) the HRhL∩L(CO)$_2$ complex was prepared using a ligand to rhodium ratio of four. The HRhL∩L(CO)$_2$ complex appeared as a diastereoisomeric mixture derived from **27** and **30** in a ratio of 2.65 : 1. No complexes derived from **29** were observed in this mixture. In summary, ligands **25a-c**, **27**, and **30** all form stable hydrides containing bisequatorially co-ordinating diphosphites.

5.2 Hydroformylation Results

Some of the results are given in tables 1 and 2. Enantiomeric excesses of 60 to 87% were found using ligand **25a**. In our hands this means that the trimethylsilyl derivatives are slightly more selective than the *t*butyl derivatives reported by Babin, for which we consistently found lower e.e.'s than those mentioned in the patent. Bulky *t*butyl groups at the *ortho* positions gave rise enantiomeric excesses up to 68% under the same reaction conditions. The rhodium catalyst derived from ligand **25b**, with bulky triethylsilyl substituents at the *ortho* positions, resulted in a very low reaction rate. Comparably low reaction rates were found at 50 °C with the *tert*-butyldimethyl analogue **25c**.

Table 1. Hydroformylation of Styrene with HRhL∩L(CO)$_2$, L∩L = **25a-c**[a]

Ligand	T (°C)	pCO	pH$_2$	% conv.[b]	% iso[c]	% n[d]	% e.e.[e]	Abs.conf.
25a	40	10	10	21	89	8	67	*(S)*
25a	25	10	10	26[f]	93	5	87	*(S)*
25a	25	10	20	69[g]	95	4	53	*(S)*
25b	50	10	10	14	85	12	25	*(S)*
25b	40	10	10	7	93	3	34	*(S)*
25b	25	10	10	7[h]	89	8	29	*(S)*
25c	50	10	10	30	67	29	11	*(S)*
25c	50	5	10	72	71	23	20	*(S)*
25c	25	10	10	8[i]	78	20	4	*(S)*

[a]Styrene catalyst molar ratio is 1000, P/Rh molar ratio of 2.2.. [b]% Conversion of styrene. [c]Selectivity to branched aldehyde [d]Selectivity to linear aldehyde [e]Enantiomeric excess. [f]After 23 h. [g]After 110 h. [h]After 24 h. [i]After 72 h.

Diastereoisomerically pure ligands **27-30** having fixed absolute configurations have also been used in the asymmetric hydroformylation of styrene. The results are given in table 2. The results obtained using **27** resemble strongly those reported for the bisphenol analogue **25a**. In the temperature range of 50-15 °C the catalytic system shows almost identical reaction rates, regio- and enantioselectivities. Hydridorhodium complexes of the enantiomeric ligands **28** and **29** show high catalytic activities, but the asymmetric inductions are very low. Hydroformylation using ligand **30** results in both low catalytic activity and low enantioselectivity.

Table 2. Hydroformylation with HRhL∩L(CO)$_2$, L∩L = 27-30[a]

ligand	T (°C)	% conv.[b]	% iso[c]	% n[d]	% e.e.[e]	Abs.conf.
27	50	43	83	13	58	*(S)*
27	25	38[f]	88	8	69	*(S)*
27	15	12[f]	92	6	86	*(S)*
28	50	98	89	10	8	*(R)*
28	40	89	91	8	18	*(R)*
28	25	21	94	5	40	*(R)*
29	50	99	87	10	12	*(S)*
29	40	99	92	6	30	*(S)*
29	25	18	95	4	38	*(S)*
30	50	36	87	11	16	*(S)*
30	40	25	88	11	18	*(S)*
30	25	2	91	8	23	*(S)*

[a]Styrene catalyst molar ratio is 1000, $p(CO) = p(H_2) = 10$ bar, P/Rh ratio 2.2. [b]% Conversion of styrene after 5 h. [c]Selectivity to branched aldehyde [d]Selectivity to linear aldehyde. [e]Enantiomeric excess. [f]After 24 h.

The introduction of larger substituents at the *ortho* position showed the expected trend of steric bulkiness on the reaction rate. The reaction rates decreased with an increase in steric bulk of the *ortho* substituents *i.e.*: $C(CH_3)_3 < Si(CH_3)_3 < Si(tert\text{-}Bu)(CH_3)_2 < Si(CH_2CH_3)_3$. However, the bulky $Si(tert\text{-}Bu)(CH_3)_2$ and $Si(CH_2CH_3)_3$ containing ligands **25b,c** did not result in an improvement of the enantiomeric excess. The optimal steric bulk in the *ortho* position seems to be obtained with trimethylsilyl substituents. Therefore, the trimethylsilyl group was also used for the optically pure bisnaphthol substituents in ligands **27-30**.

The different behavior of ligands **25** and **27-30** in the formation of the HRhL∩L(CO)$_2$ complexes clearly demonstrates that both the absolute configuration of the 2,4-pentanediol ligand backbone and the chiral bisnaphthol substituents determine the stability and catalytic performance of the rhodium complexes. Well defined stable complexes could be prepared with ligands **25**, **27** and **30**, whereas ligands **28** and **29** lead to undefined mixtures of complexes and ligand decomposition. From the results in table 2 it can be concluded that the absolute configuration around the binaphthyl axis also has a dramatic effect on the catalyst performance. When both the pentane backbone and the binaphthyl substituents have all *S* (**28**) or all *R* (**29**) configuration a mismatching diastereoisomer is formed resulting in low enantiomeric excesses. *In situ* NMR studies showed that the HRhL∩L(CO)$_2$ complexes of **28** and **29** could not selectively be synthesized. The coexistence of highly active rhodium species, in which the ligands coordinate as monodentates, is held responsible for the low enantioselectivity and relatively high reaction rates observed.

Ligands **27** and **30** both form stable rhodium complexes, but only **27** gives the matching diastereoisomer with high e.e. The results obtained for **27** strongly resemble those for **25a**. These results strongly support the conclusion that fast interconverting atropisomers of **25a** adopt predominantly the *(S,2R,4R,S)* conformation in the HRh(**25a**)(CO)$_2$ complex which is in agreement with the observed low temperature NMR spectrum; *i.e.* resonances belonging to competitive diastereoisomeric hydridorhodium complexes could not be detected in any appreciable amounts.

Further evidence for the existence of single diastereoisomeric HRhL∩L(CO)$_2$ complexes containing ligands **25a-c** can be derived from the results found for **26a** and **27**. Hydroformylation using **26a**, which consists of three diastereoisomers, resulted in lower asymmetric induction compared with the results obtained with the single isomer **27**. The observed enantiomeric excess corresponds to the average value, when the different reaction rates of the diastereoisomeric complexes are taken into consideration. Therefore, it seemed reasonable to assume that ligand **25a**, with adaptable conformations in the diaryl substituents, preferentially forms the most stable catalyst complex that also induces the highest enantioselectivity *(S,2R,4R,S)*. Unfortunately, if this is true, "fixation" of the atropisomerism cannot lead to a further increase of e.e.'s.

6 New Diphosphine Ligands: the Effect of the Bite Angle

Casey and coworkers were the first to report that the bite angle of bidentate diphosphines can have a dramatic influence on the regioselectivity of the rhodium catalyzed hydroformylation of 1-alkenes.[25,41] For the bisequatorially coordinated 2,2'-bis((diphenylphosphino)methyl)-1,1'-biphenyl (BISBI, **31**), a linear to branched aldehyde ratio as high as 66:1 was reported, while equatorially-axially coordinating (i.e. ∠P-Rh-P = 90°) 1,2-bis(diphenylphosphino)ethane (dppe) gave a linear to branched ratio of only 2.1.

BISBI was discovered and applied by workers at Eastman Kodak.[24,42] The observed selectivity is likely to be due to the bite angles of the ligands, but until recently no detailed study has been done on the effect of subtle changes of the bite angle in a series of ligands with similar electronic properties and steric size, thus solely examining the influence of the bite angle. We started the present study on the effect of the bite angle in a series of new bidentate diphosphines, based on xanthene-like backbones, on the regioselectivity in the rhodium catalyzed hydroformylation reaction. The bite angles of these ligands are fine-tuned by subtle alterations in the backbone of the ligands.

6.1 Molecular Mechanics

We used molecular mechanics in our development of new bidentate diphosphines. The natural bite angle (β_n) and flexibility range of new candidates were calculated using molecular mechanics analogously to the method used by Casey and Whiteker. The natural bite angle is defined as the preferred chelation angle determined only by ligand backbone constraints and not by metal valence angles. The flexibility range is defined as the accessible range of bite angles within less than 3 kcal mol^{-1} excess strain energy from the calculated natural bite angle.

To examine the effect of the bite angle on the selectivity in the rhodium catalyzed hydroformylation we developed a series of new bidentate ligands based on rigid heterocyclic xanthene-like aromatics. By varying the bridge in the 10-position we were able to induce small variations in the bite angle (structures **32-36**). According our Molecular Mechanics calculations, these ligands have natural bite angles ranging from 102° to 131°, and a flexibility range of ca. 35°. Ligand **36** had been synthesised before.[44]

DPEphos

Sixantphos

Thixantphos

Ph₂P   PPh₂

32   **33**   **34**

Xantphos

DBFphos

Ph₂P   PPh₂

35   **36**

6.2 Ligand Synthesis

The ligands were easily obtained in good yields (typically 70-80%) by deprotonation of the backbones with 3 equivalents of *sec*-butyllithium/TMEDA in ether. Due to the presence of the ether-oxygen, selective dilithiation at the positions *ortho* to the ether-bridge takes place. The dilithiated species is then reacted with chlorodiphenylphosphine. Washing with water to remove lithium salts, followed by washing with hexanes to remove *sec*-butyldiphenylphosphine yielded the pure product as a powder. No purification by column chromatography is necessary, but all ligands were recrystallized before utilization in catalysis. The ligands are very stable, and insensitive towards oxidation by air, both in solution and in the solid phase.

The X-ray crystal structure of the free Xantphos ligand shows that only very little adjustment of the structure is necessary to form a chelate; the orientation of the diphenyl-phosphine-moieties is nearly ideal. The observed P··P distance in the free ligand is 4.080 Å, while MM studies indicate that a decrease of the P··P distance to 3.84 Å is necessary for chelation with a P-Rh-P angle of 111.7°, a decrease of only 0.24 Å. The P atoms are brought together by means of a slight decrease of the angle between the two phenyl planes in the backbone of the ligand from ca. 166° to 158°.

The ^{13}C NMR spectra of free Xantphos, Thixantphos and Sixantphos all show a virtual triplet signal for the C-*ipso* on the backbone. This resonance is the result of a through space coupling of the two phosphorus atoms. The orientation of the lone pairs of the two phosphorus atoms is such that it causes degeneracy of the magnetic resonances of the P-nuclei. In the ^{13}C NMR the degenerate magnetic resonance of the phosphorus causes the *ipso*-carbon atom to couple with the two phosphorus atoms, with observed $^{1}J_{13C-P}$ coupling constants of 40 to 60 Hz. The appearance of this P-P coupling is a strong indication that the structure in solution of the free ligands is similar to the crystal structure of Xantphos. This through space coupling is not observed for DPEphos (due to the absence of such a rigid backbone) and DBFphos (in which the P··P-distance is apparently too large: MM calculations gave P··P= 5.760 Å, PM3 calculations gave P··P= 5.956 Å).

6.3 Hydroformylation of 1-Octene

We tested the selectivity of our ligands in the rhodium catalyzed hydroformylation of 1-octene (table 3). DPEphos, with a calculated natural bite angle of 102.2°, induced an enhanced, though moderate selectivity (compared to most diphosphines), but no isomerisation was detected. The ligands with a one-atom bridge between C(11) and C(14) (Sixantphos: X= Si(CH$_3$)$_2$; Thixantphos: X= S; Xantphos: X= C(CH$_3$)$_2$) having a calculated natural bite angle near 110° showed a very high regioselectivity and a very low rate of isomerization to internal alkenes. DBFphos, with a calculated natural bite angle of 131.1°, proved not to be very selective.

Under these mild reaction conditions, the selectivities toward the linear aldehyde observed for Sixantphos, Thixantphos, and especially Xantphos are somewhat higher than that observed for BISBI. This is mainly due to the very low selectivity to isomerization of 1-octene. The normal to branched ratios of our ligands are very close to that of BISBI. Furthermore, no hydrogenation was observed. Even though the normal to branched ratio is 80.5 for BISBI, the selectivity towards the linear aldehyde amounts to only 89.6 %. This is a result of the large selectivity for isomerization of 1-octene to 2-octene.

Table 3. Hydroformylation of 1-octene[a]

ligand	calculated bite angle	flexibility range	linear/ branched	% linear aldehyde	% isomeri-zation	t.o.f.
DPEphos	102.2	86 - 120	6.7	87.0	0	250
Sixantphos	108.7	93 - 132	34	94.2	3	168
Thixantphos	109.4	94 - 130	41	93.0	4.7	445
Xantphos	111.7	97 - 135	53.5	97.7	0.5	800
DBFphos	131.1	117 - 147	3	71	5.5	125
BISBI	122.6	101 - 148	80.5	89.6	9.3	850

[a] Conditions: $CO/H_2 = 1$, total pressure 10 bar, substrate/Rh = 674, ligand/Rh = 2.2, [Rh] = 1.78 mM. In all cases the percent hydrogenation was zero.

According to MM calculations, BISBI is very flexible; we calculated a natural bite angle of 122.6° and a flexibility range of 101°-148° for a rhodium complex, while Casey and coworkers calculated a natural bite angle of 113° and a flexibility range of 92°-155°. These results are supported by studies of the fluxional behaviour of BISBI in solution, in which a rapid exchange of the two phosphine moieties was observed. Indeed, X-ray structural analyses of transition metal complexes of BISBI show a wide variety of P-M-P chelation angles: ∠P-Mo-P = 103.5° in $(BISBI)Mo(CO)_4$,[45] ∠P-Ir-P = 117.9° in $(BISBI)Ir(H)(CO)_2$, ∠P-Rh-P= 124.8° in $(BISBI)Rh(H)(CO)(PPh_3)$,[41] and ∠P-Fe-P = 152.0° in $(BISBI)Fe(CO)_3$[46] The flexibility of BISBI allows the formation of a rhodium complex with a slightly less rigidly defined geometry thus giving somewhat lower selectivity to n-alkyl intermediates.

6.4 Hydroformylation of Styrene

Hydroformylation of styrene with a (Xantphos)Rh catalyst resulted in relatively high selectivity for the linear aldehyde (a linear to branched ratio of up to 2.35 was obtained, see table 4). This is remarkable, since styrene is a substrate with a distinct preference for the formation of the branched aldehyde due to the stability of the 2-alkyl-rhodium species, induced by the formation of an η^3-benzyl complex.

Table 4. Hydroformylation of styrene

ligand	T (°C)	p (bar)	linear/ branched	% linear alde-hyde	t.o.f.
Xantphos	60	10	0.77	43.5	128
	80	10	0.88	46.8	724
	120	10	2.35	70.1	4285
PPh$_3$[33]	70	62	0.08	7.4	(a)
DIOP[36]	25	1	0.25-0.49[a]	20-32.5[a]	(a)

Conditions: $CO/H_2 = 1$, substrate/Rh= 674, [Rh]= 1.78mM. (a) Depending on the rhodium precursor used

Summarizing the distinct effect of minor changes in the ligand backbone by alterations of the bridge between C(11) and C(14) on the observed selectivity shows that the effect of the bite angle is very subtle. For the chelating ligands in this series, we found a regular increase of both the normal/branched ratios and the percentage linear aldehyde formed with an increasing calculated natural bite angle in the hydroformylation of 1-octene. This correlation is observed at both 40 °C and 80 °C. For the ligand DBFphos no chelates were observed, and its selectivity is consequently out of the range. The very high selectivity of Xantphos (calculated natural bite angle 111.7°) and BISBI (our calculated natural bite angle 122.6° indicates that the optimum is near 112-120°. Selectivity in the hydroformylation reaction increases when the bite angle of the ligand becomes larger. However, the rigidity of the ligand is also an important factor. This is strongly supported by the results obtained for the more flexible BISBI ligand which shows a lower selectivity. The importance of this ligand rigidity becomes even more pronounced at higher temperatures. Rigid ligands with a larger calculated bite angle are unable to form stable chelates, as was demonstrated with DBFphos.

6.5 Rhodium Complexes

In order to investigate the coordination behaviour of our series of ligands, (diphosphine)Rh(H)(CO)(PPh$_3$) complexes were synthesized for all ligands. Facile exchange of PPh$_3$ in (PPh$_3$)$_3$Rh(H)(CO) with the diphosphines gave quantitative yield (although some loss was observed during the work-up procedure). Detailed studies of the ^{31}P{^{1}H} NMR spectra led to the conclusion that the two phosphine-moieties of the diphosphine in the complex were equivalent for all ligands (except DBFphos). The coupling constants between the diphosphine-P and the PPh$_3$ (^2J= 114 to 129 Hz) show that these ligands are indeed coordinated bisequatorially. The attempts to synthesize (DBFphos)Rh(H)(CO)(PPh$_3$) (DBFphos has a calculated natural bite angle of 131°) did not yield any characterizable complex.

Bubbling CO through a solution of (diphosphine)Rh(H)(CO)(PPh$_3$) led to facile displacement of the remaining PPh$_3$. The complex (diphosphine)Rh(H)(CO)$_2$ formed by this exchange (presumably the catalytically active species in the hydroformylation reaction) is stable under an atmosphere of CO. The ^{31}P{^{1}H} NMR spectra of these compounds exhibit a clean doublet for the diphosphine, indicating that these ligands are also coordinated in a bisequatorial fashion in these complexes.

The ^{1}H NMR spectra of the (Xantphos)Rh(H)(CO)(PPh$_3$), (Xantphos)Rh(H)(CO)$_2$ (Sixantphos)Rh(H)(CO)(PPh$_3$), and (Sixantphos)Rh(H)(CO)$_2$ complexes show an inequivalence of the two methyl groups, indicating a rigid conformation of the ligand (formally the absence of a mirrorplane through the equatorial plane dictates the inequivalence of the methyl and phenyl groups). Furthermore, the *ortho*-hydrogens of the phenyl groups on the diphenylphosphine moieties positioned in the equatorial plane and the axial plane are inequivalent in the (diphosphine)Rh(H)(CO)(PPh$_3$) complexes. This strongly indicates that in solution the two diphenylphosphine-moieties are rigidly placed in space around the rhodium-centre.

Therefore, we think that the enhanced selectivity for the linear aldehyde in the hydroformylation of styrene and the low isomerization of 1-octene (even at higher temperatures) is induced by a well-defined "docking area" on the rhodium-centre. The size of the bite angle dictates directly the shape of this docking area. The fact that isomerization is almost absent in these catalyst systems indicates that the strict geometry is inducing the formation of the 1-alkyl (leading to the linear aldehyde) and actually inhibits the formation of 2-alkyls.

7 Conclusion

The above examples of diphosphite and diphosphine containing rhodium catalysts have shown that the bite angle plays a key role in determining the selectivity of the hydroformylation reaction. Molecular mechanics studies have contributed to the dedvelopment of new diphoshine ligands. The results for constrained diphosphines are very encouraging. Diphosphites are less accessible for molecular mechanics because the rotation around the P-O and O-C bonds leads to many more conformations. Usually a large number of conformation within a few kcal mol^{-1} were found. Also, the substituents at the aryl groups have a large effect on the bite angle and the catalytic properties of the complexes formed.

References

1. P.W.N.M. van Leeuwen and C.F. Roobeek, Brit. Pat. 2 068 377, **1980**, to Shell.
2. P.W.N.M. van Leeuwen and C.F. Roobeek, Eur. Pat. Appl. EP 54986, **1982**, GB Appl. 41098, **1980**) to Shell and *J. Organometal. Chem.* **1983**, *258*, 343.
3. P.W.N.M. van Leeuwen and C.F. Roobeek, Eur. Pat. Appl. EP 33554, **1981**, to Shell and *J. Mol. Catal.* **1985**, *31*, 345.
4. R.L. Pruett and J.A. Smith, S. African Patent 68 4937, 1968 to UCC and *J. Org. Chem.* **1969**, *34*, 327.
5. A.M. Treciak and J. Ziólkowski, *J. Mol. Catal.* **1987**, *43*, 13 and references therein.
6. D.R. Bryant, 203rd ACS Meeting, San Francisco, **1992**, on the occasion of the Monsanto Award address by Bryant.
7. E. Billig, A.G. Abatjoglou, D.R. Bryant, Eur. Pat. 86/112257 and 214622, **1986** to UCC.
8. N. Yoshinura and Y. Tokito, Eur. Pat. 223 103, 1987 and T. Omatsu, Eur. Pat. 0 303 060, **1989** to Kuraray.
9. A. Polo, J. Real, C. Claver, S. Castillón, and J.C. Bayón, *J. Chem. Soc. Chem. Commun.* **1990**, 600. A. Polo, E. Fernandez, C. Claver, and S. Castillón, *J. Chem. Soc. Chem. Commun.* **1992**, 639.

10. E. Billig, A.G. Abatjoglou, D.R. Bryant, U.S. Pat. 4 769 498, **1987** to UCC.

11. P.W.N.M. van Leeuwen and C.F. Roobeek, unpublished.

12. (a) P.W.N.M. van Leeuwen, C.F. Roobeek, R.L. Wife, and J.H.G. Frijns, *J. Chem. Soc. Chem. Commun.* **1986**, 31; (b) P.W.N.M. van Leeuwen and C.F. Roobeek, *New J. Chem.* **1990**, *14*, 487; (c) P.W.N.M. van Leeuwen, C.F. Roobeek, J.H.G. Frijns and A.G. Orpen, *Organometallics* **1990**, *9*, 1211; (d) P.W.N.M. van Leeuwen and C.F. Roobeek, in *Homogeneous transition metal catalyzed reactions*, W.R. Moser and D.W. Slocum Ed. Advances in Chemistry Series 230, Washington **1992**, 367; (e) P.W.N.M. van Leeuwen and C.F. Roobeek, Eur. Pat. Appl. EP 82576, **1983** to Shell.

13. (a) T. Jongsma, P.Kimkes, G. Challa and P.W.N.M. van Leeuwen, *Polymer* **1992**, *33* 161; (b) T. Jongsma, M. Fossen, G. Challa, P.W.N.M. van Leeuwen, *J. Mol. Catal.* **1993**, *83*, 17; (c) T. Jongsma, H. van Aert, M. Fossen, G. Challa, P.W.N.M. van Leeuwen, *J. Mol. Catal.* **1993**, *83*, 37.

14. H.K.A.C. Coolen, P.W.N.M. van Leeuwen, R.J.M. Nolte, *J. Organomet. Chem.* **1995**, *496*, 159.

15. (a) A. van Rooy, E.N. Orij, P.C.J. Kamer, F. van den Aardweg, and P.W.N.M. van Leeuwen, J. *Chem. Soc. Chem. Commun.* **1991**, 1096. (b) T. Jongsma, G. Challa and P.W.N.M. van Leeuwen, *J. Organometal. Chem.* **1991**, *421* 121.

16. A. van Rooy, J.N.H. de Bruijn, C.F. Roobeek, P.C.J. Kamer, and P.W.N.M. van Leeuwen, *J. Organomet. Chem.* **1996**, *507*, 69.

17. C. Muilwijk, P.C.J. Kamer, and P.W.N.M. van Leeuwen, *J. Am. Oil Chem. Soc.* **1996**, in press.

18. A. van Rooy, E.N. Orij, P.C.J. Kamer, P.W.N.M. van Leeuwen, *Organometallics* **1995**, *14*, 34.

19. A. van Rooy, P.C.J. Kamer, P.W.N.M. van Leeuwen, to be published (Thesis A. van Rooy, **1995**).

20. E. Billig, A.G. Abatjoglou, D.R. Bryant, Eur. Pat. Appl. EP 213,639, **1987**, to UCC.

21. P.W.N.M. van Leeuwen, P.C.J. Kamer, G.J.H. Buisman, and A. van Rooy, 203rd ACS Meeting, San Francisco, **1992**.

22. (a) A. van Rooy, P.C.J. Kamer, P.W.N.M. van Leeuwen, K. Goubitz, J. Fraanje, N. Veldman, and A.L. Spek, *Organometallics,* **1996**, *15*, 835. (b) P.W.N.M. van Leeuwen, G.J.H. Buisman, A. van Rooy, P.C.J. Kamer, *Rec. Trav. Chim. Pays-Bas* **1994**, *113*, 61.

23. B. Moasser, W.L. Gladfelter, and D.C. Roe, *Organometallics* **1995**, *14*, 3832.

24. T.J. Devon, G.W. Phillips, T.A. Puckette, J.L. Stavinoha, and J.J. Vanderbilt, US Appl. 873918, **1986** to Eastman Kodak.

25. C.P. Casey, and G.T. Whiteker, *Isr. J. Chem.* **1990**, *30*, 299.

26. A. van Rooy, P.C.J. Kamer, P.W.N.M. van Leeuwen, N. Veldman, and A.L. Spek, J. *Organomet. Chem.* **1995**, *494*, C15.

27. For recent advances in enantioselective hydroformylation see: S. Gladiali, J.C. Bayón, and C. Claver, *Tetrahedron: Asymmetry* **1995**, *6*, 1453 and references cited therein

28. D.J. Wink, T.J. Kwok, and Y. Yee,., *Inorg. Chem.* **1990**, *29*, 5006.

29. N. Sakai, K. Nozaki, K. Mashima, and H. Takaya, *Tetrahedron: Asymmetry* **1992**, *3*, 583

30. D. Seebach, A.K. Beck, R. Imwinkelried, S. Roggio and A. Wonnacott, *Helv. Chim. Acta* **1987**, *70*, 954.

31. G.J.H. Buisman, P.C.J. Kamer, P.W.N.M. van Leeuwen, *Tetrahedron: Asymmetry* **1993**, *4*, 1625.

32. G.J.H. Buisman, E.J. Vos, P.C.J. Kamer, P.W.N.M. van Leeuwen, *J. Chem. Soc. Dalton Trans.* **1995**, 409.

33. G.J.H. Buisman, M.E. Martin, E.J. Vos, A. Klootwijk, P.C.J. Kamer, P.W.N.M. van Leeuwen, *Tetrahedron: Asymmetry* **1995**, *6*, 719.

34. J.E. Babin and G.T. Whiteker, *WO 93/03839 US 911,518,* **1992** to Union Carbide Corporation.

35. N. Sakai, S. Mano, K. Nozaki, and H. Takaya, *J. Am. Chem. Soc.* **1993**, *115*, 7033.

36. (a) G.T. Whiteker, A.M. Harrison, and A.G. Abatjoglou, *J. Chem. Soc., Chem. Commun.* **1995**, 1805. (b) T.K. Prakasha, R.O. Day, and R.R. Holmes, *Inorg. Chem.* **1992**, *31*, 1913. (c) M.J. Baker, K.N. Harrison A.G. Orpen, P.G. Pringle,and G. Shaw, *J. Chem. Soc., Chem. Commun.* **1991**, 803.

37. S.D. Pastor, S.P. Shum, R.K. Rodebaugh, A.D. Debellis, and F.H. Clarke, *Helv. Chim. Acta* **1993**, *76*, 900.

38. S. Masamune, W. Choy, J.S. Petersen, and L.R. Sita, *Angew. Chem., Int. Ed. Engl.* **1985**, *24*, 1

39. K. Burgess, M.J. Ohlmeyer, and K.H. Whitmire, *Organometallics* **1992**, *11*, 3588

40. G.J.H. Buisman, L.A. van der Veen, A. Klootwijk, W.G.J. de Lange, P.C.J. Kamer, P.W.N.M. van Leeuwen, D. Vogt, submitted for publication.

41. C.P. Casey, G.T. Whiteker, M.G. Melville, L.M. Petrovich, J.A. Gavney, and D.R. Powell, *J. Am. Chem. Soc.* **1992**, *114*, 5535.

42. T.J. Devon, G.W. Phillips, T.A. Puckette, J.L. Stavinoha, and J.J. Vanderbilt, U.S. Patent 4,694,109 (to Eastman Kodak) **1989**.

43. M. Kranenburg, Y.E.M. van der Burgt, P.C.J. Kamer, P.W.N.M. van Leeuwen, K. Goubitz, and J. Fraanje, *Organometallics* **1995**, *14*, 3081.

44. M.W. Haenel, D. Jakubik, E. Rothenberger, and G. Schroth,. *Chem. Ber.* **1991**, *124*, 1705.

45. W. A. Herrmann, C. W. Kohlpaintner, E. Herdtweck, and P. Kiprof,. *Inorg. Chem.* **1991**, *30*, 4271.

46. C.P. Casey, G.T. Whiteker, C.F. Campana, and D.R. Powell, *Inorg. Chem.* **1990**, *29*, 3376.

Development of Transition Metal Complexes as Redox Reagents and Redox Catalysts

Eberhard Steckhan*, Gerhard Hilt, Rainer Kempf and Masahiro Sadakane

Institut für Organische Chemie und Biochemie der Universität Bonn, Gerhard-Domagk-Straße 1, D-53121 Bonn, Germany

Summary

Transition metal complexes are prototypes of redox reagents and catalysts. Their behavior can be tuned to the requirements of the desired reaction by proper selection of either the central metal ion or the ligand. We are presenting three cases of fine-tuned transition metal complexes which can be applied as chemically or electrochemically regenerable oxidizing agents. 1. 1,10-Phenanthroline-5,6-dione complexes of ruthenium or cobalt have been developed as efficient catalysts for the chemical aerobic and indirect electrochemical in-situ generation and regeneration of the enzymatic cofactors NAD(P) $^+$ in enzymatic syntheses under very mild conditions. 2. Manganese(III)trispicolinate has been applied in radical C-C coupling recations between CH-acidic β-keto esters or 1,3-diketones and acid labile alkenes like silylenolethers, enolacetates, enamines, and allylsilanes in organic non-acidic media. In addition, manganese(III)bispicolinate was developed as an in-situ electrochemically regenerable redox catalyst for the same types of reactions in DMF or t-butanol/water. 3. The α-Keggin heteropolyanion $K_6SiW_{11}O_{39}Mn(H_2O)$ was studied as a possible stable electrochemically regenerable redox catalyst for oxidation reactions in aqueous media. The formation and regeneration of an intermediate Mn(IV)(OH) species was demonstrated and applied for the indirect electrochemical oxidation of different alcohols.

1 Introduction

The availability of highly selective regenerable redox reagents and redox catalysts is very often the key to successful complex organic syntheses. Therefore, their development is a very active area of chemical research. Frequently, transition metal complexes are the solution for problems in selective redox reactions. However, one has to distinguish between two types of reagents: 1. Outer-sphere electron transfer agents, and 2. inner-sphere electron transfer agents. While in the case of outer-sphere electron transfer agents, the selectivity can only be influenced by the potential difference between the reagent and the substrate and by the speed of chemical follow-up reactions, in the case of inner-sphere electron transfer agents the selectivity is determined by a chemical step like a ligandation, ligand exchange, oxygen, hydrogen or hydride ion transfer. However, in the latter case the regeneration process for the active redox reagent either electrochemically or chemically, is more difficult to develop because the rate of all the different reaction steps involved in the catalytic cyle have to be fast and brought into accordance. The advantage is that there are many possibilities to tune the transition metal reagent to the desired process either by changing the central metal ion or by modification of the ligands. One can distinguish four types of inner-sphere redox reactions using transition metal complexes: 1. metal-centered redox reactions; 2. ligand-centered redox reactions in which only the ligand is redox active changing its redox state; 3. ligand respective atom transfer reactions, like oxygen transfer or hydride transfer from central metal under change of ist redox state; 4. metal ion assisted atom transfer reactions within the ligand sphere without change of the metal redox state. In this communication, we want to present redox catalytic processes of the first, second and third type i.e. metal-centered redox reactions, ligand-centered redox reactions and atom transfer reactions from the central metal ion.

2 Transition Metal Complexes of 1,10-Phenanthroline-5,6-dione as Efficient Catalyst for the NAD(P)$^+$ Cofactor Regeneration in Enzymatic Synthesis

Enzymatic oxidation reactions can be of great value in organic synthesis[1]. Nature has developed a large number of highly selective oxidizing enzymes among which mono-oxygenases like cytochrome P-450 depending enzymes[2], dioxygenases like the arene dioxygenases[3], and oxidases like glycerol 3-phosphate oxidase[4], have found important applications in synthesis. The most widely employed class of oxidizing enzymes, however, are dehydrogenases like horse liver alcohol dehydrogenase (HLADH)[5]. These dehydrogenases are depending on NAD(P)$^+$ as the electron accepting cofactor. Thus, for the synthetic application of these enzymes an efficient and mild system for the cofactor regeneration is necessary[1,6]. Apart from enzymatic cofactor recycling systems, chemical and electrochemical regeneration procedures have attracted strong interest because of their greater flexibility. Chemical regeneration systems usually consist of o-quinoid mediators acting as hydride ion abstracting agents and oxygen as final electron acceptor. Catalase is added to destroy the hydrogen peroxide produced. Effective chemical or electrochemical regeneration systems must meet the following requirements: 1. To obtain a high space-time yield the turnover frequencies of the quinoid mediators must be as high as possible. 2. The mediators must be stable in both redox states within the regenerating cycle over a long period of time. 3. They should be easily soluble in aqueous buffer systems which are necessary for enzymatic conversions. 4. They should react selectively only with the reduced cofactor.

All chemical systems described thus far have limited use because they are either reacting too slowly (FMN[5, 7]: 1.8 turnovers/h; 4,7-phenanthrolin-5,6-dione[8]: 4-6 turnovers/h) or are chemically not stable enough under basic conditions[9], which usually are favourable for the enzymatic oxidations. Electrochemical regeneration of the oxidized cofactors is either possible by direct anodic oxidation[10] or *via* redox catalysts also mainly applying *o*-quinonoid systems as mediators. While for the mediators the same disadvantages are observed as for the chemical systems, the direct electrochemical regeneration is problematic because electrode fouling has been observed quite often. Moreover, the anodic potential is too positive for selective oxidations[10] because of the electrochemically highly irreversible process.

To solve the above mentioned problems for the efficient chemical and electrochemical regeneration of NAD(P)$^+$, we used the 1,10-phenanthroline-5,6-dione system as a starting point. However, the electron density within the o-quinoid structure should be diminished to accelerate the hydride transfer from NAD(P)H and to lower the oxidation potential for the electrochemical regeneration to about 0 V vs. Ag/AgCl to exclude side-reactions of the substrates at the electrode surface. At the same time, the solubility should be enhanced by the use of charged species, and the formation of insoluble precipitates by oligomerization of the reduced forms *via* hydrogen bonds should be avoided by blocking the nitrogen atoms. All these problems could finally be solved by complexation of 1,10-phenanthroline-5,6-dione (PD) with a transition metal ion. We have shown previously[11], that homoleptic complexes of Co^{II}, Ni^{II} and Cu^{II} with 1,10-phenanthroline-5,6-dione are indeed undergoing fast hydride transfers from NADH. However, the reduced forms of the complexes are still slowly precipitating *via* oligomerization. In the case of Co^{II} and Ru^{II} we could avoid the precipitation by using heteroleptic complexes in which four of the coordination sites are blocked by one *N,N,N*-tris(aminoethyl)amine (tren) or *N,N,N*,-tris(pyridylmethyl)amine (TPA) ligand. In addition, the homoleptic tris-PD complex of Ru^{II} showed an extremely high stability even in the reduced form. Therefore, we concentrate in this publication on the following transition metal complexes: the homoleptic Ru^{II}(PD)$_3$ and the heteroleptic Co^{II}(TPA)(PD) and Ru^{II}(TPA)(PD) complexes We focused our investigations on the study of those properties which are important for their efficiency as redox catalysts for the NAD(P)H oxidation .

We studied the properties of the complexes using mainly electroanalytical methods like cyclic voltammetry, rotating disc electrode measurements, spectroelectrochemical studies using an optically transparent thinlayer electrochemical cell (OTTLE), exhaustive controlled potential electrolysis, together with chemical reductions by NADH and NaBH$_3$CN. While we are presenting a condensation of the results here, the details of the studies are published elsewhere[12].

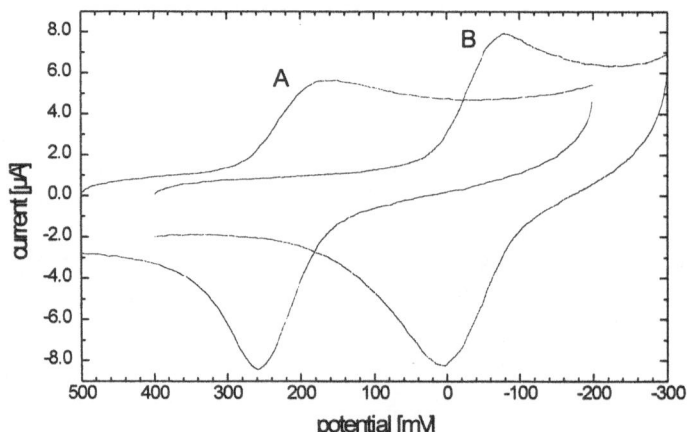

Structures of the 1,10-phenanthroline-5,6-dione transition metal complexes studied

heteroleptic
M(TPA)(PD) - complexes
with M = Co(II) and Ru(II)

homoleptic
RuPD$_3$ - complex

In the cyclic voltammetric studies all compounds show the reduction in the potential region of - 0.2 to + 0.3 V vs. the Ag/AgCl reference electrode in phosphate buffer. The shapes and potentials are strongly depending on the pH values of the solution. The reduction potentials are shifting to more negative potentials with increasing pH. The voltammograms in general show the behavior of a system with a preceding chemical reaction (CrEr) under intermediate kinetic control in the more acidic region (Fig. 1, curve A) and more Nernstian behavior in the neutral and basic region (Fig. 1, curve B).

Fig. 1. Cyclic voltammograms of [Ru(PD)$_3$](ClO$_4$)$_2$ at pH 2.5 (A) and pH 7.4 (B) at a scan rate of 100 mV/s vs Ag/AgCl reference electrode (working electrode: glassy carbon)

Because the Nernstian behavior dominates in the neutral to basic region the peak potentials are reported for pH 7 (Table 1).

Table 1. Reductive ($E_{p,\,red}$) and corresponding oxidative ($E_{p,\,ox.}$) peak potentials at pH 7.0 and formal potentials (E'°) in pH 2.5 and pH 7.0 or 8.0 phosphate buffer

compound	E_p,red / E_p,ox, V[a] (pH)	E'°, V[b] (pH)
[Ru(PD)$_3$](ClO$_4$)$_2$	-0.060 / +0.020 (7.0)	+0.242 (2.2) / -0.055 (8.0)
[Co(TPA)(PD)](BF$_4$)$_2$	-0.150 / 0 (7.0)	-0.040 (8.0)
[Ru(TPA)(PD)](Cl)$_2$	-0.110 / -0.040 (7.0)	+0.217 (2.5) / -0.015 (7.0)

[a] peak potentials from cyclic voltammograms at 20 to 100 mV/s scan rates; [b] formal potentials taken from spectroelectrochemistry at OTTLE electrodes

Except for the homoleptic cobalt complex [Co(PD)$_3$](BF$_4$)$_2$, which shows two overlapping waves indicating that the ligands are interacting within the complex, the corresponding ruthenium complex only shows one redox wave. This demonstrates that in this case the ligands are acting independently. The kinetic control of the cyclic voltammetric behavior can be interpreted by a preceeding equilibrium between the electrochemically active PD compound and its electrochemically inactive hydrated form as has recently been described by *Anson et al.*[13] (see Scheme 1).

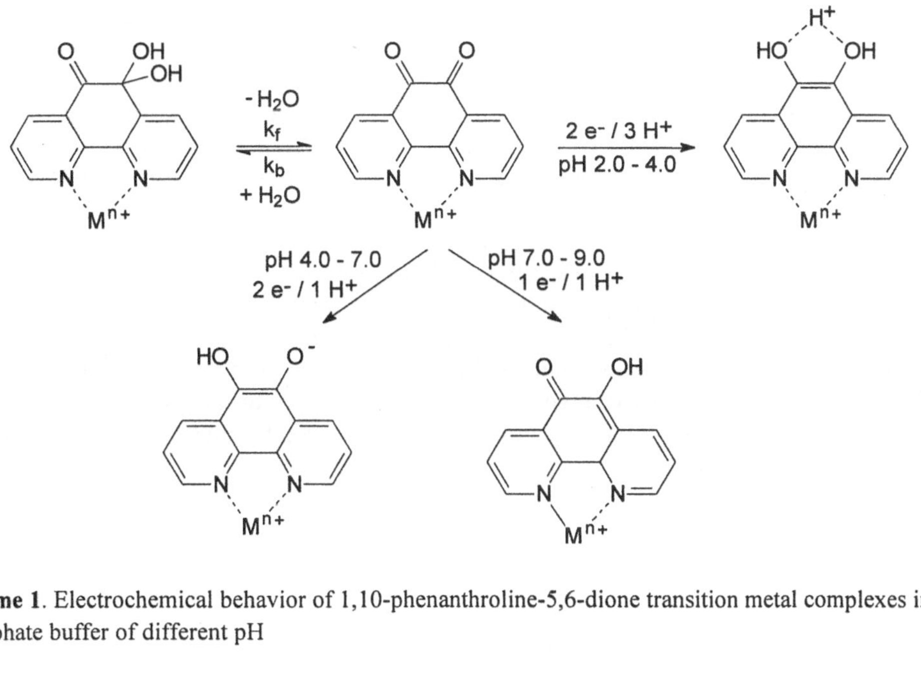

Scheme 1. Electrochemical behavior of 1,10-phenanthroline-5,6-dione transition metal complexes in phosphate buffer of different pH

257

The hydration is favored by the complex formation because thus the electron density at the quinone groups is lowered. The rate of the establishment of the equilibrium is obviously influenced by the pH leading to more Nerstian behavior of the cyclic yvoltammograms at higher pH values. Therefore, by changing the pH the peak potentials are not only influenced by the number of protons which are involved in the electrochemical step but also to a certain extent by the pH influence on the rate constants of the preequilibrium. The reoxidation peaks, however, are less influenced by the preequilibrium. The cyclic voltammetric behavior can thus best be expalined by the reactions given in Scheme 1.

The total number of transferred electrons during the reduction of the complexes has been determined by exhaustive controlled potential electrolysis (Table 2) and a spectro-electrochemical investigation according to *Heineman et al.*[14] Both measurements give the same results. That is, under acidic conditions two electrons for each PD ligand are transferred while at pH 7 or 8 only one electron for each ligand is exchanged.

Table 2. Number of transferred electrons during exhaustive controlled potential electrolysis at different pH values

compound	number of electrons (pH)
$[Co(PD)_3](BF_4)_2$	5.99 (2.2) / 2.96 (8.0)
$[Ru(PD)_3](ClO_4)_2$	5.89 (2.2) / 2.91 (8.0)
$[Co(TPA)(PD)](BF_4)_2$	1.96 (2.2) / 0.91 (7.0)
$[Ru(TPA)(PD)](Cl)_2$	2.14 (2.2) / 1.03 (7.0)

These results could be rationalized in two ways. The first possibility is the formation of a hydroquinone-quinone dimer of the transition metal complexes either by a charge transfer complex or by the attack of hydroxy groups of the hydroquinone on the carbonyl groups of the quinone leading to a semiacetal dimer structure. In the case of the homoleptic tris-phenanthrolinedione complexes such interactions would lead to oligomers or dimers which should precipitate. However, even after total electrolysis under basic conditions the systems stayed soluble. The second possibility is a stabilization of the semiquinone complex by the transition metal ion leading to an increase of the coordinative binding order of the metal under basic conditions. In an acidic medium the protonation of the semiquinone complex at the carbonyl group would favor the further reduction to the hydroquinone complex (Scheme 2). A stabilization of protonated semiquinones by metal ions has already been proposed previously.

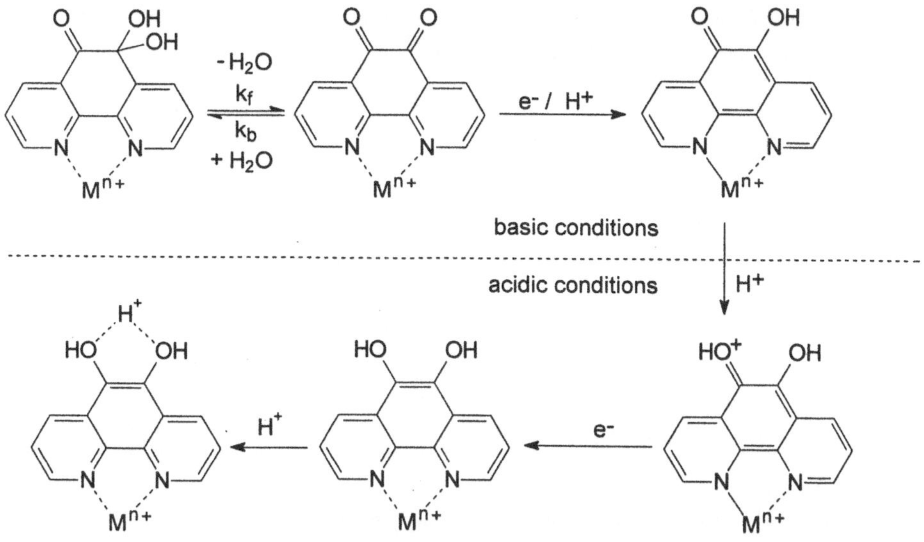

Scheme 2. Possible one electron/one proton process under basic conditions and two electron/three proton process under acidic conditions

As it was our aim to use the PD transition metal complexes as effective hydride ion abstracting agents towards the chemical or indirect electrochemical regeneration of NAD(P)$^+$ from NAD(P)H in enzymatic syntheses we studied the rate of the aerobic PD transition metal complex catalyzed generation of NAD$^+$ from NADH by UV spectroscopy using oxygen as end electron acceptor. The turnover frequencies are strongly dependent on the pH of the solution, We found that our previously reported values for some of the complexes[11] can even be increased considerably by changing the pH from 6.0 to pH 8.0 (Table 3). This is advantageous because most of the enzymatic oxidations catalyzed by alcohol dehydrogenases are favored by basic conditions. The turnover frequencies for these new redox catalysts with up to 900 turnovers per hour are more than two orders of magnitude higher than the best results reported up to now.

Table 3. Turnover frequencies for the indirect aerobic generation of NAD[+] from NADH followed by UV spectroscopy at 340 nm using PD based catalysts[a]

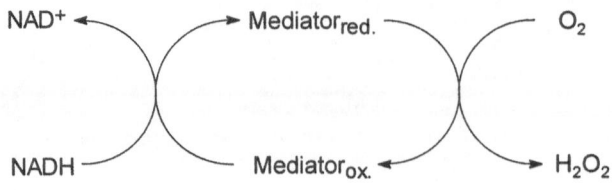

compound	TN/h[b] pH 6.0	TN/h[b] pH 8.0
[Ru(PD)$_3$](ClO$_4$)$_2$	55	900
[Ru(TPA)(PD)](Cl)$_2$	31	737
Co(TPA)(PD)](BF$_4$)$_2$	23	96

[a] Conditions: NADH : mediator = between 50 and 200 : 1 in buffered solutions; mediator < 5x10^{-6} mol/L; [b] Turnover number for the PD catalyst per hour

Table 4. pH dependence of the turnover frequencies for the indirect electrochemical generation of NAD[+] from NADH followed by UV spectroscopy at 340 nm using PD based catalysts[a]

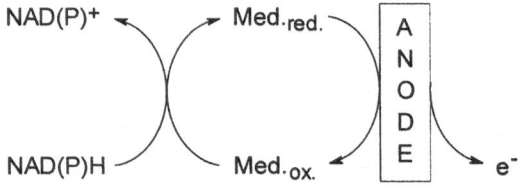

	turnover numbers/h		
compound	pH = 6.0	pH = 7.0	pH = 7.9
[Ru(PD)$_3$]	30	35	39
[Ru(TPA)(PD)]	23	21	21
PDMe[b]	34	36	35
[Co(TPA)(PD)]	24	20	26

[a] Conditions: batch electrolysis with a ratio of NADH ro mediator = 60 : 1; mediator = 3 - 5x 10^{-4} mol/l; [b] PDMe is N-methyl 1,10-phenthrolinium-5,6-dione tetrafluoroborate

The indirect electrochemical regeneration also works very well which is demonstrated by the strong catalytic oxidative peak current enhencement in cyclic voltammetry at pH 8.2. However the turnover frequencies (Table 4) are practically independent of the pH of the solution. The lower turnover freuquencies and the independence of the pH indicates that in the case of the indirect electrochemical process the diffusion of the reduced mediator to the electrode surface seems to be the limiting factor. Instead in the case of the indirect aerobic process the homogeneous reaction between oxygen and the reduced mediator is rate determining.

The results of both the indirect aerobic and the indirect electrochemical anaerobic cofactor regenrations were so convincing that we coupled both regeneration systems with the horse liver alcohol dehydrogenase (HLADH) catalyzed enzymatic oxidation of *meso*-diols to form chiral γ-lactones.

These lactones are valuable starting materials for complex syntheses of natural products.

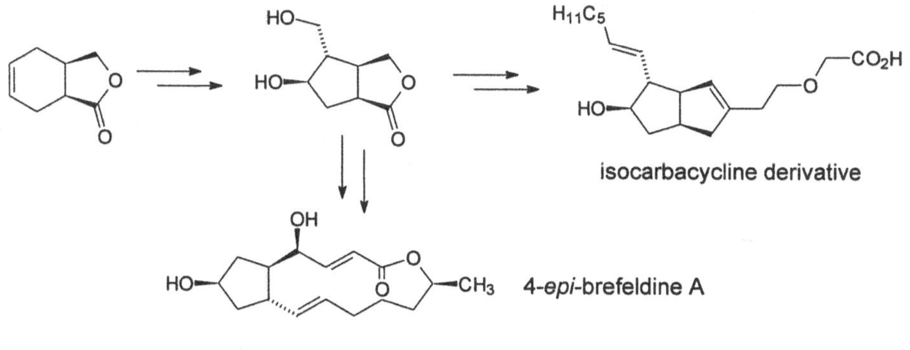

isocarbacycline derivative

4-*epi*-brefeldine A

The enzymatic syntheses using our newly developed cofactor regeneration systems are summarized below.

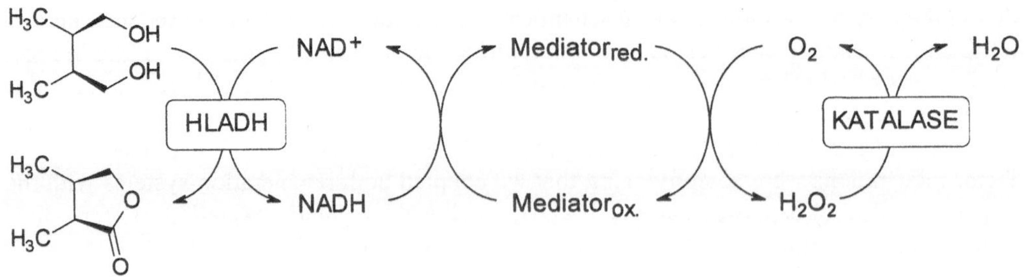

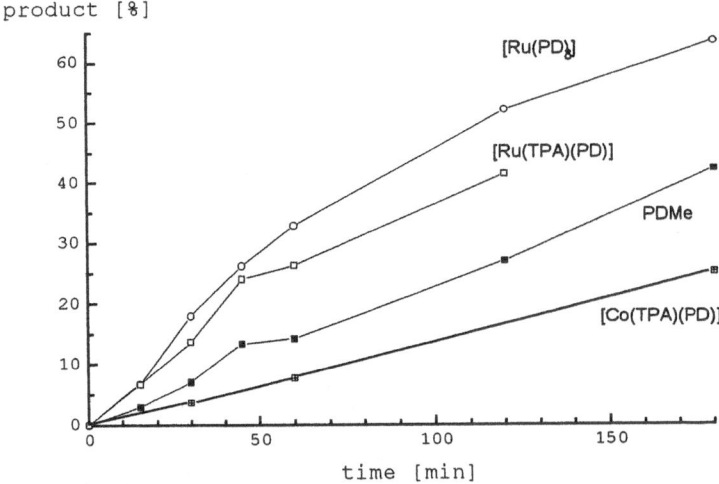

Scheme 3. HLADH catalyzed oxidation of a *meso*-diol using the indirect aerobic regeneration of the cofactor for the determination of the turnover rates

Fig. 3. Reaction progress during the HLADH catalyzed oxidation with indirect aerobic cofactor regeneration using different redox catalysts at pH 8.0. PDMe is *N*-methyl 1,10-phenthrolinium-5,6-dione tetrafluoroborate

In a batch process using 200 mg of diol the best results were obtained for Ru(PD)$_3$ which reached 131 turnovers per hour.

Table 5. Turnover frequencies for different redox catalysts for the HLADH oxidation of a *meso*-diol

mediator	turnover frequency [h^{-1}]
[Ru(PD)$_3$]	131
[Ru(TPA)(PD)]	107
PDMe	65
[Co(TPA)(PD)]	30

with indirect aerobic cofactor regeneration at pH 8.0

Conditions: ratio of mediator: NAD: diol = 1 : 10.3 : 220 (mediator 7.8×10^{-5} mol/L; diol $1.6 \times^{-2}$ mol/L)

Under similar conditions the enzyme catalyzed oxidation under indirect electrochemical cofactor regeneration was studied (Scheme 4).

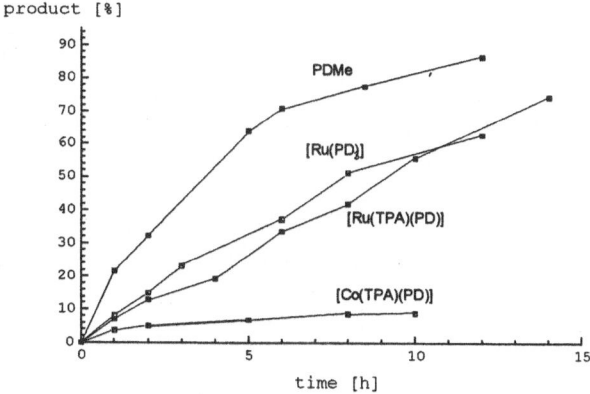

Scheme 4. HLADH catalyzed oxidation of *meso*-diols using the indirect electrochemical regeneration of the cofactor for the determination of the turnover rates

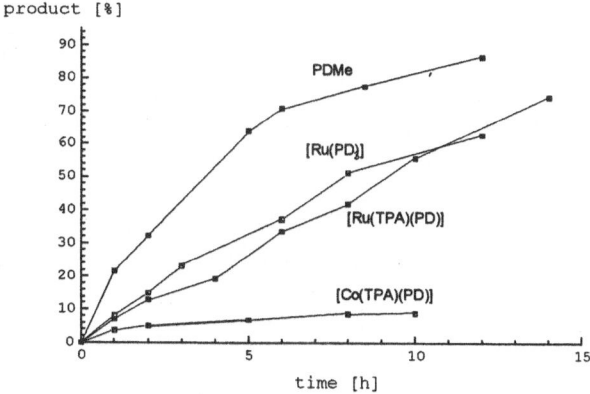

Fig. 4. Reaction progress during the HLADH catalyzed oxidation with indirect electrochemical anaerobic cofactor regeneration using different redox catalysts at pH 8.0

Table 6. Turnover frequencies for different redox catalysts for the HLADH oxidation of a *meso*-diol with indirect anerobic electrochemical cofactor regeneration at pH 8.0

mediator	turnover frequency [h^{-1}]
PDMe	31
[Ru(PD)$_3$]	15
[Ru(TPA)(PD)]	10
[Co(TPA)(PD)]	5

Conditions: ratio of mediator : NAD : diol=1 : 6 : 90 (mediator1.6×10^{-4} mol/L; diol 1.44×10^{-2} mol/L)

In the indirect electrochemical process up to 31 turnovers/h could be reached with the present cell construction in batch electrolyses of 200 mg of diol. Finally, we performed batch experiments and compared them with the published results by *Jones*[15] of the same substrate according to the following reaction.

Table 7. Comparison of the enzymatic oxidation of a *meso*-diol 1. under indirect aerobic regeneration of NAD using the system FMN/O$_2$ by *Jones*[15], 2. under indirect aerobic regeneration using Ru(PD)$_3$](ClO$_4$)$_2$ as catalyst, and 3. under indirect electrochemical regeneration using PDMe (N-methyl 1,10-phenthrolinium-5,6-dione tetrafluoroborate) as catalyst

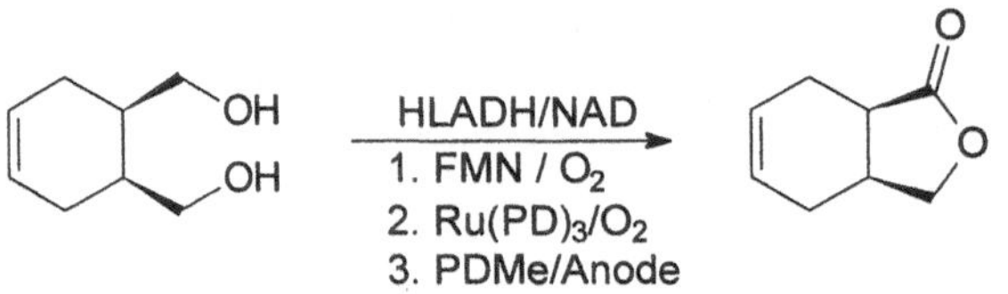

Conditions	1. FMN/O$_2$ (pH 9)[a]	Ru(PD)$_3$/O$_2$ (pH 8)	PDMe/Anode (pH 8)
catalyst	9.72 g (20.3 mmol)	2.6 mg (0.8×10^{-5} mol)	3.2 mg (1×10^{-5} mol)
NAD	720 mg (1.1 mmol)	53 mg (8×10^{-5} mol)	70 mg (1×10^{-4} mol)
substrate	2 g (13.9 mmol)	1 g (7 mmol)	1 g (7 mmol)
time	24 h	86 h	49 h
conversion	83 %	99 %	63 %
turnover frequency	0.06 TN/h	21 TN/h	18 TN/h[a]
ratio of substrate : NAD : catalyst	0.64 : 0.05 : 1	900 : 10 : 1	700 : 10 : 1

[a] After 4.5 h a turnover of 27% was reached corresponding to a turnover frequency of 82 TN/h

Our proposed systems for the NAD(P)$^+$ regeneration during enzymatic syntheses are not yet optimized. However, even now they prove to be advantageous over present methods.

3 Development of an Electrochemically Regenerable Manganese(III) Picolinate as Oxidation Mediator for Radical C-C Bond Formations

The construction of complex molecules by using efficient and selective C-C bond formations is a key objective in modern organic synthesis. Besides the more classical polar reactions via carbanions and carbocations the field of radical C-C bond formations has expanded considerably recently.[16] In addition to a high reactivity carbon radicals exhibit good chemo-, regio- and stereoselectivities. Normally, the radical forming reactions are mild enough to tolerate highly functionalized complex molecular structures.[17]

In addition to the frequently used reductive radical forming reactions for example via tributyl tin hydride[18], recently oxidative procedures have found more interest because higher functionalized products could be obtained. Common oxidation reagents are complexes of transition metals in higher oxidation states like copper(II) chloride[19] or cerium(IV) ammoniumnitrate[20]. Additionally, very often the one-electron oxidation reagent manganese(III) acetate in acetic acid as solvent is applied. It is able to oxidize enol forms of CH-acidic compounds like carbonyl compounds to the α-oxoalkyl radicals. The oxidation ocurs in the form of an *inner-sphere* mechanism after complexation of the substrate by the manganese salt. The thus formed radical behaves as an electrophile and adds easily to nucleophilic π-systems. The generated product radical can either abstract a hydrogen atom from the solvent to give the saturated product, be further oxidized to a carbocation and add a nucleophile, or it can add to another π-system like in tandem cyclizations.

Oxidative radical C-C bond forming reactions have been applied and reviewed intensively.[21] In contrast to manganese(III) acetate, which requires acetic acid as solvent, manganese(III) trispicolinate Mn(pic)$_3$ is active in DMF and some other organic solvents and therefore extends the process to the application of acid sensitive substrates[22]. In contrast to manganese(III) acetate, this complex has a monomeric octahedral structure[23]. It also acts as inner-sphere electron transfer agent so that highly selective reactions can be expected. Until now, Mn(pic)$_3$ has only been applied to β-keto carboxylic acids as carbonyl component. According to *Narasaka*[24] they add to olefins under decarboxylation.

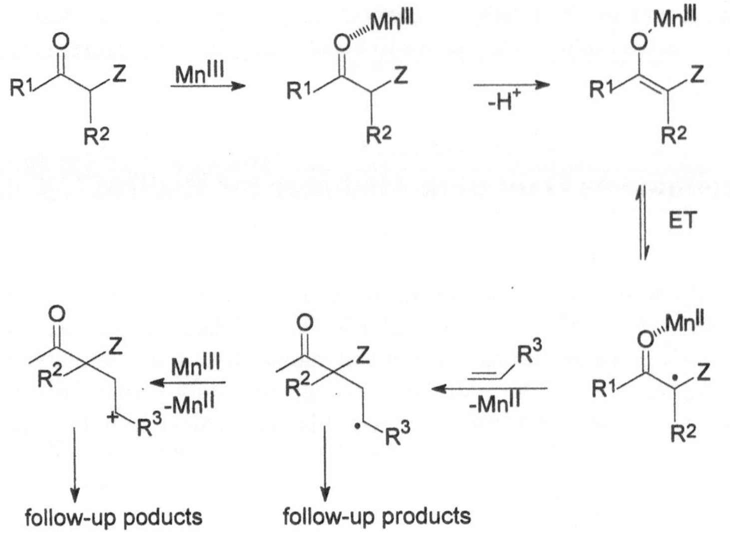

R^1 = H, Alkyl, Aryl, OR, OH   Z = H, Alkyl, Aryl, COR, COOR, CN, SꞆR,
R^2 = H, Alkyl, Aryl, Cl

Scheme 5. Inner-sphere mechanism for the manganese(III) acetate induced C-C bond formation between an α-oxoalkyl radical and an electron-rich olefin

3.1 Reactions Induced by Stoichiometric Amounts of Manganese(III) Trispicolinate [Mn(pic)₃]

Addition reactions in acetic acid

As of our knowledge, nothing was known about the reaction of diesters, ketoesters and diketones with Mn(pic)₃ in the presence of alkenes. In a test reaction, we found that 2-methyl malonates can only be added to 1-hexene using acetic acid as solvent at elevated temperatures (115°C). The expected unsaturated addition products are obtained in a 6 to one ratio. As another product, a picolinate could be identified which must be formed by the addition of picolinate to the intermediate carbocation. This is quite unusual because acetate is present in large excess. Therefore, this result might indicate that also the oxidation of the adduct radical to the cation and its further addition to a nucleophile takes place within the ligand sphere of the manganese complex.

266

However, further studies are necessary to prove this assumption. Under identical conditions, the addition to 2,4,4-trimethyl-1-pentene was also successful yielding the expected three regioisomers. Surprisingly, the thermodynamically unfavorable terminal olefin was the major product.

Addition reactions in non-acidic organic solvents

Because Mn(III)(pic)$_3$ can be used in non-acidic organic solvents like for example DMF, acid-labile substrates can be applied. Thus, we were successful in the addition of β-keto esters and β-diketones of type 1 to silylenolethers 3 - 5, enolacetate 6, enamine 7 and allylsilane 8 using two equivalents of manganese(III) trispicolinate at 60°C. Malonates do not react under these conditions. The results are summarized in Table 8.

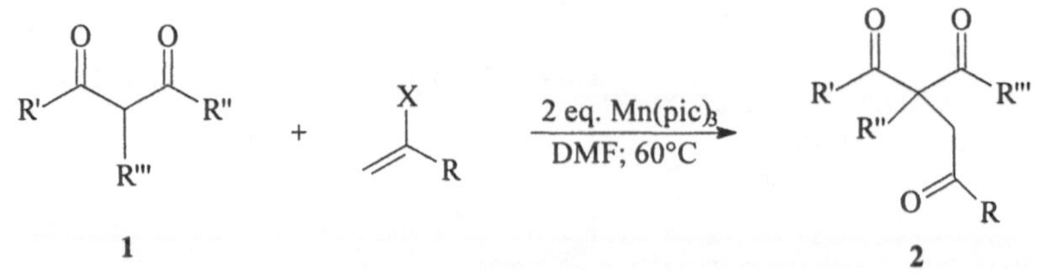

R, R', R'' = alkyl, aryl; R''' = alkyl, OEt; X = OTMS, OTBDMS, OAc, N(alkyl)

Scheme 6: General reaction scheme for the radical addition of β-keto esters and β-diketones to acid sensitive alkenes using Mn(III)(pic)₃ as stoichiometric oxidant

The method is characterized by the following points: 1. oxidant: 2 equivalents Mn(pic)₃; no other co-oxidant needed; 2. solvents: DMF, N-methyl pyrrolidone, acetonitrile, DMSO, methanol, ethanol are also possible; 3. temperature: 60°C; higher temperatures lead to by-products; 4. easy flash chromatographic work-up; 5. selectivity: no hydrogen atom transfer was observed in any solvent; fast outer-sphere oxidation of the adduct radical and loss of the leaving group is observed; 6. reaction rate is proportional to carbonyl reactivity and enolization rate of the substrate: at least one keto function is necessary; malonates do not react; 7. steric effects: fast addition to terminal olefines; trisubstituted olefins provide poor yields.

Table 8. Results of the radical addition of β-keto esters and β-diketones to acid sensitive alkenes using Mn(III)(pic)₃ as stoichiometric oxidant (conditions: 2 eq. Mn(pic)₃, 1 eq. olefine, DMF, 60°C)

reaction	olefin	reaction time	isolated yield
	3	7 h	58 %
	3	2 h	57 %
	6	1 h	86 %
	3	< 1 h	64 %
	8	2 h	[a]
	3	< 1 h	76 %
	4	< 1 h	58 %
	6	< 1 h	26 %
	7	2,5 h	43 %
	5	2 h	8 %

[a] Decomposition during chromatographic workup

3.2 Reactions Induced by Electrochemically Generated and Regenerated Manganese(III) Bispicolinate

The reaction process implies two oxidation steps, so two equivalents of Mn(pic)$_3$ are required. Therefore, it would be advantageous if only catalytic amounts of the oxidant, would be necessary. In principle, this should be possible if anodic in-situ generation and regeneration of the oxidant could be achieved. The regeneration of manganese(III) tris-picolinate in non-acid organic solvents fails due to the poor solubility of the manganese(II) complex formed during the reaction. However, electrocatalysis is successfully achieved if the manganese(III) reagent is electrogenerated from manganese(II) **bis**picolinate bishydrate instead of using the **tris**picolinate. This complex is octahedral with the two aquo ligands occupying the axial positions. Continous in-situ generation of this oxidant by anodic oxidation of manganese(II) bispicolinate Mn(pic)$_2$ is easily possible in different solvents like DMF or *tert*-butanol/water (3:1). The cyclic voltammogram of Mn(pic)$_2$ in DMF (Fig. 5) shows a quasi reversible redox wave for Mn(II)/Mn(III) at + 700 mV vs. Ag/AgCl and a second irreversible oxidation step for Mn(III)/Mn(IV) at + 1.5 V. Controlled potential electrolysis of a solution of manganese(II) bispicolinate in DMF in presence of a dicarbonyl compound and silyl enol ether yields the same products as obtained in the stoichometric reaction. This reaction is the first example of an indirect electrochemical oxidative C-C-bond formation mediated by manganese picolinate. A process of this type combines the advantages of selective homogeneous reactions with those of a reagent-free heterogenous electron transfer.

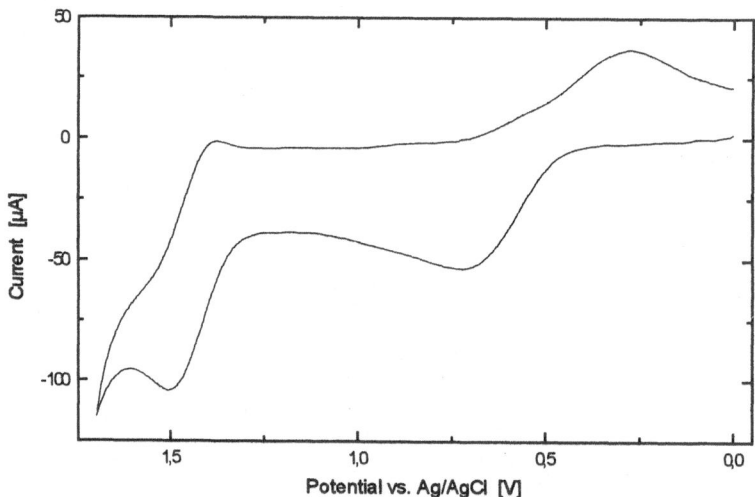

Fig. 5. Cyclic voltammogram of manganese(II) bispicolinate in DMF vs. Ag/AgCl

Controlled potential electrolyses

Controlled potential electrolyses were performed in a quasi-divided cell using a Sigraflex®-anode, a Pt-wire cathode, and a 2.5 mol% Mn(pic)₂ solution in DMF:

isol. yield: 38 %
current efficiency: 55 %

Constant current electrolyses

In order to simplify the process and to achieve a higher reaction speed, we have investigated constant current electrolyses. Water-containing solvents provide a suitable cathode reaction and allow higher current densities. Now it is possible to conduct the reaction within the same time and with the same or even higher yield as obtained in the stoichiometric reaction using Mn(pic)₃. The amount of oxidant was reduced by 95 %. Current densities are in a range from 5 - 12 mA/cm². The reactions are not yet optimized.

isol. yield: 36 %
current efficiency: 36 %

As a conclusion it can be said that radical C-C-bond formation induced by manganese(III) oxidants in non-acidic media is an advantageous reaction in synthetic organic chemistry. The use of stoichiometric amounts of manganese(III) trispicolinate extends the process to a variety of attractive acid-labile alkenes as trapping agents which formerly could not be employed. Reactions proceed with high product selectivities. However, the most important progress was that we were able to achieve a redox catalytic process with in-

271

situ electrochemical regeneration of the oxidant by using the soluble manganese(II) bispicolinate. Employing this technique the process can be conducted catalytically with reference to the metal oxidant.

4 Electrochemical Oxidations using the Manganese-substituted α-Keggin-Heteropolyanion $K_6SiW_{11}O_{39}Mn(H_2O)$ as a Mediator

Homogeneous catalytic oxidations have been one of the most interesting areas of chemical research. Numerous metalloporphyrin-based systems including iron, manganese, cobalt, chromium and ruthenium have been described[25]. Recently, considerable attention has been directed towards metal substituted α-Keggin heteropolyanions as analogues of metalloporphyrin[26].

From α-Keggin heteropolyanions ($AB_{12}O_{40}$), so-called "lacunary heteropolyanions" ($AB_{11}O_{39}$) are derived by removing one BO_6 unit. Many transition metals (Co, Cr, Fe, Mn, Ru etc.) can fill this hole, giving rise to metal substituted α-Keggin heteropolyanions such as $PW_{11}O_{39}H_2O$). Similar to metalloporphyrins, metal substituted α-Keggin heteropoly-anions can work as oxygen carriers. These transition metals M^{n+} accept oxygen from different oxygen donors (e. g. iodosylarenes, peroxides, etc.) and the thus formed oxometal species $Mn^{n+2}=O$ are able to oxidize various organic substrates like hydrocarbons, alcohols and olefins[26]. The heteropolyanion ligands are robust under strongly oxidative conditions and thus have an important advantage over metalloporphyrin systems, which decompose under these conditions. The transition metal incorporated in the heteropolyanion resides in an octahedral environment with one coordination site occupied by a labile water molecule. In this work, we have focused our attention on manganese substituted α-Keggin silicon tungstoheteropolyanion for the application as an electrochemically regenerable mono-mediator system. This complex ($SiW_{11}O_{39}Mn(II)$) can catalyze the oxidation of phenols, olefins and hydrocarbons in combination with some stoichiometric oxidants[26]. Our goal for systems of this type is the development of an electrochemically regenrable redox catalyst with long-time stability in aqueous media to avoid the use of stoichiometric oxygen donors as they were mentioned above. On electrochemical oxidation in aqueous media the aquo ligand of the transition metal may work as oxygen source. To be able to act as an efficient redox catalyst, the active form of the complex must be generable electrochemically from the transition metal aquo complex in aqueous media. In addition, after the homogeneous oxidation reaction with the substrate the transition metal aquo complex must be regenerated easily. If these conditions are fullfilled can be determined by using electroanalytical measurements.

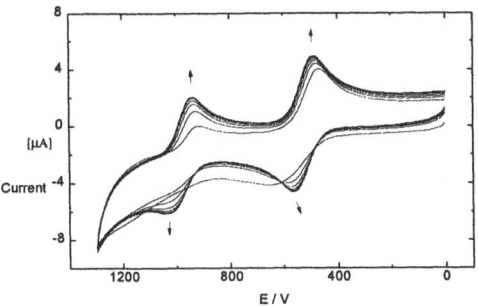

Fig. 6 Continuous cyclic voltammograms at a glassy carbon electrode of a 1 mM solution of SiW$_{11}$O$_{39}$Mn in pH 6.0 phosphate buffer vs Ag/AgCl reference electrode. Supporting electrolyte: 0.5 M phosphate solution. Scan rate = 20 mV s^{-1}. The initial potential was 0 V and the initial scan direction was towards more positive potential. The arrows are indicating the direction of peak changes during continuous scanning

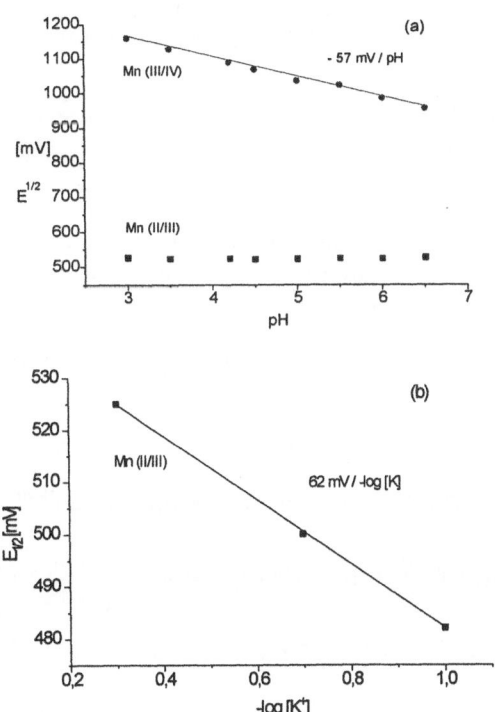

Fig. 7 (a) pH dependence of E$_{1/2}$ for the Mn(II/III) and Mn(III/IV) and (b) effect of potassium cation concentration of E$_{1/2}$ of Mn(II/III)

The cyclic voltammogram shows two redox waves at + 0.53 V vs. the Ag/AgCl reference electrode for the Mn(II)/Mn(III) pair and at + 0.99 for the Mn(III)/Mn(IV) pair. As shown in Fig. 6, with continous scanning of the potential region from 0 to + 1.3 V, the first oxidative peak shifted to a more negative potential and the second oxidation peak developed with increasing number of scans. With continous scanning all peaks finally reach a stable value (consolidated peaks). Each of the two consolidated reversibel redox pairs correspond to a single electron transfer. Fig. 7 (a) shows the dependence of $E_{1/2}$ on the pH value with constant potassium concentration for the two anodic waves. The $E_{1/2}$ of $Mn^{III/IV}$ show a clear linear dependence on the pH with a slope of 57 mV per pH value. This means that one proton is involved in the redox step. The consolidated redox couple $Mn^{II/III}$, however, shows no pH dependence but depends on the concentration of the potassium cation. As shown in Fig. 7 (b), the $E_{1/2}$ of $Mn^{II/III}$ changed ca. 62 mV per -log $[K^+]$. This suggests that the potassium cation is involved in this redox step.

The consolidated peaks in the cyclovoltammogram of $SiW_{11}O_{39}Mn(II)(H_2O)^{6-}$ in the potential region between 0 V and 1.3 V vs. Ag/AgCl can be explained quite easily by the following two reactions.

$$K_nSiW_{11}O_{39}Mn(II)(H_2O)^{(6-n)-} = K_{n-1}SiW_{11}O_{39}Mn(III)(H_2O)^{(5-n)-} + e^- + K^+$$
$$(E_{1/2} = 0.53 \text{ V at pH 6.0})$$
$$K_{n-1}SiW_{11}O_{39}Mn(III)(H_2O)^{(5-n)-} = K_{n-1}{}_1SiW_{11}O_{39}Mn(IV)(OH)^{(5-n)-} + e^- + H^+$$
$$(E_{1/2} = 0.99 \text{ V at pH 6.0})$$

The change of the cyclic voltammograms during continous scanning can be explained in the following way: At the start the complex contains an ion from the supporting electrolyte which in our case is phosphate instead of a water molecule. At the hard Mn(IV) oxidation state, the phosphate anion (L) is easily replaced by the hard hydroxy group thus transforming the redox system back to the hydroxy (aquo) complex which shows the consolidated peaks.

$$K_nSiW_{11}O_{39}Mn(II)(H_2O)^{(6-n)-} = K_{n-1}SiW_{11}O_{39}Mn(III)(H_2O)^{(5-n)-} + e^- + K^+$$

$$K_{n-1}SiW_{11}O_{39}Mn(III)(L)^{(5-n)-} = K_{n-1}SiW_{11}O_{39}Mn(IV)(L)^{(4-n)-} + e^-$$

$$K_{n-1}SiW_{11}O_{39}Mn(IV)(L)^{(4-n)-} + HO^- = K_{n-1}{}_1SiW_{11}O_{39}Mn(IV)(OH)^{(5-n)-} + L$$

In fact, if the cyclic voltammogram is recorded from an initial potential, where the Mn(IV)(OH) complex is produced at the electrode surface (1.25 V vs Ag/AgCl in pH 6.0 solution), the consolidated peaks were obtained immediately, instead of after multiple cycling. Therefore, by controlled potential electrolysis (at 1.25 V vs Ag/AgCl) the Mn(IV)(OH) complex is generated. The retransformation of the manganese (IV) phosphate system into the manganese (IV) hydroxy complex at the second oxidation step makes the $SiW_{11}O_{39}Mn(IV)(OH)^{5-}$ system a promising candidate for the application as a mediator in indirect electrochemical oxidation processes. The Mn(IV)(OH) complex which can not be generated chemically by hydrogenperoxide could be generated electrochemically at + 1.25 V and total turnover was obtained. This complex is slowly reduced to the Mn(III) complex at room temperature in the solution If alcohols are added, the reaction is considerably accelerated which could be followed by UV spectroscopy. In this way, rate constants for the reaction betweeen the Mn(IV)(OH) complex and different alcohols could be obtaiend.

Under pseudo-first order condition ([Mn] << [alcohol]), the appropriate logarithmic kinetic plots were linear (r > 0.9995), establishing first oder kinetics in the concentration of Mn. The observed rate (k_{obs}) increased linearly with increasing concentration of iso-propanol and was independent on the pH of the solution. We compared the oxidation rate of some other alcohols in the pH 6.0 phosphate buffer at constant concentration of alcohols ([alcohol] = 0.1 M). The alcohols reacted with the Mn(IV)OH) complex in the following order: 1-phenyl-ethanol ($k_{obs} = 1.9 * 10^{-3}[s^{-1}]$), allyl alcohol ($k_{obs} = 1.65 * 10^{-3}[s^{-1}]$), benzylalcohol (k $k_{obs} = 1.3 * 10^{-3}[s^{-1}]$) > iso-propanol ($k_{obs} = 2.5 * 10^{-4}[s^{-1}]$) > ethanol (k $k_{obs} = 7.5 * 10^{-5}[s^{-1}]$).

Continous indirect electrochemical oxidations of 1-phenyl ethanol, p-methoxy benzylalcohol, p-chloro benzylalcohol, p-nitro benzylalcohol, and 2-cyclohexene-1-ol using catalytic (0.5 to 5 mol %) of the $SiW_{11}O_{39}Mn(IV)(OH)^{5-}$ mediator have been performed (Table 9). The redox catalyst underwent up to 160 cycles without decomposition. Thus, it could be demonstrated, that the manganese heteropolytungstate can act as redox catalyst for the indirect electrochemical oxidation of certain alchols. Further studies with regard to the oxidation of other functional groups are under way.

Table 9. Electrochemical oxidations of alcohols using $SiW_{11}O_{39}Mn(IV)(OH)^{5-}$ as redox catalyst

$$R \overset{OH}{\underset{R'}{\diagdown}} \xrightarrow[\text{pH 6.0, phosphate buffer (8 mL)}]{\substack{\text{C-anode; Pt-cathode; + 1.25 V vs Ag/AgCl} \\ K_6SiW_{11}O_{39}Mn(II)(H_2O)\ (0.5 - 5\ mol\%)}} R \overset{O}{\underset{R'}{\diagdown}}$$

Alcohols (mM)	Catalyst (mM)	Products	Time (Charge)	Yield (%)	Current Yield (%)
1-phenylethanol (50)	5 mol% (2.5)	acetophenone	4.5 h (100 As)	61.4 (37.4 % alcohol)	48
1-phenylethanol (100)	0.5 mol% (0.5)	acetophenone	8 (147 As)	42 (57 % alcohol)	27
2-cyclohexene-1-ol (50)	5 mol% (2.5)	2-cyclohex-enone	5.5 h (100 As)	29 (69.7 % alcohol)	22
p-methoxy benzyl-alcohol (50)	5 mol% (2.5)	p-methoxy benzaldehyde	3 h (80 As)	64	100
		p-methoxy benzoic acid		15.7 (0 % alcohol)	
p-chloro benzyl-alcohol (50)	5 mol% (2.5)	p-chloro benzaldehyde	5.5 (80 As)	13	64
		p-chloro benzoic acid		24 (44 % alcohol)	
p-nitro benzyl-alcohol (50)	5 mol% (2.5)	p-nitro benzaldehyde	3 (80 As)	12	66
		p-nitro benzoic acid		26 (34 % alcohol)	

Acknowledgement

Financial support by the Volkswagen-Stiftung (AZ. : I/68384), the European Community under the program Human Capital and Mobility, CRX-CT92-0073, the Fonds der Chemischen Industrie, and the BASF AG is gratefully acknowledged.

References

1. V. Alphand, A. Archelas, M. Frede, M. Hofbauer, K.-H. van Pée, T. Pohl, E. Steckhan, in *Enzyme Catalysis in Organic Synthesis - A Comprehensive Handbook*, K. Drauz, H. Waldmann, Eds., Vol. II, Chapter B.6, VCH, Weinheim **1995**.

2. F. P. Guengerich, CRC Press, Boca Raton, Fl, *Mammalian Cytochromes P-450*, **1987**, Vols 1 and 2; F.S. Sariaslani, *Critical Reviews in Biotechnology*, **1989**, *9*, 171-257; Cytochrome P-450: Advances and prospects, *Faseb J.*, **1992**, *6*, 666792.

3. D. T. Gibson, M. Hensley, H. Yoshioka, T. S. Mabry, *Biochemistry* **1970**, *9*, 1626-1630; T. Hudlicky, H. Luna, H.F. Olivo, C. Andersen, T. Nugent, J.D. Price, *J. Chem. Soc., Perkin Trans. 1* **1991**, 2907-2917; S. Ley, M. Parra, A.J. Redgrave, F. Sternfeld, *Tetrahedron* **1990**, *46*, 4995-5026.

4. W.-D. Fessner, G. Sinerius, *Angew. Chem.* **1994**, *106*, 217-220; *Angew. Chem. Int. Ed. Engl.* **1994**, *33*,209; E. Steckhan, *Topics Curr. Chem.* **1994**, *170*, 99.

5. J. B. Jones, E. Jakovac, *Org. Synth.* **1985**, *63*, 10-17.

6. H. G. Davies, R. H. Green, D. R. Kelly, in *Biotransformations in Preparative Organic Chemistry*, Chapter 4, Academic Press, London **1989**; J. B. Jones, C. J. Sih, D. Perlman (eds), in *Applications of Biochemical Systems in Organic Synthesis*, in *Techniques of Chemistry*, Vol. X, Part II, Wiley, New York **1976**; K. Nakamura, M. Aizawa, O, Miyawaki, in *Electroenzymology, Coenzyme Regeneration*, Springer-Verlag, Berlin **1988**.

7. J. B. Jones, K. E. Taylor, *J. Chem. Soc., Chem. Commun.* **1973**, 205; J. B. Jones, K. E. Taylor, *Can. J. Chem.* **1976**, *54*, 2969, 2974.

8. S. Itoh, M. Kunigawa, N. Mita, Y. Ohshiro, *J. Chem. Soc., Chem. Commun.* **1989**, 694; S. Itoh, N. Mita, Y. Ohshiro, *Chem. Lett.* **1990**, 1949; M. Frede, Ph.D. thesis, Bonn **1993**.

9. H. Huck, H. L. Schmidt, *Angew. Chem.* **1981**, *93*, 421; *Angew. Chem. Int. Ed. Engl.* **1981**, *20*, 402; C. Degrand, L. L. Miller, *J. Am. Chem. Soc.* **1980**, *102*, 5728; N. K. Lau, L. L. Miller, *J. Am. Chem. Soc.* **1983**, *105*, 5271.

10. A. Fassoune, J.M. Laval, J. Moiroux, *Biotechnol. Bioeng.* **1990**, *35*, 935; J. M. Laval, C. Bourdillon, J. Moiroux, *Biotechnol. Bioeng.* **1987**, *30*, 157; J. Bonnefoy, J. Moiroux, J. M. Laval, C. Bourdillon, *J. Chem. Soc., Faraday Trans 1* **1988**, *84*, 941.

11. Preliminary results, see: G. Hilt, E. Steckhan, *J. Chem. Soc., Chem. Commun.* **1993**, 1706.

12. G. Hilt, T. Jarbawi, W.R. Heineman, E. Steckhan, *Chem. Eur. J.*, in print.

13. Y. Lei, F.C. Anson, *J. Am. Chem. Soc.* **1995**, *105*, 5271; *Inorg. Chem.* **1996**, *35*, 3044.

14. T.P. DeAngelis, W.R. Heineman, *J. Chem. Edu.* **1976**, *9*, 594; W.R. Heineman, F.M. Hawkridge, H.N. Blount, in *Electroanalytical Chemistry*, Vol. 7 (A.J. Bard, Ed.), Marcel Dekker, New York 1974, 1.

15. H.P Lok, I.J. Jakovac, J.B. Jones, *J. Am. Chem. Soc.* **1985**, *107*, 2521.

16. Reviews: a) B. Giese, *Radicals in Organic Synthesis: Formation of Carbon-Carbon-Bonds*, Pergamon Press, Oxford 1986; b) B. Giese, *Angew.Chem.* **1989**, *101*, 993; c) D.P. Curran, *Synlett* **1991**, 63; d) D.P. Curran, *Synthesis* **1989**, 417, 489; e) D.P. Curran,

in: *Comprehensive Organic Synthesis*, B.M. Trost, I. Fleming, Eds., Pergamon Press, Oxford 1991; f) H.M.R. Hoffmann, *Angew.Chem.* **1992**, *104*, 1361; g) W. Smadja, *Synlett* **1994**, 1; h) B. Giese, W. Damme, R. Batra, *Chemtracts Org.Chem.* **1994**, *7* (6), 355.

17. a) D. P. Curran, *Synlett* **1991**, 63; b) D. J. Hart, *Science* **1984,** 223, 883; c) M. Ramaiah, *Tetrahedron* **1987**, *43*, 3541; d) D. P. Curran, in: *Comprehensive Organic Chemistry*, Vol. 4, B. M. Trost, I. Fleming (Eds.), Pergamon Press, Oxford 1991.

18. W. P. Neumann, *Synthesis* **1987**, 665.

19. M. G. Vinogradov, A. E. Kondorsky, G. I. Nikishin, *Synthesis* **1988**, 10.

20. a) G. A. Molander, *Chem. Rev.* **1992,** *92*, 595; b) A. Citterio, R. Sebastianio, A. Marion, *J. Org. Chem.* **1991**, *56*, 5328.

21. a) W. J. de Klein in: *Organic Synthesis by Oxidation with Metal Compounds*, W. J. Mijs, C. R. H. de Jonge (Eds.), Plenum Press, New York 1986; b) R. Warsinsky, *Diplomarbeit*, Universität Bonn, 1990; c) G. G. Melikyan, *Synthesis* **1993**, 3185; d) B.B. Snider, J.J. Patricia, S.A. Kates, *J. Org. Chem.* **1988**, *53*, 2137; e) B.B. Snider, R. Mohan, S.A. Kates, *Tetrahedron Lett.* **1987**, *28*. 841; f) R. Mohan, S.A. Kates, M.A. Dombroski, B.B. Snider, *ibid.* 845; g) M.A. Dombroski, S.A. Kates, B.B. Snider, *J. Am. Chem. Soc.* **1990**, *112*, 2759; h) J.M. Mellor, S. Mohammed, *Tetrahedron Lett.* **1991**, *32*, 7107, 7111, *Tetrahedron* **1993**, 7547, 7557, 7567; C. Scolastico, A. Bernardi, *Chemtracts Org. Chem.* **1994**, *7*, 252; i) W. E. Fristad, S. S. Hershberger, *J. Org. Chem.* **1985**, *50*, 1026; j) J. P. Coleman, R. C. Hallcher, D. E. McMackins, T. E. Rogers, J. H. Wagenknecht, *Tetrahedron* **1985**, *41*, 831.

22. K. Narasaka, N. Miyoshi, K. Iwakura, T. Okauchi, *Chem. Lett.* **1989**, *12,* 2169.

23. B.N. Figgis, C.L. Raston, R.P. Sharma, A.H. White, *Aust. J. Chem.* **1978**, *31*, 2545-8.

24. K. Narasaka, N. Miyoshi, K. Iwakura, T. Okauchi, *Chem. Lett.* **1989**, 2169.

25. B. Meunier, *Chem. Rev.* **1992**, *92*, 1411.

26. C. L. Hill and C. M. Prosser-McCartha, *Coord. Chem. Rev.* **1995**, *143*, 407.

Synthesis of Highly Enantioenriched Compounds *via* Iron Mediated Allylic Substitutions

Dieter Enders*, Bernd Jandeleit, Stefan von Berg

Institut für Organische Chemie, Rheinisch-Westfälische Technische Hochschule, Professor-Pirlet-Straße 1, D-52074 Aachen, Germany

Summary

In this account we describe our efforts over more than a decade to develop a synthetically useful and practical methodology for the synthesis of highly enantioenriched compounds *via* iron promoted allylic substitutions. After first attempts based on a kinetic resolution of planar chiral tetracarbonyl(η^3-allyl)iron(1+) complexes and an auxiliary controlled diastereoselective formation of the iron complexes, the "chirality transfer" approach turned out to be an efficient solution. Starting from enantiopure easily accessible and cheap acceptor substituted allylic substrates, the corresponding (η^3-allyl)tetracarbonyl iron cation complexes are formed and trapped with a variety of carbon and heteroatom nucleophiles including silyl enol ethers, electronrich arenes and heteroarenes, functionalized organozinc compounds and amines. The iron mediated allylic substitutions proceed with virtually no loss of chirality information from central (C-O) over planar (C-Fe) back to central chirality (C-C or C-X) affording products of high enantiomeric purity with overall retention (double inversion). In addition, complete γ-regioselectivity and conservation of the double bond geometry is achieved. First applications in the synthesis of virtually enantiopure natural products are described.

1 Introduction

In recent years cationic metal π-complexes of odd and even numbered unsaturated polyenic ligands have received considerable attention as useful synthetic tools in organic synthesis. They can be regarded as stabilized carbocation equivalents coordinated to a transition metal. One advantage is their enhanced reactivity towards a multitude of soft carbon and hetero atom nucleophiles[1,2]. In addition, they are of increasing significance in enantiomerically pure form for the development of various new synthetic methods and as valuable building blocks for the convergent regio- and stereocontrolled construction of organic target molecules. In general, transition metal π-complexes of unsymmetrically substituted ligands in which the

279

metal fragment distinguishes the two enantiotopic faces of the ligand are planar chiral[3]. Nucleophilic addition reactions are subject to complete stereocontrol by the metal-ligand moiety while regiochemical outcomes are mostly controlled by substitution patterns and/or the influence of substituents.

Among the various carbon-carbon and carbon-heteroatom bond forming reactions promoted or catalysed by transition metals, allylic substitution *via* electrophilic π-allyl complexes have been one of the most intensively investigated[4-9]. Current knowledge about the stereochemical course of the formation and reactivity of cationic tetracarbonyl-(π-allyl)iron complexes is limited. Studies devoted to the synthetic potential of alkyl and aryl substituted tetracarbonyl(η^3-allyl)iron(1+) complexes have demonstrated that these species undergo regioselective nucleophilic attack by a multitude of soft carbon and heteroatom nucleophiles preferentially at the less substituted or at the *syn* substituted allyl termini affording addition products of (Z)-configuration[10]. Likewise, polar effects on the regioselectivity of nucleophilic addition reactions to tetracarbonyl(η^3-allyl)iron(1+) complexes caused by electron withdrawing functionalities (e.g. CO_2R, $CONR_2$) have been examined recently and independently to our work by *Green et al.*[11] and *Speckamp et al.*[12] to give allyl coupled addition products with complete γ-regioselectivity after oxidative removal of the stabilizing $Fe(CO)_4$-fragment.

Since the early eighties, our group has been engaged in the investigation of the influence of planar chirality of complexes of the forementioned type on the design of synthetic strategies and their use in the preparation of highly enantiomerically enriched organic target molecules[13-22]. In this course, we developed an efficient approach to acceptor-functionalized highly diastereo- and enantiomerically enriched planar chiral tetracarbonyl(η^3-allyl)-iron(1+) complexes starting from easily accessible enantiopure precursors based on natural products. A general use of these iron complexes as synthetic equivalents of a^4-synthons for the homologous (1,5)-*Michael* addition[23] is of interest both as a form of umpolung[24] of classical d^4-chemistry and as a synthetic method of considerable potential. Now we have completed a series of preliminary studies towards mechanistic details and the stereochemical outcome of our iron mediated "chirality transfer" methodology. Furthermore, we have investigated the synthetic potential of this new process and have found some useful applications in natural product synthesis. In order to give an overview, we wish to summarize and to discuss our results obtained so far in this aerea of research.

2 Development of Basic Concepts

The following three basic concepts have been examined to reach our goals:

1) Nucleophilic addition of enantiopure d^2-nucleophiles to racemic planar chiral electrophilic tetracarbonyl(η^3-allyl)iron(1+) complexes ("kinetic resolution").
2) Preparation of diastereomerically enriched tetracarbonyl(η^3-allyl)iron(1+) complexes by means of an auxiliary controlled complexation of (*E*)-configurated enoates and subsequent addition of achiral nucleophiles.

3) Diastereoselective complexation of enantiopure (E)-configurated acceptor substituted olefins, subsequent generation of enantiomerically enriched acceptor substituted tetra-carbonyl(η^3-allyl)iron(1+) complexes and addition of achiral nucleophiles ("chirality transfer").

In the following discussion each of the concepts will be presented in greater detail taking account to its impact in understanding the stereochemical pathways of the complete reaction sequences as well as to its synthetic potential.

2.1 Kinetic Resolution / Chiral Nucleophile Approach

Scheme 1 displays the first concept in which the cationic (η^3-allyl)iron complex represents a planar chiral but racemic a^4-synthon and the nucleophile a chiral d^2-synthon offering an access to chiral 1,6-dicarbonyl compounds, which are proved to be versatile intermediates in the synthesis of more complex target molecules.

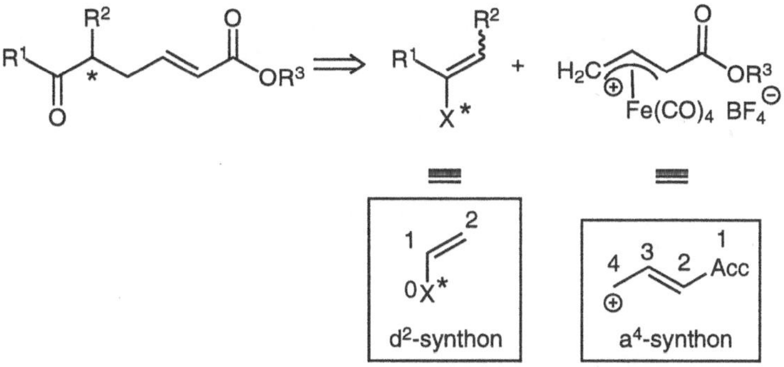

Scheme 1

The realisation of the concept is illustrated in scheme 2. The reaction of the ester **1** with diironnonacarbonyl [$Fe_2(CO)_9$] initially yields neutral methoxycarbonyl-substituted tetra-carbonyl(η^2-olefin)iron(0) complexes, which are directly converted without isolation to the corresponding tetracarbonyl(η^3-allyl)iron(1+) salts **2** by treatment with excess anhydrous HBF$_4$ or anhydrous HPF$_6$ in diethylether, respectively[25,26]. The salts were obtained as pale yellow moderately moisture- and air-stable powders in 68 - 78 % yield. The complexes thus obtained were then used in nucleophilic addition reactions with chiral nucleophiles, such as enamines **3**, metalated imines **4** and metalated α-silyl ketones **5**[27-29] (Scheme 2). The first test series were carried out by the addition of the enantiopure (S)-proline derived en-amines **3** to the complexes **2**. The enantiomeric excesses of the reaction products **7** are in line with the growing steric bulk of the substituents of the pyrrolidine ring in **3** increasing in the order R = H (ee = 34 %), R = Me (ee = 64 %), R = Et (ee = 72 %). In addition, the reaction of various metalated *Schiff* bases was examined. The usual optimisation tests showed that best results were obtained when the imines were transformed to the α-lithio derivatives with t-BuLi and the reaction was carried out in THF at low temperature (-78 °C) (yield: 17 - 71 %, *ee* = 40 - 67 %). Furthermore, the reaction of lithiated acyclic α-silylated ketones furnished after workup diastereomeric mixtures (*de* = 49 - > 96 %) of α-silyl α'-allylated ketone derivatives, which could be desilylated furnishing the 6-oxoenoates **7** (yield: 35 - 95 %). Here, racemisation occured in some cases (*ee* = < 5 - 92 %) during the desilylation step.

In compilation, the resulting enantiomeric excesses obtained with the reaction of the metalated α-silyl ketones **5** are only slightly higher than those obtained by the alternative routes employing the enamines **3** or metalated imines **4**. The enantiomeric excesses of the 6-oxoenoates **7** thus obtained are generally in line with increase in steric bulk of the nucleophile, as for instance in the case of the enamines **3**. The results support the assumption that the planar chiral tetracarbonyl(η^3-allyl)iron(1+) complexes are effectively kinetically resolved by the predominant reaction of the enantiopure carbon nucleophiles **3**, **4** and **5** with one of the enantiomers of the racemic iron complexes **2**, as has been reported in the literature for similar systems[30].

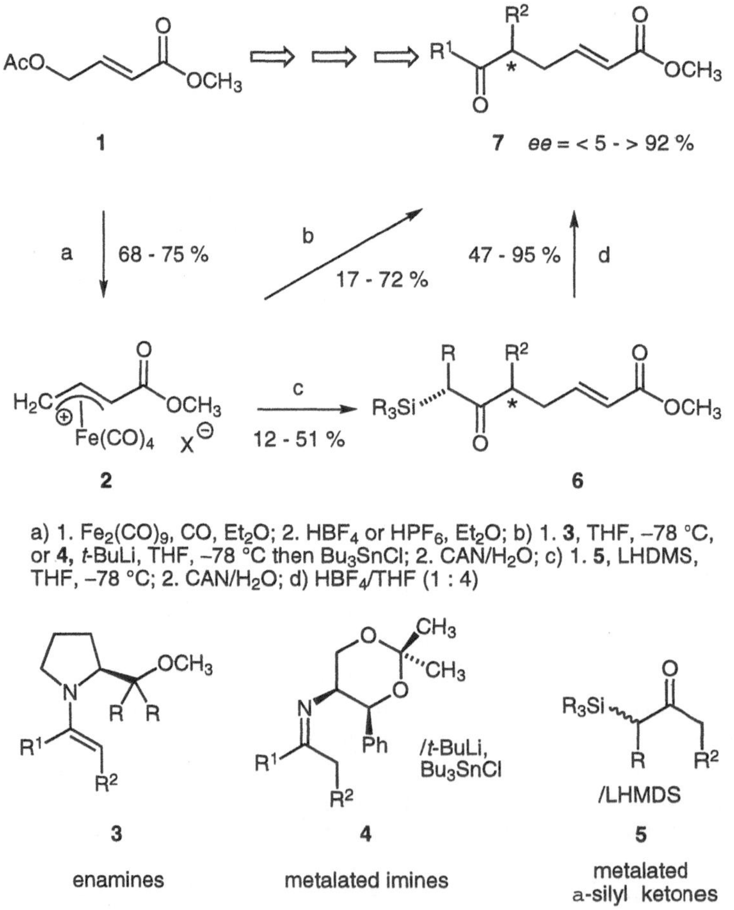

a) 1. $Fe_2(CO)_9$, CO, Et_2O; 2. HBF_4 or HPF_6, Et_2O; b) 1. **3**, THF, −78 °C, or **4**, *t*-BuLi, THF, −78 °C then Bu_3SnCl; 2. CAN/H_2O; c) 1. **5**, LHDMS, THF, −78 °C; 2. CAN/H_2O; d) HBF_4/THF (1 : 4)

Scheme 2

In summary, we have shown that the nucleophilic addition of various enantiopure d^2-carbon nucleophiles (chiral enamines **3** or metalated imines **4** and silylketone derivatives **5**) to racemic planar chiral electrophilic tetracarbonyl(η^3-allyl)iron(1+) complex **2** proceeds with complete γ-regioselectivity and with kinetic resolution of the iron complex. Subsequent oxidative demetalation opens an access to enantiomerically enriched (*ee* = < 5 > 92 %) 1,6-dicarbonyl compounds **7** in moderate overall yields (14 - 53 %, 4 steps or 5 - 21 %, 5 steps) and with retention of the (*E*)-double bond geometry with respect to the starting material **1**[20].

2.2 Auxiliary Controlled Complexation / Achiral Nucleophile Approach

The second concept, namely the synthesis of highly diastereomerically enriched tetracarbonyl(η^3-allyl)iron(1+) complexes *via* auxiliary controlled complexation is illustrated in scheme 3.

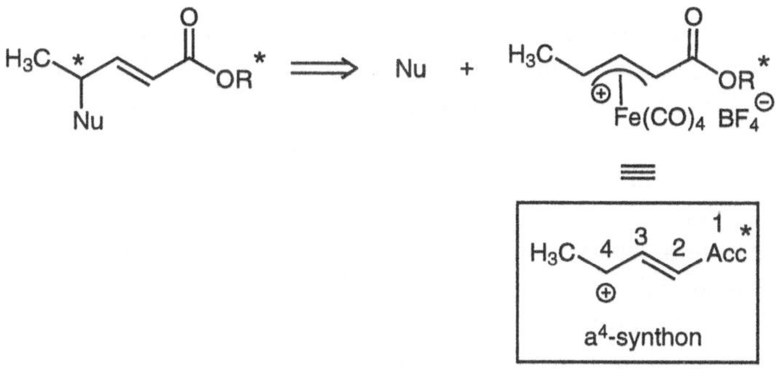

Scheme 3

To reach this goal, first we tried to find an approach to alkoxy carbonyl functionalized tetracarbonyl(η^3-allyl)iron(1+) complexes by complexation of the epimeric acetyl protected 8-phenylmenthylester (4*R/S*)-**8a**[31] as a model system employing the *Corey-Ensley* alcohol as chiral auxiliary.

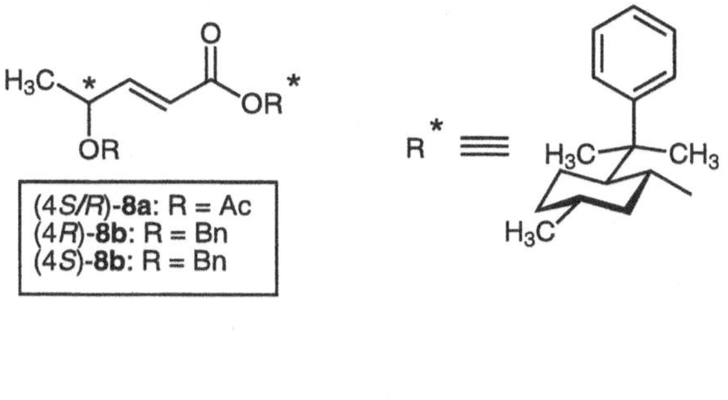

The (*E*)-enoate **8a** was transformed to the tetracarbonyl(η^3-allyl)iron(1+) complexes by initial complexation with $Fe_2(CO)_9$ to the neutral tetracarbonyl(η^2-alkene)iron(0) species, followed by subsequent treatment with anhydrous HPF_6 or HBF_4 in diethyl ether yielding the cationic complex **9** with a moderate diastereomeric excess (80%, *de* ≈ 40 %). Repeated precipitation of **9** from a solution in nitromethane with cold ether afforded the complex in virtually diastereomerically pure form (20 %, *de* ≥ 98 %) as could easily be determined by ¹H-NMR spectroscopy. In addition, ¹H-NMR spectroscopic analysis showed that both the alkoxy carbonyl functionality and the methyl group of **9** are placed in a *syn* relationship with respect to the β-hydrogen atom of the allylic subunit. Unfortunately, the exact position of the $Fe(CO)_4$ fragment could not unambiguously be determined. Variations of the chiral auxiliary based on alternative chiral pool precursors [e.g. R^* = (−)-menthol, (−)-borneol, (-)-*p*-anisidylmenthol][32] proved to be less diastereomerically discriminating during the complexation step (*de* ≈ 0 %). In addition, no synthetical attractive enrichment could be observed by precipitation. In this context, the influence of the stereochemical uniformity of the carbon atom bearing the leaving group on the trajectory of the incoming $Fe(CO)_4$ moiety was examined next.

Starting from the diastereomerically pure epimeric benzyl-protected enoates (4*S*)-**8b** and (4*R*)-**8b**, which were readily obtained in acceptable yields from the corresponding protected enantiomeric lactaldehydes after conventional olefination procedures, the iron complexes (2*R*,3*R*,4*S*)-**9** and (2*S*,3*S*,4*R*)-**9** were obtained in highly diastereomerically enriched form (*de* > 90 %) and in acceptable to good yields [(2*R*,3*R*,4*S*)-**9**: (30 %); (2*S*,3*S*,4*R*)-**9** (53 %)] as single *syn*,*syn*-isomers as determined by ¹H-NMR spectroscopy. Furthermore, ¹H-NMR spectroscopy unambiguously demonstrated that the complex (2*S*,3*S*,4*R*)-**9** is, in contrast to (2*R*,3*R*,4*S*)-**9** not identical with the diastereomer **9** obtained from complexation of the epimeric acetates (4*R/S*)-**8a**. From these results it seems reasonable that the trajectory of complexation by the $Fe(CO)_4$-moiety is mainly determined by the configuration of the carbon atom bearing the leaving group, with less influence of the chiral auxiliary.

The electrophilic tetracarbonyl(η^3-allyl)iron(1+) complexes **9** thus obtained were subjected towards nucleophilic addition reactions by various achiral silyl enol ethers or silyl ketene acetals **10**, which in turn are readily accessible from their corresponding carbonyl precursors according to established procedures[33]. 6-Oxoenoates **11** of excellent diastereomeric purity (*de* > 90 % - ≥ 95 %) were obtained in fair to excellent yield (25 - 90 %) after oxidative removal of the $Fe(CO)_4$-moiety. The results showed that the reaction products of the iron complexes obtained from (*R/S*)-**8a** and (*R*)-**8b** with the silyl enol ether **10** are identical, whereas the reaction of iron complex obtained from (*S*)-**8b** with **10** generates a diastereomeric 6-oxoenoate differing in its absolute configuration at the carbon atom where the nucleophilic attack of the corresponding nucleophile **10** occured (Scheme 4)[21].

284

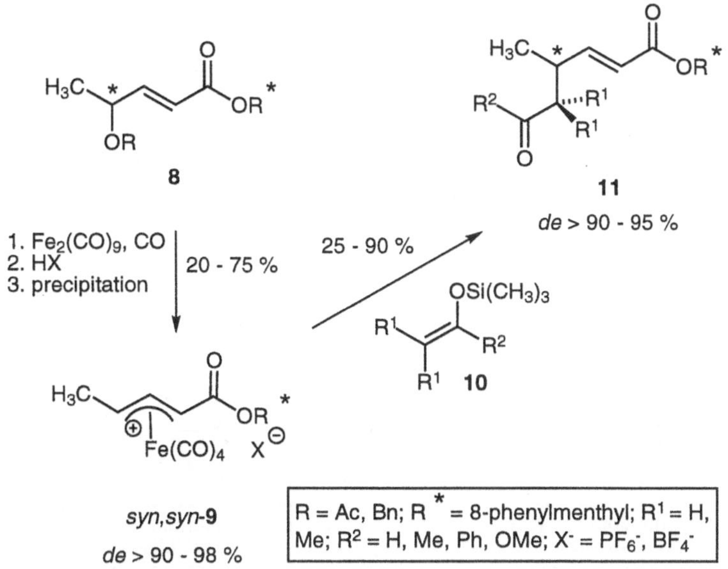

8

1. Fe₂(CO)₉, CO
2. HX
3. precipitation

$|$ 20 - 75 %

25 - 90 %

11

de > 90 - 95 %

OSi(CH₃)₃

R^1 R^2

R^1 **10**

H₃C

O

OR *

Fe(CO)₄   X⊖

syn,syn-**9**

de > 90 - 98 %

| R = Ac, Bn; R * = 8-phenylmenthyl; R^1 = H, |
| Me; R^2 = H, Me, Ph, OMe; X^- = PF_6^-, BF_4^- |

Scheme   4

From the results obtained it is obvious that the facial selectivity of complexation of the double bonds in the starting enoates **8** by the Fe(CO)₄-moiety as well as the absolute configuration of the resulting nucleophilic addition products is predominantly controled by the absolute configuration of the stereogenic center bearing the leaving group in **8**. Based on these results we examined the following third concept which is discribed in chapter 2.3.

2.3 "Chirality Transfer" Approach

Starting from enantiopure protected α-hydroxy aldehydes, easily available from natural sources such as lactic acid or *via* asymmetric synthesis, are transformed into enantiopure, acceptor substituted allylic substrates, which in turn are converted into planar chiral iron-tetracarbonyl cation complexes. Reaction with carbon or heteroatom nucleophiles and removal of the iron moiety should then lead to the final allylic substitution products. The question was wether the whole procedure from central (C-O) over planar (C-Fe) back to central chirality (C-C or C-X) would be possible without any loss of chirality information.

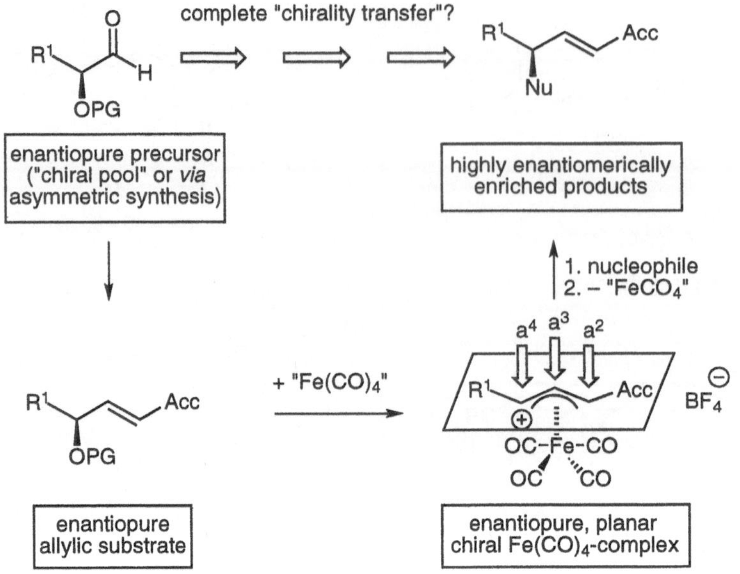

Scheme 5

The transformation of the enantiopure easily accessible enoates (4S)- and (4R)-**12**[34] to the planar chiral tetracarbonyl(η^3-allyl)iron(1+) complexes (1S,2S,3R)-**13** and (1R,2R,3S)-**13**, following the reaction procedure described above yielded the electrophilic complexes as single *syn,syn*-configurated diastereomers (75 %, *de* ≥ 95 %) after repeated precipitation. Unfortunately it can not be excluded that a selective enrichment of the *syn,syn*-configurated isomers during the precipitation procedure has taken place. Furthermore, their enantiomeric purity could only be indirectly determined from the enantiomeric excesses of the addition products **14** obtained from nucleophilic addition reactions of various silyl enol ethers, silyl ketene acetals or amines (*ee* ≥ 95 %) (Scheme 6).

The absolute configuration of the stereogenic centers in the addition products **14** were unambiguously determined to be (*S*) by comparing the sign of optical rotation of derivatives of reaction products with those data given for authentic samples with known absolute configuration in the literature.

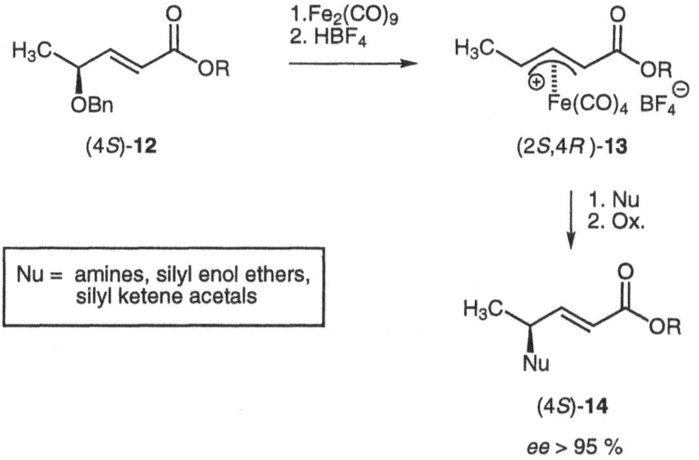

Scheme 6

In general, variation of the substituents at the allylic termini of the iron complexes should extend the method to make it more flexible and valuable and hence more attractive for synthetic chemists. Considering the possibilities, one could vary the acceptor group and the substituent R simply by starting from different enantiopure precursors from cheap sources (Scheme 7).

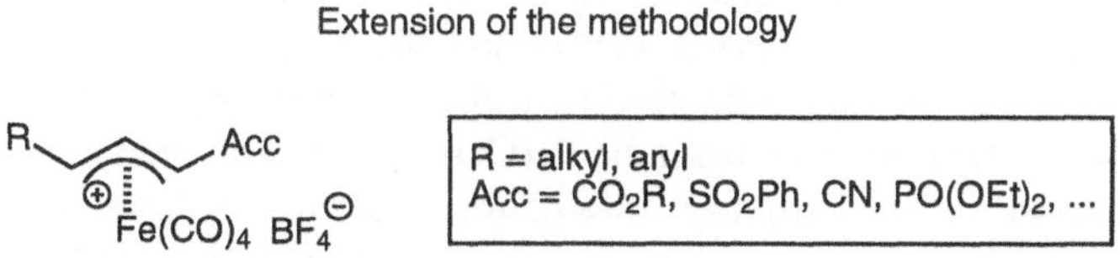

Extension of the methodology

Scheme 7

Based on the results obtained and in order to gain more insight into the overall stereochemical pathway of the formation and reactivity of allylic complexes of the type discussed, further investigations were emphasized on a phenylsulfonyl substituted model

287

system. Due to their enhanced stability and easy accessibilty reported so far in the literature[35], we were inspired to try to isolate the initially formed neutral tetracarbonyl(η^2-alkene)iron(0) complexes and planned to examine their stereochemistry with respect to the outcoming stereochemistry of the resulting tetracarbonyl(η^3-allyl)iron(1+) complexes. Starting from (S)-ethyllactate the required enantiomerically pure sulfone (3S)-**15** is easily accessible[36]. Its reaction with nonacarbonyldiiron [Fe$_2$CO$_9$] initially yielded a mixture of the diastereomeric tetracarbonyl(η^2-alkene)iron(0) complexes (1S,2R,3S)-**16** and (1R,2S,3S)-**16** in a ratio of about 15 : 85 which was easily determined by ^{1}H-NMR spectroscopy (Scheme 8).

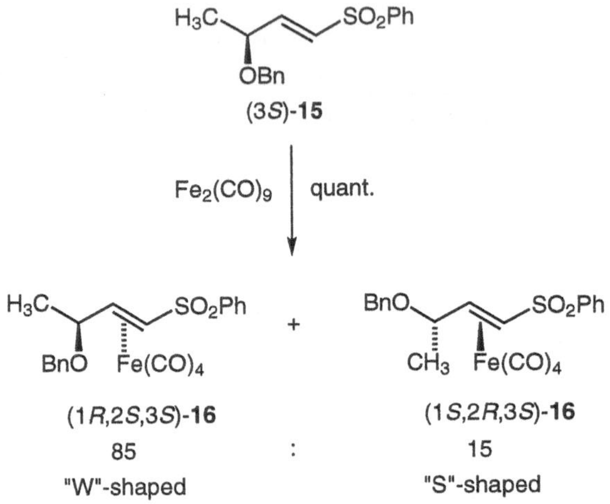

Scheme 8

The diastereomers (1S,2R,3S)-**16** and (1R,2S,3S)-**16** could be separated by fractional crystallisation and (1R,2S,3S)-**16** could be obtained in virtually diastereo- and enantiopure form (de > 99 %, ee > 99 %). The crystal structures of both enantiopure diastereomers could be solved[37-39]. Some structural aspects concerning the stereochemistry of the complexed ligand in (1R,2S,3S)-**16**, the major diastereomer, merit closer inspection (figure 1).

The SO$_2$Ph unit is twisted away from the Fe(CO)$_4$ moiety and the terminal methyl group is, on the other hand, inclined slightly towards the Fe-atom resulting an "W"-shaped geometry of the ligand, which in turn determins a syn,syn-substitution pattern in the resulting cationic complex **17** (vide supra). As a consequence of the single configuration at the 3-position, one side of the double bond unit is selectively blocked by the Fe(CO)$_4$ moiety, which lies trans to the OBn leaving group. These stereochemical features are very important for the transformation of (1R,2S,3S)-**16** into the stereochemically well defined syn,syn-substituted cationic complex (1R,2S,3R)-**17**, which was obtained upon addition of excess anhydrous HBF$_4$ to an etheral solution of the neutral (η^2-olefin)iron complex (1R,2S,3S)-**16** in 96 %

and in virtually isomeric pure form (*de, ee* > 98 %) as a pale yellow moderate moisture- and air stable powder.

Since all starting materials are accessible in a multigram scale, the even complex (1*R*,2*S*,3*R*)-**17** can be synthesised in scales up to 10 g and be stored for extended periods in a refrigerator. The complex (1*R*,2*S*,3*R*)-**17** thus obtained, was now used in reactions with a wide range of nucleophiles, including electronrich arenes and heteroarenes[16] and organozinc carbon nucleophiles bearing further functional groups[22] (Scheme 9).

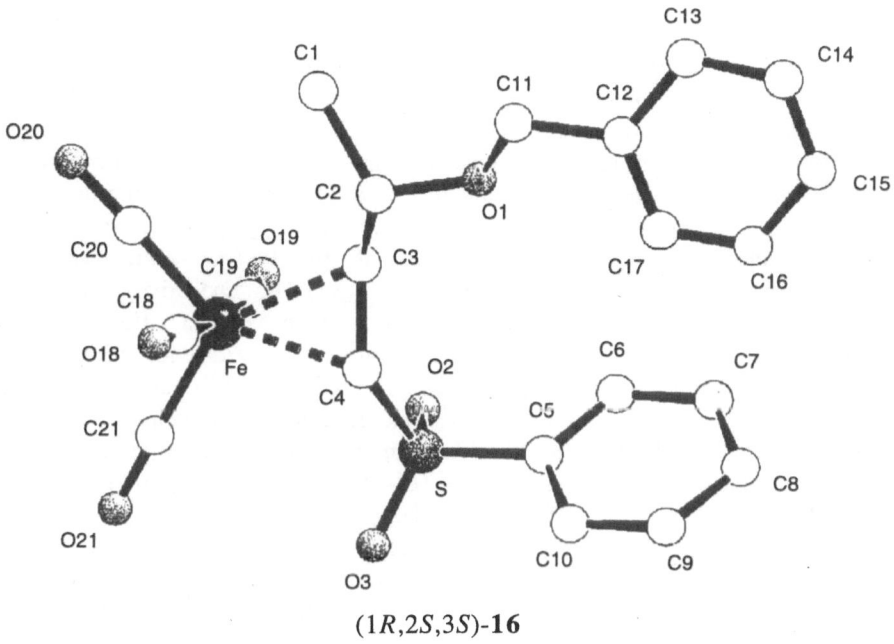

(1*R*,2*S*,3*S*)-**16**

Figure 1

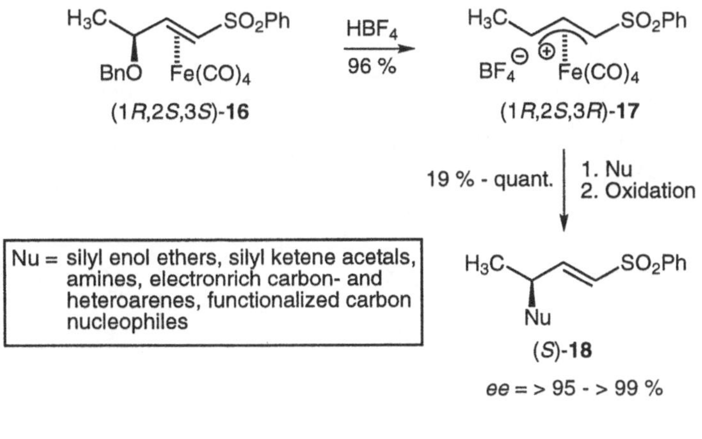

H₃C SO₂Ph
BnO Fe(CO)₄
(1*R*,2*S*,3*S*)-**16**

$\xrightarrow[\text{96 \%}]{\text{HBF}_4}$

H₃C SO₂Ph
BF₄⊖ ⊕ Fe(CO)₄
(1*R*,2*S*,3*R*)-**17**

19 % - quant. | 1. Nu 2. Oxidation

Nu = silyl enol ethers, silyl ketene acetals, amines, electronrich carbon- and heteroarenes, functionalized carbon nucleophiles

H₃C SO₂Ph
Nu
(*S*)-**18**

ee = > 95 - > 99 %

Scheme 9

289

We now had collected enough data to propose a reaction mechanism for the overall reaction sequence starting from the olefins (S)-**12** or (S)-**15**, respectively. Due to the shielding of the OBn-leaving group, complexation of the olefins (S)-**12** and (S)-**15** by an incoming Fe(CO)$_4$-moiety seems to be directed preferentially to the opposite face of the double bond with respect to the sterically demanding OBn-leaving group. Cleavage of the C-O-bond of the OPG leaving group proceeds with formation of a new Fe-C-bond. According to the relative *anti*-arrangement of the Fe(CO)$_4$-moiety and the OPG leaving group, the absolute configuration of the carbon atom C-3 which bore the OPG unit is inverted during this step. Based on the assumption of a uniform reaction mechanism for the complexes with all different types of soft nucleophiles employed and due to the overriding *anti*-directing effect of the Fe(CO)$_4$ fragment, the nucleophilic attack occurs from the opposite side with respect to the Fe(CO)$_4$ moiety as it has been described for similar nucleophilic addition reactions to numerous other transition metal complexed carbocations[40]. During the addition step a new carbon-carbon or carbon-hetero atom bond is formed with inversion of configuration at C-3 by cleavage of the C-Fe bond. The initially resulting neutral C-3 substituted tetracarbonyl(η^2-alkene)iron(0) complexes have to possess the same relative and therefore absolute stereochemistry with respect to the Fe(CO)$_4$-moiety and the new substituents as decribed for the complex $(1R,2S,3S)$-**16**. Oxidative cleavage of the tetracarbonyliron fragment completes the reaction circle without affecting the newly generated stereogenic centre. Thus, the reaction sequence proceeds with virtually complete "chirality transfer" from C-O over C-Fe to C-C or C-X and with overall retention (double inversion) of the stereochemistry of the stereogenic centre with respect to the starting materials (S)-**12** and (S)-**15**. In addition, the reaction proceeds with complete γ-regioselectivity and with conservation of the (E)-double bond geometry leading to highly functionalized molecules of well defined stereochemistry.

Due to their origin from enantiopure building blocks bearing methyl substituents (e.g. isoprenoids, alanine, lactic acid derivatives) many naturally occuring compounds possess methyl branched carbon atom skeletons containing structural fragments of the general type **A** which in turn can be retrosynthetically disconnected to the synthons **B**, **C** and **D** or **B**, **E** and **F** respectively (Scheme 10).

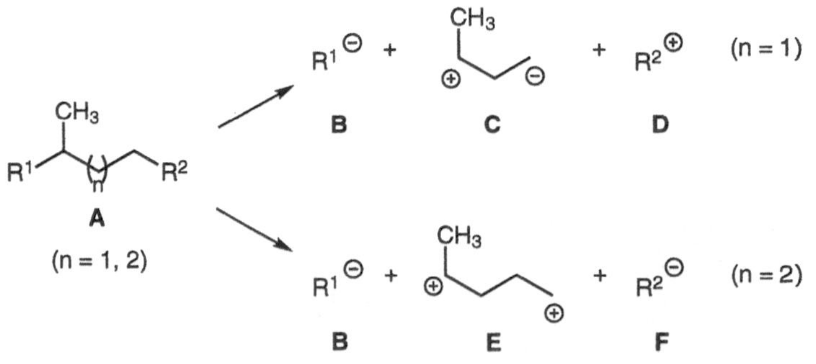

Scheme 10

290

The synthetic equivalence of the methyl ester substituted iron complex with a chiral a^1/a^4-synthon as well as the synthetic equivalence of the phenylsulfonyl substituted iron complex with a d^1/a^3-synthon can be drawn back from their potential bifunctionality opening an approach to a flexible sequential functionalisation which in turn is of great synthetic value (Scheme 11).

The following chapters demonstrate some of our efforts made so far in natural product synthesis taking advantage of our methodology.

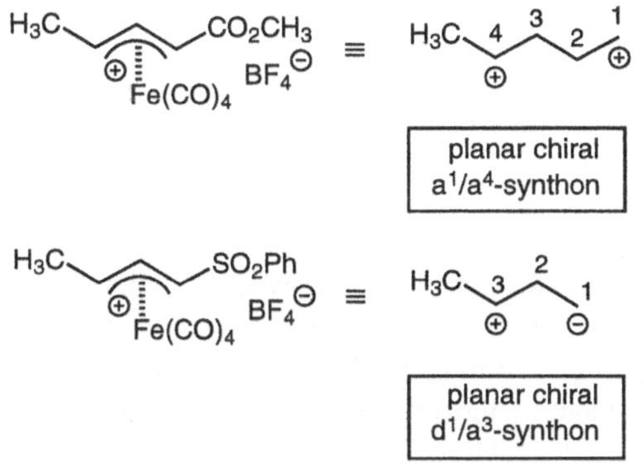

planar chiral
a^1/a^4-synthon

planar chiral
d^1/a^3-synthon

Scheme 11

3 Application in the Total Synthesis of Simple Natural Products

3.1 Synthesis of the C_{14}-Amine of the New Zealand Ascidian *Pseudodistoma Novaezelandiae*

The C_{14}-amine (2S,3E,5Z)-2-Amino-3,5,13-tetradecatrien [(S)-**19**] was first isolated and characterised in 1991 in connection with one of the large numbers of research programs directed towards the discovery of marine-derived drugs[41]. The compound was obtained as the major constituent of an extract of the New Zealand ascidian *Pseudodistoma novaezelandiae* and showed in biological assays cytotoxic activity against diverse cell lines (P 388 murine leukaemia and host mammalian cells). Furthermore, it was also active against the bacteria *Escherichia coli* and *Bacillus subtilis* and showed antifugal growth inhibiting activity against *Candida albicans*, *Cladosporium resinae* and *Trichophyton mentagrophytes*. As illustrated in scheme 12 the retrosynthetic analysis of the title compound (S)-**19** leads to two subunits, a γ-amino enal (S)-**20** and *Wittig* reagent **21**.

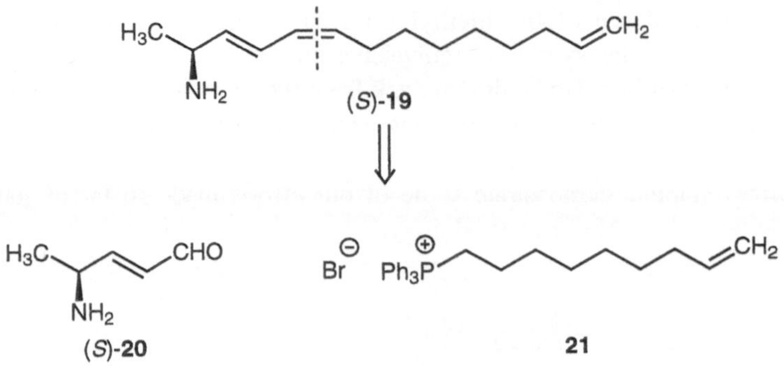

H₃C. ‖ CH₂

NH₂   (*S*)-19

H₃C. CHO   Br⁻ Ph₃P⁺ ⌁⌁⌁ CH₂

NH₂
(*S*)-20   **21**

Scheme 12

Starting from the enantiopure complex *syn,syn*-(1*S*,2*S*,3*R*)-**13** the γ-amino enal (*S*)-**24** was prepared in three steps in an overall yield of 69 % (Scheme 13). Key step in the reaction sequence is the nucleophilic addition of *t*-butyl carbamate to the cationic iron complex followed by oxidative cleavage of the tetracarbonyliron fragment yielding 88 % of the pure (*E*)-enoate (*S*)-**22** (*ee* ≥ 95 %). The reaction proceeded with complete γ-regioselectivity, conservation of the (*E*) double bond geometry and stereoselectively *trans* with respect to the tetracarbonyliron moiety. DIBAH-reduction and subsequent *Swern*-oxidation of the resulting allylic alcohol (*S*)-**23** afforded finally the protected γ-amino enal (*S*)-**24**.

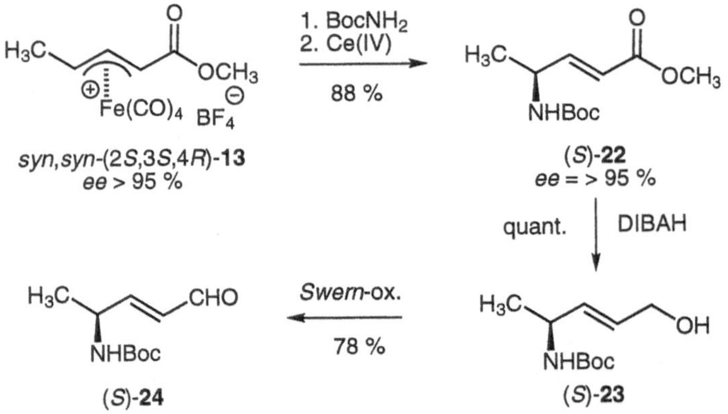

H₃C. OCH₃
Fe(CO)₄ BF₄⁻

syn,syn-(2*S*,3*S*,4*R*)-**13**
ee > 95 %

1. BocNH₂
2. Ce(IV)
→
88 %

H₃C. OCH₃

NHBoc
(*S*)-**22**
ee = > 95 %

quant. | DIBAH

H₃C. CHO

NHBoc
(*S*)-**24**

Swern-ox.
←
78 %

H₃C. OH

NHBoc
(*S*)-**23**

Scheme 13

As illustrated in scheme 14, the synthesis of the *Wittig* salt **21** was readily accomplished in 4 steps and in an overall yield of 45 % starting from the commercially available octan-1,8-diol **25** according to standard literature procedures.

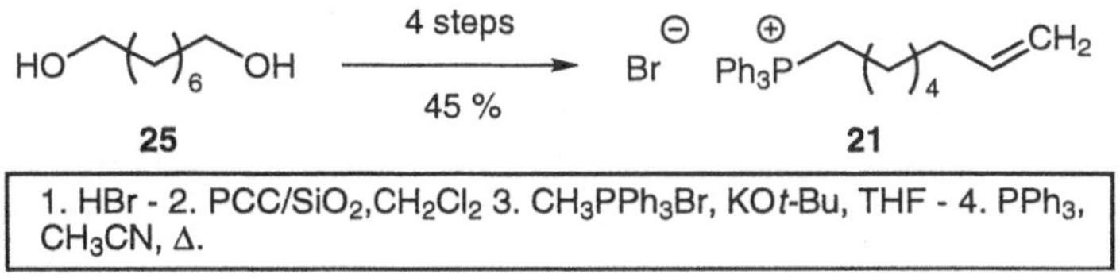

Scheme 14

1. HBr - 2. PCC/SiO$_2$,CH$_2$Cl$_2$ 3. CH$_3$PPh$_3$Br, KOt-Bu, THF - 4. PPh$_3$, CH$_3$CN, Δ.

For the final construction of the carbon backbone of the target molecule the *Wittig* reagent **21** was transformed into the corresponding ylide by deprotonation with potassium *t*-butoxide. Its subsequent reaction with the γ-amino enal (*S*)-**24** furnished the protected amino triene (*S*)-**26** in 85 % yield with high (*Z*)-selectivity with respect to the geometry of the new generated carbon double bond [(*Z*)/(*E*) = 97:3] (Scheme 15). Finally, after virtually quantitative cleavage of the carbamat protecting group from (*S*)-**26** the free amine (*S*)-**19** was obtained as a pale yellow viscous oil.

The enantiomeric purity (*ee* > 95 %) was determined by ^{13}C-NMR spectroscopy and GLC analysis of the corresponding *Mosher* amide of (*S*)-**19** and by comparison with the *Mosher* derivative of the racemic material obtained by an analogous route starting from the racemic complex *syn,syn*-(1*R/S*,2*R/S*,3*R/S*)-**13**. The absolute configuration of the stereogenic centre could be assigned as (*S*) based on our earlier investigations (*vide supra*). In conclusion, we have carried out, the first total synthesis of (2*S*,3*E*,5*Z*)-2-Amino-3,5,13-tetradecatrien [(*S*)-**19**] which in turn also confirmed its proposed structure. The compound was obtained in excellent overall yield starting from the iron complex *syn,syn*-(1*S*,2*S*,3*R*)-**13** (5 steps, 57 %) and in high enantiomeric purity (*ee* > 95 %)[15].

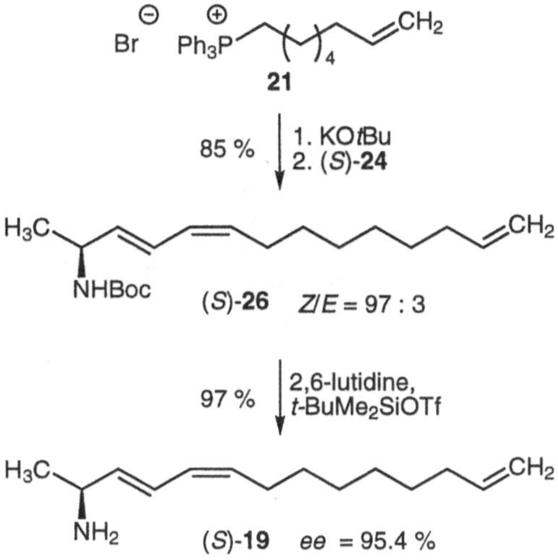

Scheme 15

3.2 Synthesis of (–)-(S)-Myoporone

Myoporone, a hepatotoxic furanosesquiterpene diketone, was first reported in 1957 as the main furanoid constituent of the essential oil of the Japanese *Myoporum bontioides* A. Gray. Later its presence was also detected in various other *Myoporum* species, *Eremophila* species, *Eumorphia serica* and *Eumorphia prostata*. Interestingly, different species of *Myoporum* and *Eremophila* yielded myoporone samples of widely differing optical purity. The *Jackson* variety of the Australian shrub *Myoporum deserti* contains on the one extreme the (–)-(S)-myoporone as a secondary metabolite and *Myoporum montanum* on the other the (+)-(R)-enantiomer[42,43]. (–)-(S)-Myoporone was also isolated in enantiopure form as a stress metabolite of sweet potatoes (*Ipomoea batatas*) during microbial infection ("black rotted disease") or under the challenge from mercuric chlorid[44]. However, only a few total syntheses of myoporone have been reported so far in the literature starting either from enantiomerically enriched natural precursors or by asymmetric synthesis[45].

As depicted in scheme 16, retrosynthetic analysis of the target molecule lead to three subunits, the silyl enol ether **28**, the cationic phenylsulfonyl substituted (η^3-allyl)tetracarbonyliron complex **17** and the aldehyde **29**.

As illustrated in scheme 17, the synthesis of the 6-oxosulfone (R)-**30** was accomplished by nucleophilic addition of the silyl enol ether **28**, easily obtained in 92 % yield as a single regioisomer from commercially available 4-methylpentane-2-one, to the enantiopure phenylsulfonyl substituted (η^3-allyl)tetracarbonyliron(1+) complex *syn,syn*-(1R,2S,3R)-**17** in dichloromethane after subsequent oxidative demetalation with aqueous ceric ammonium nitrate solution.

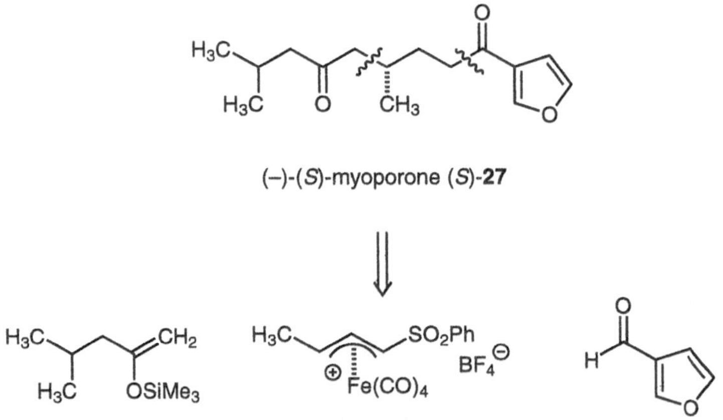

(–)-(S)-myoporone (S)-**27**

28   *syn,syn*-(1R,2S,3R)-**17**   **29**

Scheme 16

294

This key step proceeded as expected with complete regioselectivity and stereoselectively *anti* with respect to the metal carbonyl moiety. Thus, the reaction sequence resulted in an overall retention (double inversion) of configuration with respect to the starting material and the stereogenic centre at C-3 could be assigned the (*R*)-configuration as can be concluded from the sign of optical rotation of the final product (*S*)-**27** (*vide infra*). The enantiomeric purity of (*R*)-**30** (*ee* > 99 %) was determined by HPLC employing a chiral stationary phase (Daicel-OD) and by comparison with the racemic material, which was synthesized by an analogous route using the racemic iron complex (1*R/S*,2*R/S*,3*R/S*)-**17**. Palladium catalyzed hydrogenation of the double bond of (*R*)-**30** furnished the saturated sulfone (*R*)-**31** in quantitative yield. Although difficult, the keto group was quantitatively transformed to the dioxolane derivative according to *Noyori*'s acetalisation method. The β-keto sulfone (*R/S,R*)-**33** was synthesised as a mixture of diastereomers in a two step procedure with an overall yield of 89 %. Aldol-type reaction of furan-3-aldehyde **29** with the α-lithio derivative of the protected sulfone of (*R*)-**32** at −78 °C furnished a mixture of diastereomeric alcohols with 96 % conversion with respect to the starting material (*R*)-**32** and completed the construction of the carbon skeleton of the target molecule. Subsequent *Swern* oxidation of the diastereomeric secondary alcohols yielded the β-keto sulfone (*R/S,R*)-**33** in 93 % yield and with a moderate diastereomeric excess of only 23 % of one of the α-epimeric sulfones. Simultaneous removal of both the acetal protecting group and the sulfone group of (*R/S,R*)-**33** was achieved by refluxing (*R/S,R*)-**33** in a mixture of ethyl acetate, acetic acid and ethanol (3:1:1) in the presence of zinc dust and a few drops of water to afford the target compound as a pale yellow liquid in 93 % yield, solidifying to a colourless solid in the refrigerator. The enantiomeric excess of (*S*)-**19** (*ee* > 99 %) was determined by HPLC on a chiral stationary phase (Daicel OD) and by comparison with a racemic sample. As clearly shown by the HPLC-plots, the stereogenic centre bearing the methyl branch does not suffer from any significant racemisation during the reaction sequence under the described conditions.

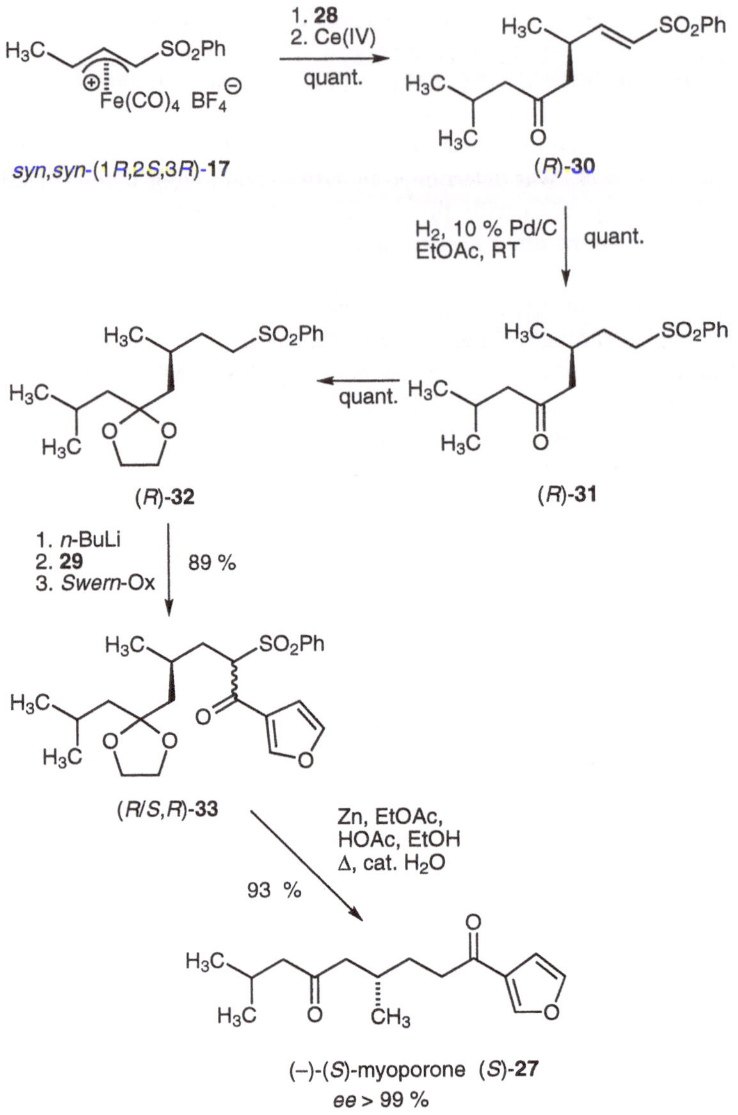

Scheme 17

In summary, the naturally occuring hepatotoxic furanosesquiterpene (–)-(*S*)-myoporone has been synthesised straightforward and efficiently in excellent enantiomeric purity (*ee* > 99 %) and in a very high overall yield of 82 % (six steps) from readily available starting materials[17]. This reaction sequence examplifies very nicely the great synthetic value of the phenylsulfonyl-substituted (π-allyl) complex *syn,syn*-(1*R*,2*S*,3*R*)-**17** due to its potential bifunctionality.

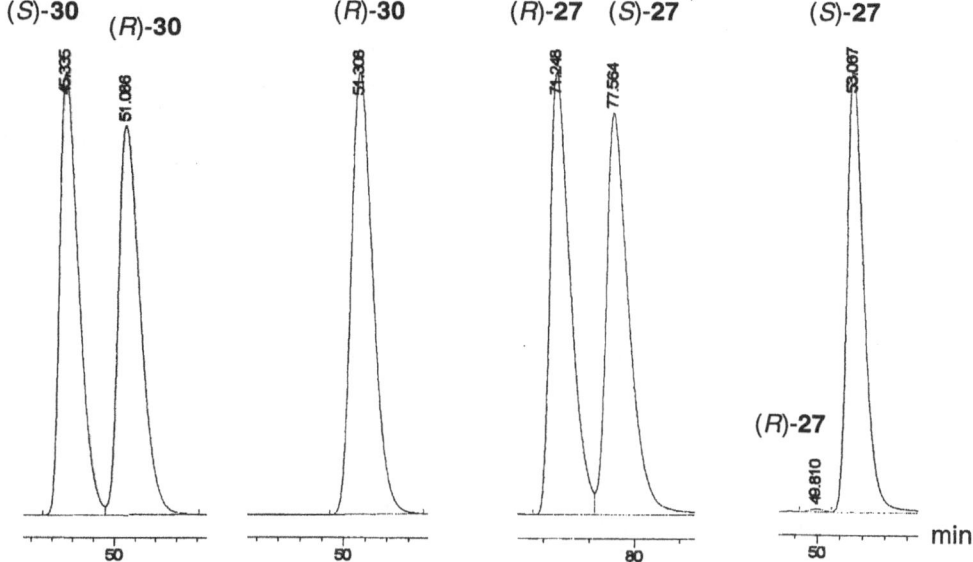

Figure 2

3.3 Synthesis of the Sex Pheromones of the Spotted and Banded Cucumber Beetle

The Spotted Cucumber Beetle (SCB) *Diabrotica undecimpunctata howardi* Barber as well as the Banded Cucumber Beetle (BCB) *Diabrotica balteata* LeConte are polyphagous insects (*Coleoptera: Chrysomelidae*) belonging to the *fucata* species group of the genus *Diabrotica*. In general, both species are confined to North America from Canada to Mexico but the Banded Cucumber Beetle is found more often in the warmer southern regions down to Costa Rica and Cuba[46,47]. These beetles are known to cause severe damage to several crop plants due to their action as vectors of a number of plant diseases. In contrast, the larvae of these beetles [e.g. Southern Corn Rootworm (SCR)] are most damaging pests to e.g. seedling curcubits, groundnut, corn and sweet potatoe[48]. (–)-(*R*)-10-Methyltridecan-2-one [(*R*)-**34**] was identified as the sex pheromone secreted by virgin females of *Diabrotica undecimpunctata howardi* Barber[48], while the female produced sex pheromone of *Diabrotica balteata* LeConte has been isolated and identified as (–)-(*R,R*)-6,12-Dimethylpentadecan-2-one [(6*R*,12*R*)-**35**][49]. The chemical structures of sex pheromones of this species included in the *fucata* group are characterised by the methyl ketone functionality and a methyl branch on the fourth carbon from the hydrocarbon end of the chain (Scheme 18)[47,49,50]. Field trapping experiments directed towards sex pheromones as biotechnical weapons with regard to integrated future pest management programs generally demonstrated that the (*R*)- or the (*R,R*)-stereoisomers, respectively, show the highest attractive bioactivity towards males of the SCB or the BCB and closely related species[48,49]. However, only few total syntheses of all possible stereoisomers of **34** and **35** in high enantiomeric purity have been reported so far in the literature, mostly starting from enantiopure methyl-branched building blocks from nature e.g. citronellol[50-52].

With our method we were able to realize efficient and convergent syntheses of highly enantiomerically enriched (6*R*)-**34** as well as (6*R*,12*R*)-**35** taking advantage of the bifunctionality of the phenylsulfonyl substituted iron complex (1*R*,2*S*,3*R*)-**17**. In the following chapters the straightforward syntheses of both sex pheromones in their naturally occuring absolute configuration are described.

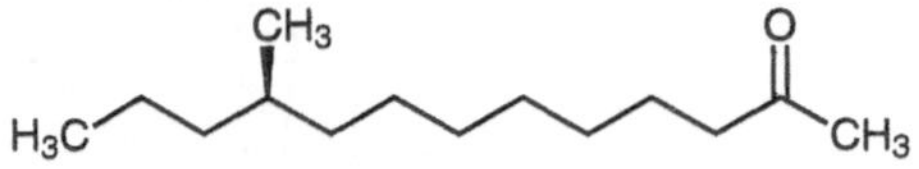

(−)-(*R*)-10-methyltridecan-2-one (*R*)-**34**

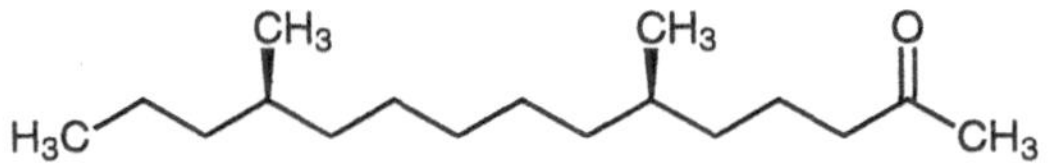

(−)-(*R*,*R*)-6,12-dimethylpentadecan-2-one (*R*,*R*)-**35**

Scheme 18

3.3.1 Synthesis of (−)-(*R*)-10-Methyltridecan-2-one

The retrosynthetic analysis of the spotted cucumber beatle sex pheromone is illustrated in scheme 19. The target molecule (*R*)-**34** can be devided into three subunits, the allyl silane **36**, the planar chiral, cationic phenylsulfonyl substituted (η^3-allyl)tetracarbonyl iron complex *syn,syn*-(1*R*,2*S*,3*R*)-**17** and the bromide **37**.

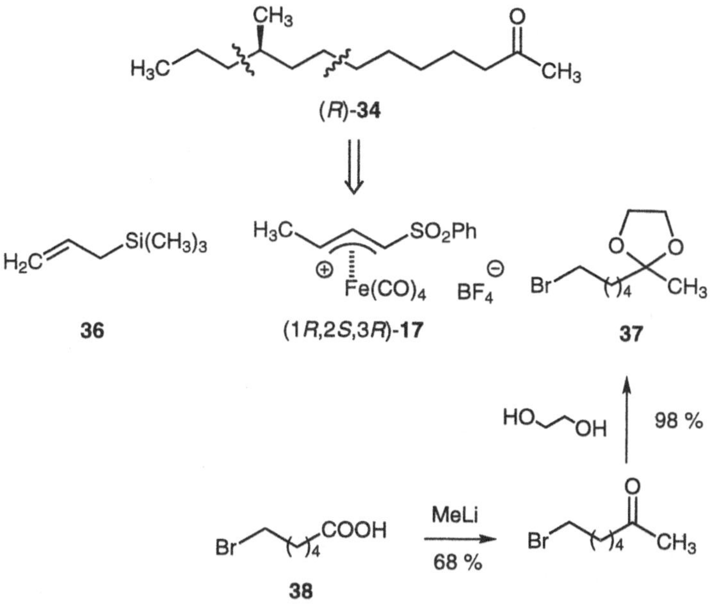

Scheme 19

The synthesis of the protected electrophile **37** was easily accomplished in two steps starting from commercially available 6-bromo hexanoic acid **38** in 67 % overall yield. Again, key step of the synthesis is a nucleophilic addition, in this case of allyltrimethylsilane, to the electrophilic iron complex *syn,syn*-(1R,2S,3R)-**17**, followed by oxidative cleavage of the tetracarbonyl iron fragment furnishing the γ-propenyl substituted alkenyl sulfone (R)-**39** in quantitative yield. The enantiomeric excess of (R)-**39** (*ee* = 99.4 %) was determined by GC employing a chiral stationary phase (Lipodex E) and by comparison with the appropriate racemic material. Catalytic hydrogenation of (R)-**39** yielded (R)-**40** virtually without any racemisation as it has been proven again by GC analysis on a chiral stationary phase (Lipodex E). α-Lithiation of the saturated sulfone (R)-**40** with *n*-butyllithium and subsequent alkylation of the metalated species with bromide **37** completed the final construction of the carbon skeleton to furnish (R/S,R)-**41** in 95 % yield as a mixture of diastereomers (*de* < 2 %). The sulfonyl group was reductively removed by treatment with an excess of sodium amalgam. Cleavage of the acetal protecting group was accomplished by acid hydrolysis to yield finally the pure pheromone (R)-**34** in 79 % yield (Scheme 20). As it has been shown for very similar systems no synthetic operation should had effected the stereogenic centre at C-10 during the reaction course and the enantiomeric purity of the pheromone (R)-**34** should be identical with that determined for (R)-**39** or (R)-**40** (*ee* ≥ 99 %).

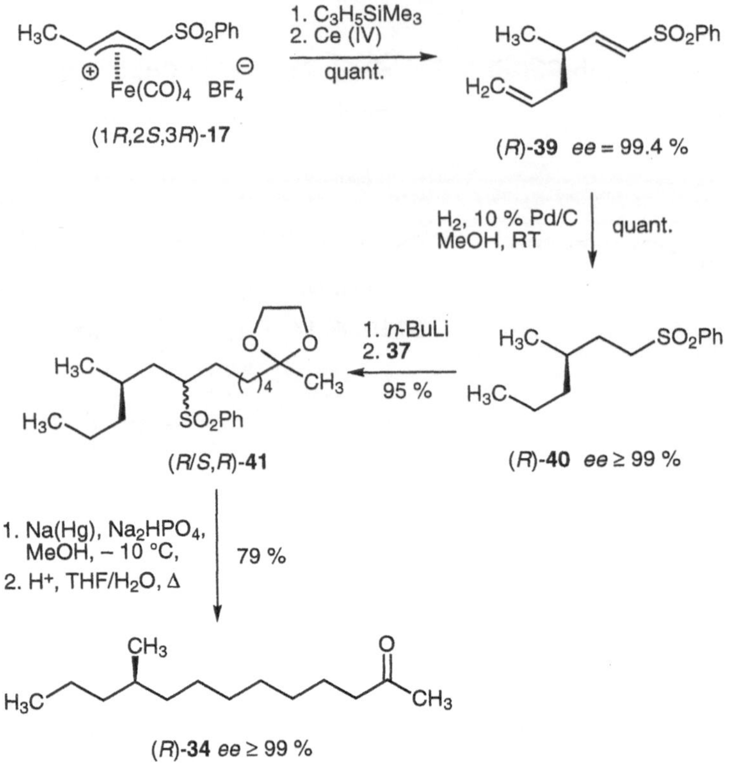

Scheme 20

In conclusion, an efficient and highly convergent synthesis of the sex pheromone of *Diabrotica undecimpunctata howardi* Barber, (–)-(*R*)-10-methyltridecan-2-one, in excellent overall yield (five steps, 75 %) and in high enantiomeric purity (*ee* ≥ 99 %) was realized[19].

3.3.2 Synthesis of (–)-(*R*,*R*)-6,12-Dimethylpentadecan-2-one

Retrosynthetic analysis of the target molecule by C-9/C-10 disconnection leads to the sulfone (*R*)-**40** and the bromide (*S*)-**42**, which can be subsequently further disconnected to five readily accessible allylic subunits as illustrated in scheme 21.

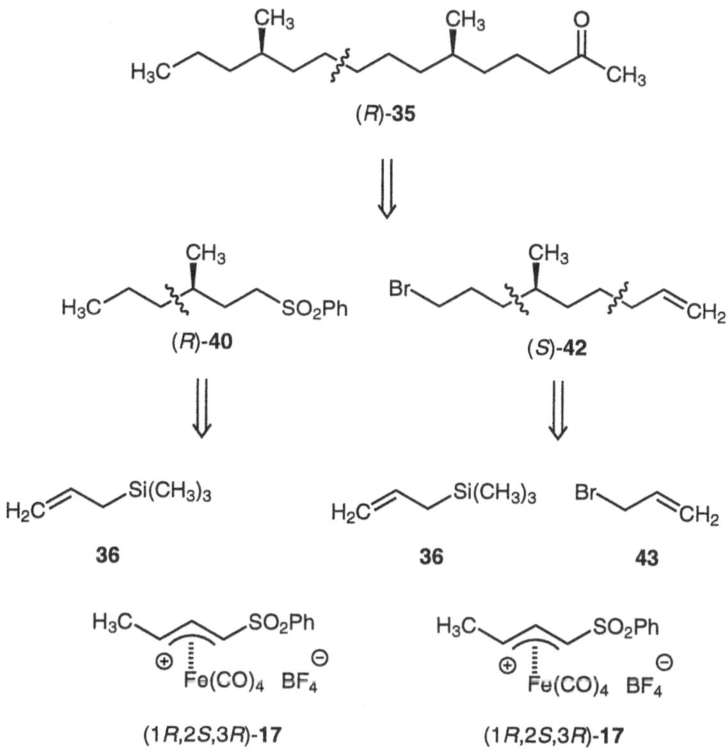

Scheme 21

The synthesis of the sulfone (R)-**40** has already been described in the previous chapter (*vide supra*), while the bromide (S)-**42** can be synthezised from allyltrimethylsilane **36**, the iron complex *syn,syn*-(1R,2S,3R)-**17** and allyl bromide. The crucial step for the construction of the carbon chain of the title compound was the alkylation of the sulfone (R)-**40** with the bromide (S)-**42**. The building block (R)-**40** was prepared in three steps and practically quantitative yield starting from iron complex *syn,syn*-(1R,2S,3R)-**17** as described earlier (*ee* ≥ 99 %). For the preparation of the subunit (S)-**42** the regioselective hydroboration of the terminal double bond of (R)-**39** was achieved by treatment with 9-borabicyclo[3.3.1]nonane (9-BBN) followed by oxidative workup with H_2O_2 furnishing the alcohol (R)-**43** in 90 % yield. Palladium catalyzed hydrogenation of the remaining double bond and protection of the free hydroxy functionality as the corresponding *t*-butyldimethylsilyl-ether furnished the sulfone (R)-**44** in 94 % overall yield. The α-lithio derivative of (R)-**44** was alkylated with allyl bromide affording the alkylated sulfone. The crude alkylation product was directly deprotected with aqueous hydrofluoric acid/acetonitrile furnishing the hydroxy sulfone (R/S,R)-**45** in 99 % yield over three steps.

The reductive removal of the sulfonyl group of (R/S,R)-**45** was achieved by treatment with an excess of sodium amalgam. The resulting alcohol (S)-**46** was converted in two steps to the corresponding bromide (S)-**42** according to literature procedures (Scheme 22).

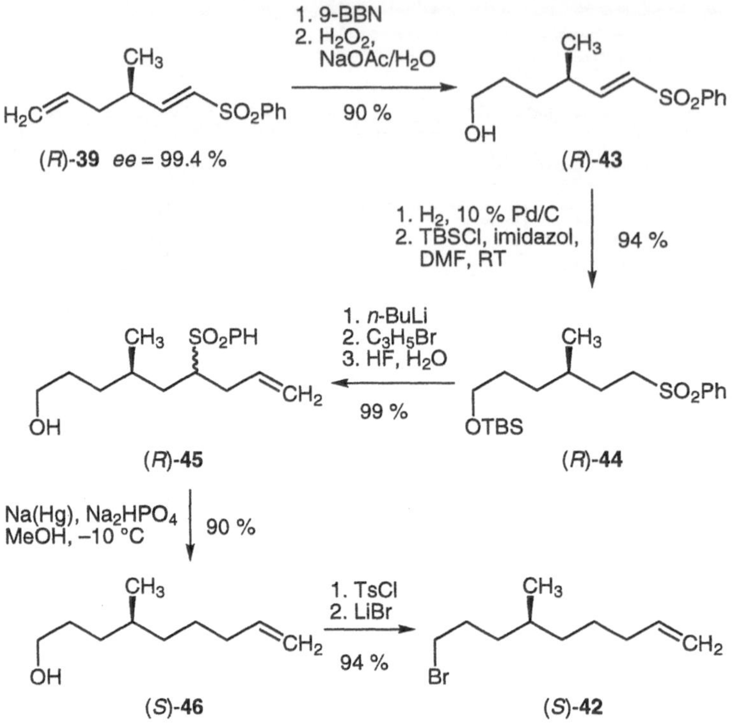

Scheme 22

The final construction of the carbon skeleton was achieved by alkylative connection of the two enantiopure building blocks (R)-**40** and (S)-**42** under the standard conditions described before (Scheme 20). The reductive removal of the phenylsulfonyl group of the resulting alkylation product (6S,10R/S,12R)-**47** and the Wacker-type conversion of the terminal double bond into the methyl ketone functionality furnished the pure pheromone (6R,12R)-**35** in 68 % yield (Scheme 23) and in excellent diastereo- and enantiomerical purity (de ≥ 98 %, ee ≥ 99 %). In conclusion, starting from readily available, inexpensive starting materials, the first stereocontrolled, efficient and highly convergent synthesis of the sex phereomone (–)-(R,R)-6,10-dimethylpentadecan-2-one was accomplished in good overall yield (13 steps, 39 %) and in high diastereo- and enantiomeric purity (de ≥ 98 %, ee ≥ 99 %)[18]. There remains to be mentioned that the sign of the optical rotations of the final products (*S*)-**27** myoporone, (*R*)-**34** and (6*R*,12*R*)-**35** were identical with those reported for the authentical natural samples in the literature, which demonstrates that we have synthesised all compounds in their naturally occuring configurations.

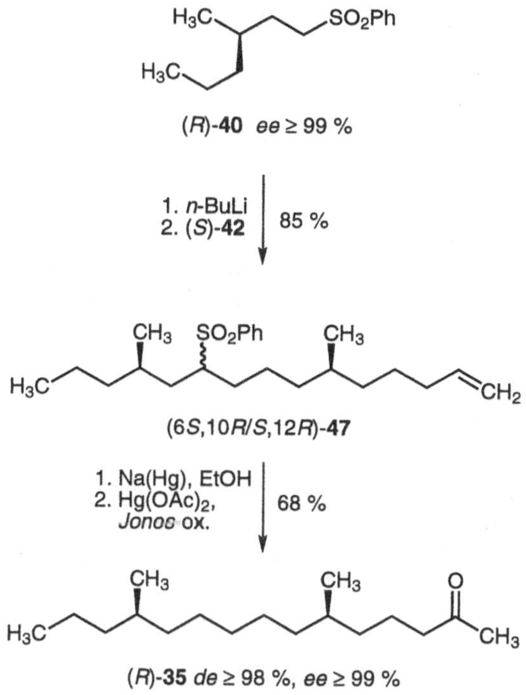
H3C
SO2Ph
H3C
(R)-40 ee ≥ 99 %
1. n-BuLi
2. (S)-42
85 %
CH3 SO2Ph CH3
H3C
CH2
(6S,10R/S,12R)-47
1. Na(Hg), EtOH
2. Hg(OAc)2,
Jones ox.
68 %
CH3 CH3 O
H3C
CH3
(R)-35 de ≥ 98 %, ee ≥ 99 %

Scheme 23

Noteworthy to say that these results also verify the overall stereochemical outcome of our "chirality transfer" process according to the proposed mechanism with double inversion (net retention) at the stereogenic centre with respect to the starting materials. The synthesis of the corresponding optical antipodes or diastereomers may be easily performed in the same manner by employing the other enantiomer of the iron complex **17**.

4 Conclusion

We presented in this paper a "chirality transfer process" in iron mediated allylic substitution reactions. Starting from enantiopure precursors from the chiral pool, the formation of the enantiopure tetracarbonyl(η^3-allyl)iron(1+) complexes proceeds with virtually complete chirality transfer from the allylic starting material with central chirality to the corresponding tetracarbonyl(η^3-allyl)iron(1+) complexes owing planar chirality. During the subsequent nucleophilic addition to the electrophilic iron complexes a new stereogenic center is formed with inversion of the configuration at the involved carbon atom. Thus, the whole reaction sequence proceeds with overall retention (double inversion) of the stereochemistry of the stereogenic centre with respect to the starting materials (S)-**12** and (S)-**15**. In addition, the reaction proceeds with complete γ-regioselectivity and with conservation of the (E)-double bond geometry. Although this "chirality transfer process" in allylic substitution reactions proceeds *via* stoichiometric tetracarbonyl(η^3-allyl)iron(1+) complexes, the great variability in the substitution patterns of the allylic subunit, the broad range of employable nucleophiles in this sequence and the possibility of preparing both enantiomers of a target molecule display the great synthetic potential and value of this protocol. Highly functionalized and stereochemically well defined compounds of high enantiomeric purity are of increasing significance and represent valuable building blocks in the synthesis of bioactive compounds. The presented methodology should be regarded as a useful supplement to alternative (catalytical) variants in allylic subsitution reactions, which are often more limited in their range of nucleophilic components and allylic substrates.

Acknowledgement

D. E. would like to express his sincere thanks to the following former coworkers, whose experimental skill and enthusiasm have contributed to the successful development of the iron mediated allylic substitution project: Dr. Peter Fey, Dr. Thomas Schmitz, Dr. Udo Frank, Dr. Braj Bhushan Lohray and Dr. Michael Finkam. This work was supported by the Volkswagen-Stiftung, the Fonds der Chemischen Industrie, the European Community (Human Capital and Mobility Network: Metal Mediated and Catalyzed Organic Synthesis) and the Deutsche Forschungsgemeinschaft (Leibniz award). We thank BASF AG, Bayer AG, Boehringer Mannheim AG, Degussa AG, Hoechst AG, Schering AG and Wacker Chemie for generously providing us with chemicals.

References

1. General use of transition metals in organic synthesis: Hegedus, L. S. *Organische Synthese mit Übergangsmetallen*, VCH, Weinheim, Germany, **1995**.
2. (a) Pearson, A. J. in Trost, B. M.; Fleming, I. (ed.) *Comprehensive Organic Synthesis, Vol. 4*, Pergamon, Oxford, **1991**, p. 663. Semmelhack, M. F. in Trost, B. M.; Fleming, I. (ed.), *Comprehensive Organic Synthesis, Vol. 4*, Pergamon, Oxford, **1991**, p. 517. Pike, R. D.; Sweigart, D. A. *Synlett*, **1990**, 565. Blystone, S. C. *Chem. Rev.*, **1989**, *89*, 1663. Consiglio, G.; Waymouth, R. W. *ibid.*, **1989**, *89*, 257. Pearson, A. J. in Hartley, F. R.; Patai, S. (ed.), *The Chemistry of the Metal-Carbon Bond, Vol. 4.*, Wiley, Chichester, **1987**, p. 889.
3. Alexander, R. P.; Morley, C.; Stephenson, G. R. *J. Chem. Soc. Perkin Trans. I* **1988**, 2069. Stephenson, G. R.; Alexander, R. P.; Morley, C.; Howard, P. W. *Phil. Trans. R. Soc. Lond.* **1988**, *A326*, 545.

4. Palladium, reviews: Trost, B. M. *Pure & Appl. Chem.* **1996**, *68*, 779. Trost, B. M.; *Acc. Chem. Res.* **1996**, *29*, 355. Williams, J. M. J. *Synlett*, **1996**, 705. Trost, B. M. *Pure & Appl. Chem.* **1996**, *68*, 779. Trost, B. M. *Chem. Rev.* **1996**, *96*, 395. Reiser, O. *Angew. Chem.* **1993**, *105*, 576. *Angew. Chem. Int. Ed. Engl.* **1993**, *32*, 547. Frost, C. G., Howarth, J.; Williams, J. M. J. *Tetrahedron: Asymmetry* **1992**, *3*, 1089. Hayashi, T. in Ojima, I. (ed.), *Catalytic Asymmetric Synthesis*, VCH, Weinheim, Germany, **1993**, p. 325.

5. Nickel: Bricout, H.; Carpentier and A. Mortreux, J.-F. *J. Chem. Soc., Chem. Commun.* **1995**, 1863. Indolese, A. F.; Consiglio, G. *Organometallics* **1994**, *13*, 2230. Consiglio, G.; Indolese, A. F. *ibid.* **1991**, *10*, 3425. Kobayashi, Y.; Ikeda, E. *J. Chem. Soc., Chem. Commun.* **1994**, 1789. Kang, S. K., Cho, D. G.; Park, C. H., Namkoong, E. Y.; Shin, J. S. *Synth. Commun.* **1995**, *25*, 1659.

6. Molybdenum: Trost, B. M.; Merlic, C. A. *J. Am. Chem. Soc.* **1990**, *112*, 9590. Faller, J. W.; Mazzieri, M. R.; Nguyen, J. T.; Parr, J.; Tokunaga, M. *Pure Appl. Chem.* **1994**, *66*, 1463. Yu, R. H.; McCallum, J. S.; Liebeskind, L. S. *Organometallics* **1994**, *13*, 1476. Dvorak, D.; Stary, I.; Kocovsky, P. *J. Am. Chem. Soc.* **1995**, *117*, 6130. Dvorakova, H., Dvorak, D.; Kocovsky, P. *Tetrahedron Lett.* **1995**, *36*, 6351.

7. Tungsten: Lloyd-Jones, G. C.; Pfaltz, A. *Angew. Chem.* **1995**, *107*, 534; *Angew. Chem. Int. Ed. Engl.* **1995**, *34*, 462. Frisell, H.; Åkermark, B. *Organometallics* **1995**, *14*, 534.

8. Ruthenium: Kondo, T.; Ono, H.; Satake, N.; Mitsudo, T.; Watanabe, Y. *Organometallics* **1995**, *14*, 1945. Zhang, S. W.; Mitsudo, T.; Kondo, T.; Watanabe, Y. *J. Organomet. Chem.* **1993**, *450*, 197.

9. Copper: Persson, E. S. M.; van Klaveren, M.; Grove, D. M.; Bäckvall, J. E.; van Koten, G. *Chem. Eur. J.* **1995**, *1*, 351. van Klaveren, M.; Persson, E. S. M.; del Vilar, A.; Grove, D. M.; Bäckvall, J. E.; van Koten, G. *Tetrahedron Lett.* **1995**, *36*, 3059. Flemming, S.; Kabbara, J.; Nickisch, K.; Westermann, J.; Mohr, J. *Synlett* **1995**, 183. Ibuka, T.; Nakai, K.; Habashita, H.; Hotta, Y.; Fujii, N.; Mimura N.; Yamamoto, Y. *Angew. Chem.* **1994**, *106*, 693; *Angew. Chem. Int. Ed. Engl.* **1994**, *33*, 652.

10. Iron: Whitesides, T. H.; Arhart, R. W.; Slaven, R. W. *J. Am. Chem. Soc.* **1973**, *95*, 5792. Pearson, A. J. *Tetrahedron Lett.*, **1975**, 3617. Salzer, A. Hafner, A. *Helv. Chim. Acta*, **1983**, *66*, 1774. Nicholas, K. M.; Landoulis, S. J. *J. Organomet. Chem.* **1985**, *285*, C13. Hafner, A.; von Philipsborn, W.; Salzer, A. *Helv. Chim. Acta* **1986**, *69*, 1757. Silverman, G. S.; Strickland, S.; Nicholas, K. M. *Organometallics* **1986**, *5*, 2117. Dieter, J. W.; Li, Z.; Nicholas, K. M. *Tetrahedron Lett.* **1987**, *28*, 5415. Li, Z.; Nicholas, K. M. *J. Organomet. Chem.* **1991**, *402*, 105. Yeh, M.-P. C.; Tau, S.-I. *J. Chem. Soc., Chem. Commun.* **1992**, 13.

11. Green, J. R.; Carrol, M. K. *Tetrahedron Lett.* **1991**, *32*, 1141. Gadja, C.; Green, J. R. *Synlett* **1992**, 973. Zhou, T.; Green, J. R. *Tetrahedron Lett.* **1993**, *34*, 4497.

12. Koot, W.-J.; Hiemstra, H.; Speckamp, W. N. *J. Chem. Soc., Chem. Commun.* **1993**, 156. Hopman, J. C. P.; Hiemstra, H.; Speckamp, W. N. *ibid.* **1995**, 617. Hopman, J. C. P.; Hiemstra, H. Speckamp, W. N. *ibid.* **1995**, 619.

13. P. Fey, Dissertation, University of Bonn, **1985**. T. Schmitz, Dissertation, Technical University of Aachen, **1990**. U. Frank, Dissertation, Technical University of Aachen, **1990.**
14. Enders, D.; Finkam, M. *Synlett* **1993**, 401.
15. Enders, D.; Finkam, M. *Liebigs Ann. Chem.* **1993**, 551.
16. Enders, D.; Jandeleit, B.; Raabe, G. *Angew. Chem.* **1994**, *106*, 2033; *Angew. Chem. Int. Ed. Engl.* **1994**, *33*, 1949.
17. Enders, D.; Jandeleit, B. *Synthesis* **1994**, 1327.
18. Enders, D.; Jandeleit, B.; Prokopenko, O. F. *Tetrahedron* **1995**, *51*, 6273.
19. Enders, D.; Jandeleit, B. *Liebigs Ann. Chem.* **1995**, 1173.
20. Enders, D.; Fey, P.; Schmitz, T.; Lohray, B. B.; Jandeleit, B. *J. Organomet. Chem.* **1996**, *514*, 227.
21. Enders, D.; Fey, P.; Frank, U.; Jandeleit, B. *J. Organomet. Chem.* **1996**, *519*, 147.
22. Enders, D.; von Berg, S.; Jandeleit, B. *Synlett* **1996**, 18.
23. Danishefsky, S. *Acc. Chem. Res.* **1977**, *12*, 66. Wong, H. N. C.; Hon M.-Y.; Tse, C.-W.; Yip, Y.-C.; Tanko, J.; Hudlicky, T. *Chem. Rev.* **1989**, *89*, 165.
24. Seebach, D. *Angew. Chem.* **1979**, *91*, 259; *Angew. Chem. Int. Ed. Engl.* **1979**, *18*, 239. Hase, T. A. *Umpoled Synthons*, Wiley, New York, **1987**.
25. Weiss, E.; Stark, K.; Lancaster, J. E.; Murdoch, H. D. *Helv. Chim. Acta* **1963**, *46*, 288.
26. Dieter, J. W.; Nicholas, K. M. *J. Organomet. Chem.* **1981**, *212*, 107.
27. Enders, D.; Klatt, M. in Paquette, L. A. (ed) *Encyclopedia of Reagents in Organic Synthesis*, Wiley, New York, **1995**. Enders, D.; Kipphardt, H.; Gerdes, P.; Breña-Valle, L.; Bhushan, V. *Bull. Soc. Chem. Belg.* **1988**, *97*, 691.
28. Enders, D.; Schankat, J. *Helv. Chim. Acta* **1993**, *76*, 402. Enders, D.; Schankat, J. *Helv. Chim. Acta* **1995**, *78*, 970. Enders, D.; Mannes, D.; Raabe, G. *Synlett* **1992**, 10. Enders, D.; Karl, W. *Synlett* **1992**, 895. Meyers, A. I. *Acc. Chem. Res.* **1978**, *11*, 275. Weinges, K.; Brunne, G.; Droste, H. *Liebigs Ann. Chem.* **1980**, 212.
29. Enders, D.; Lohray, B. B. *Angew. Chem.* **1987**, *99*, 359; *Angew. Chem. Int. Ed. Engl.* **1987**, *26*, 351. Enders, D.; Lohray, B. B. *Angew. Chem.* **1988**, *100*, 594. *Angew. Chem. Int. Ed. Engl.* **1988**, *27*, 581. Lohray, B. B.; Enders, D. *Helv. Chim. Acta* **1989**, *72*, 980.
30. Pearson, A. J.; Khetani, V. D.; Roden, B. A. *J. Org. Chem.* **1989**, *54*, 5141. Pearson, A. J.; Blystone, S. L.; Nar, H.; Pinkerton, A. A.; Roden, B. A.; Yoon, J. *J. Am. Chem. Soc.* **1989**, *111*, 134.; Donaldson, W. A.; Shang, L.; Rogers, R. D. *Organometallics* **1994**, *13*, 6.
31. Reviews:. Kipphardt, H; Enders, D. *Kontakte (Darmstadt)*, 2 **1985**, 37. Ort, O. *Org. Synth.* **1987**, *65*, 203. Herzog, H.; Scharf, H.-D. *Synthesis* **1986**, 420. Corey, E. J.; Ensley, H. *J. Am. Chem. Soc.* **1975**, *97*, 6907.
32. R. Pelzer, diploma thesis, RWTH Aachen, **1987**.

33. Revis, A.; Hilty, T. K. *J. Org. Chem.* **1990**, *55*, 2972; Stang, P. J.; Mangum, M. G.; Fox, D. P.; Haak, P. *J. Am. Chem. Soc.* **1974**, *96*, 4562. Ainsworth, C.; Chen F.; Kuo, Y.-N. *J. Organomet. Chem.* **1972**, *46*, 59. House, H. O.; Czuba, L. J.; Gall, M.; Olmstead, H. D. *J. Org. Chem.* **1969**, *34*, 2324.

34. Scolastico, C.; Bernardi, A.; Cardani, S.; Villa, R. *Tetrahedron* **1988**, *44*, 491. Annunziata, R.; Cinquini, M.; Cozzi, F.; Raimondi, C.; Pilati, T. *Tetrahedron: Asymmetry* **1991**, *2*, 1329.

35. Ibbotson, A.; Reduto dos Reis, A. C.; Saberi, S. P.; Slawin, A. M. Z.; Thomas, S. E.; Tustin, G. J.; Williams, D. J. *J. Chem. Soc. Perkin Trans. 1* **1992**, 1251.

36. Ito, Y.; Kobayashi, Y.; Kawabata, T.; Takase, M.; Terashima, S. *Tetrahedron* **1989**, *45*, 5767. Shahak, I.; Almog, J. *Synthesis* **1970**, 145. Rathke, M. W.; Nowak, M. *J. Org. Chem.* **1985**, *50*, 2624.

37. W. -J. Koot, H. Hiemstra, W. N. Speckamp, *J. Chem. Soc. Chem. Commun.* **1993**, 156.

38. Hsiou, Y.; Wang, Y.; Liu, L.-K. *Acta Crystallogr. Sect. C* **1989**, *45*, 721. Guilard, R.; Dusausoy, Y. *J. Organomet. Chem.* **1974**, *77*, 393. Luxmoore, A. R.; Truter, M. R. *Acta. Crystallogr.* **1962**, *15*, 1117.

39. Enders, D.; Schmitz, T.; Raabe, G.; Krüger, C. *Acta Crystallogr. Sect. C* **1991**, *47*, 37. Whitesides, T. H.; Slaven, R. W.; Calabrese, J. C. *Inorg. Chem.* **1974**, *13*, 1895.

40. Zhang, W.-Y.; Jakiela, D. J.; Maul, A.; Knors, C.; Lauher, J. W.; Helquist, P.; Enders, D. *J. Am. Chem. Soc.* **1988**, *110*, 4652. Rubio, A.; Liebeskind, L. S. *J. Am. Chem. Soc.* **1993**, *115*, 891.

41. Perry, N. B.; Blunt, J. W.; Munro, M. H. G. *Aust. J. Chem.* **1991**, *44*, 627.

42. For identification and isolation of myoporone from different plant species see: Blackburne, I. D.; Park, R. J.; Sutherland, M. D. Aust. J. Chem. **1972**, *25*, 1787. Dastlik, K. A.; Forster, P. G.; Ghisalberti, E. L.; Jefferies, P. R. *Phytochemistry*, **1989**, *28*, 1425. Bohlmann, F.; Zdero, C. *Phytochemistry* **1978**, *17*, 1155. Blackburne, I. D.; Park, R. J.; Sutherland, M. D. *Aust. J. Chem.* **1971**, *24*, 995. Métra, P. L.; Sutherland, M. D. *Tetrahedron Lett.* **1983**, *24*, 1749.

43. For a recent review on the phytochemistry of the Myoporaceae see: Ghisalberti, E. L. *Phytochemistry* **1994**, *1*, 7. For a review on natural sesquiterpenes see: Fraga, B. M. *Nat. Prod. Rep.* **1992**, *9*, 557. For reviews on the synthesis of furanosesquiterpenes and sesquiterpenes see: Allen, A. J.; Vaillancourt, V, ; Albiziati, K. F. *Org. Prep. Proced. Int.* **1994**, *26*, 85. Erman, W. F., in *Chemistry of the Monoterpenes*, part A & B, (Gassman, P. G.; Ed.), Marcel Dekker, New York, **1985**. Vanderwalle, M; De Clercq, P., *Tetrahedron* **1985**, *41*, 1767. S. L. Graham, C. H. Heathcock, M. C. Pirrung, F. Plavac, C. T. White in *The Total Synthesis of Natural Products*, Vol. 5; ApSimon, J. (Ed.), John Wiley & Sons, New York, **1983**.

44. Burka, L. T.; Iles, J. *Phytochemistry* **1979**, *18*, 873. Burka, L. T.; Kahnert, L.; Wilson, B. J.; Harris, T. M. *J. Am. Chem. Soc.* **1977**, *99*, 2302. Oba, K.; Uritani, I. *Plant Cell Physiol.* **1979**, *20*, 819.

45. For syntheses of racemic and enantiomerically enriched myoporone see: Kubota, T.; Matsuura, T. *Chem. Ind.* **1957**, 491. Kubota, T.; Matsuura, T. *Bull. Chem. Soc. Jpn.* **1958**, *31*, 491. Kubota, T. *Tetrahedron* **1958**, *4*, 68. Roussis, V.; Hubert, T. D. *Liebigs Ann Chem.* **1992**, 539. Hefl, T.; Zdero, C.; Bohlmann, F. *Tetrahedron Lett.* **1987**, *28*, 5643. Anand, R. C.; Singh, V. *Tetrahedron* **1993**, *49*, 6515.

46. For general information regarding Diabrotica species see: Jackson, J. J. in *Handbook of Insect Rearing*; Singh, P.; Moore, R. F. (Eds.), Elsevier, Amsterdam, **1985**, p. 237. Krysan, J. L.; Miller, T. A. in *Methods for the Study of Pest Diabrotica*, Springer Verlag, New York, **1986**, p. 1. Hill, D. S. in *Agricultural Insects Pests of the Tropics and their Control*, 2nd Ed., Cambridge, University Press, **1975**, p. 460. Booth, R. G.; Cox, M. L.; Madge, R. B. in *Guides to Insects of Importance to Man*; 3. Coleoptera, International Institute of Entomology, London, **1990**, p. 147. Dillon, E. S.; Dillon, L. S. in *A Manual of Common Beetles of Eastern North America*, Vol. II, Dover Publ., York, **1961**, p. 711.

47. For pheromones see: Bestmann, H. J.; Vostrowsky, O. *Chem. in uns. Zeit.* **1993**, *3*, 123, and literature cited therein.

48. Hummel, H. E. Med. Fac. Landbouww. Rikjsuniv. Gent **1989**, *54/3a*, 945. Guss, P. L.; Tumlinson, H. J.; Sonnet, P. E.; McLaughlin, J. R. J. *J. Chem. Ecol.* **1983**, *9*, 1363.

49. McLaughlin, J. R.; Tumlinson, J. H.; Mori, K. *J. Econ. Entomol.* **1991**, *84*, 99. Chuman, T.; Guss, P. L.; Doolitle, R. E.; McLaughlin, J. R.; Krysan, J. L.; Schalk, J. M.; Tumlinson, J. H. *J. Chem. Ecol.* **1987**, *13*, 1601.

50. For syntheses of the SCB pheromone see: Sharma, A.; Pawar, A.; Chattopadhyay, S.; Mamdapur, V. R. *Org. Prep. Proced. Int.* **1993**, *25*, 330. Nguyen Cong Hao, Mavrov, M. V.; Serebryakov, E. P. *Izv. Akad. Nauk. SSSR, Ser. Khim* **1987**, *9*, 2083. Oppolzer, W.; Dudfield, P.; Stevenson, T.; Godell, T., *Helv. Chim. Acta* **1985**, *68*, 212. Rossi, R.; Carpita, A.; Chini, M. *Tetrahedron* **1985**, *68*, 212. Guss, P. L.; Tumlinson, H. J.; Sonnet, P. E.; McLaughlin, J. R. J. *J. Chem. Ecol.* **1983**, *9*, 1363. Senda, S.; Mori, K. *Agric. Biol. Chem.* **1983**, *47*, 795. Sonnet, P. E., *J. Org. Chem.* **1982**, *41*, 627.

51. For a synthesis of the BCB pheromone see: Mori, K.; Igarashi, Y. *Liebigs Ann. Chem.* **1988**, 717.

52. For general syntheses of insect pheromones see: Mori, K. *The Synthesis of Insect Pheromones*, 1979-1989. In *The Total Synthesis of Natural Products*; ApSimon, J. (Ed.); Wiley, Inc.; New York, **1992**; V. 9. Mori, K. *Tetrahedron* **1989**, *45*, 3233, and literature cited therein.

Syntheses of Biologically Relevant Target Molecules by Transition Metal-Induced C-C-Bond Formation

Alois Fürstner

Max-Planck-Institut für Kohlenforschung, D-45470 Mülheim/Ruhr, Germany

1 New Conceptions in Low-Valent Titanium Chemistry

Some twenty years ago *Mukaiyama*, *Tyrlik* and *McMurry* made the independent and almost simultaneous discovery that low valent titanium [Ti] induces the reductive coupling of aldehydes or ketones to alkenes[1]. The strong reducing ability together with the pronounced oxophilicity of this element provide sufficient driving force even for the formation of strained olefins which are difficult to access otherwise. Additionally, intramolecular carbonyl couplings are strongly biased by the template effect of the reagent which facilitates the cyclization of medium or large ring systems. For these striking features, such „McMurry reactions" have found many applications in organic synthesis ever since[2, 3].

However, conceptual advancements were scarce. Thus, with respect to the substrates, the reaction remained essentially limited to aldehydes and ketones. One of the few exceptions to that rule are *type I* cyclizations of oxo-esters affording cyclanones after a hydrolytic work-up[4]. This literature precedence encouraged us to explore intramolecular *type II* reactions which were found to exhibit a much broader scope.

Specifically, they allow for the first time to cross-couple aldehydes or ketones with functional groups of substantially lower redox potentials which have previously been considered as hardly reactive or even as inert. In addition to esters, this includes amides, carbonates, urethanes, and urea derivatives.

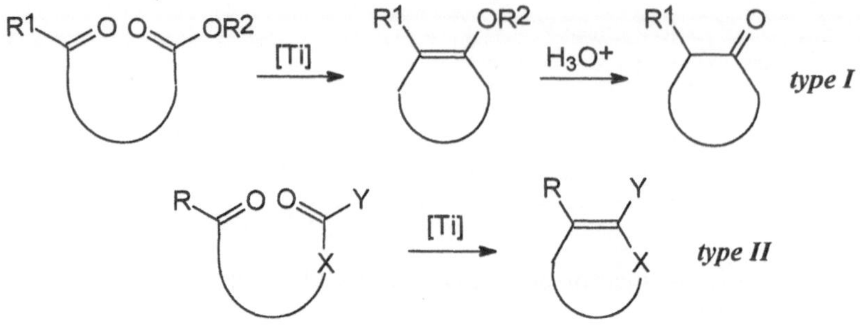

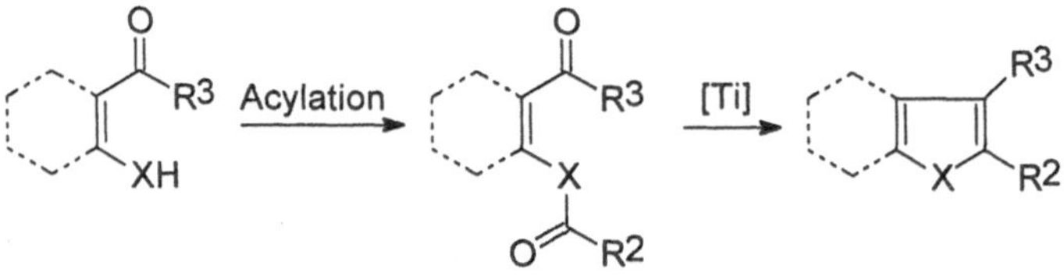

This opened up a quite general entry into several important classes of aromatic heterocycles such as furans, benzo[b]furans, pyrroles, indoles, and others which were far beyond the scope of the conventional McMurry process[5-14]. Gratifyingly, our approach turned out to be highly flexible with regard to the substituents R^2 and R^3 in the enol- (X = O) or enamine (X = NR^1) part of the product formed. Moreover, it is compatible with a wide range of functional groups and with pre-existing chiral centers in the substrates. When applied to polycarbonyl compounds, the reaction displays a striking chemo- and regioselectivity in favor of the 5-membered heterocycle. This pattern becomes particularly evident from our synthesis of the simple indole alkaloid salvadoricine (2) isolated from *Salvadora persica*, a plant which is used in traditional medicine in Pakistan[6].

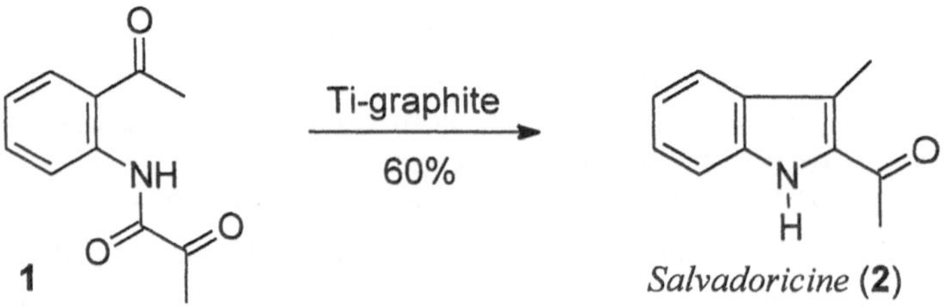

Salvadoricine (2)

310

This specific example does not only disprove the alleged inertness of amides towards low-valent titanium, but clearly shows that the driving force of the oxo-amide cyclization is even higher than that of a conventional McMurry coupling of the two ketone groups which may also be envisaged with this particular substrate. Such a seemingly reversed chemo-selectivity is unprecedented in the literature.

Initially, we relied upon the use of titanium-graphite to effect such heterocycle synthe-ses. This reagent is conveniently prepared by reduction of $TiCl_3$ with potassium-graphite laminate (C_8K) in an ethereal solvent and has proved on several occasions to be one of the most efficient McMurry coupling agents described so far[15, 16]. Its unique reactivity stems from nanosized „titanium" particles uniformly covering the extended surface of the graphite flakes. These morphological features explain both the unparalleled activity and the reasonable stability of this specific metal-graphite combination[17].

However, we are well aware that the use of the pyrophoric C_8K as the reducing agent may intimidate potential users. Although this beautifully bronze-colored intercalation com-pound can be readily prepared in a batchwise manner which minimizes any risks, we were aiming at the development of a less hazardous and hence more practical procedure. Several titanium samples were screened for this very purpose (Table 1.1)[13].

Table 1.1. Screening of various titanium species for reductive indole synthesis[13]

Entry	[Ti]	Formal Oxidation State	Isolated Yield
1	$TiCl_3 + 3\ C_8K$	0	90
2	$TiCl_3 + 2\ C_8K$	+1	90
3	$TiCl_4 + 4\ K[BEt_3H]$	0	67
4	$Ti(toluene)_2$	0	75
5	$Ti(biphenyl)_2$	0	70
6	$[HTiCl \cdot (THF)_{0.5}]$	+2	85
7	$[TiH_2(MgCl_2)_n(THF)_2]$	+2	69
8	$Cp_2Ti(PMe_3)_2$	+2	79

Although none of them rivaled the efficiency of titanium-graphite, this study finally led to the revision of some major conceptions in titanium chemistry and triggered the devel-opment of an entirely new and highly convenient preparative set-up.

In the previous literature, *metallic* titanium was claimed to be necessary for reductive carbonyl coupling reactions[2]. Although this assumption was never proved beyond doubt and has already been queried some time ago[17], this opinion seems to persist up to date[2b]. The distinct differences in reactivity of „titanium" samples prepared by various preparative routes have been explained by differences in the size, morphology and texture of the particles formed. The preparative results summarized in Table 1.1, however, clearly disprove this view. They provide the first unequivocal experimental proof that *various low-valent titanium species differing in their formal oxidation state, ligand sphere and solubility may well induce the same reductive coupling process.* Moreover, it is obvious that *Ti(0) is neither irreplacable nor necessarily the best choice*[13].

Based on this insight we devised an alternative way for performing reactions of this type: if - *in clear contrast to current practice* - TiCl$_x$ (x = 3, 4) is added to the substrate *prior* to reduction, this Lewis acid will first coordinate to the carbonyl groups leading to a pre-organization of the reactants. If reduction of the complex formed in situ can then be achieved by means of a mild reducing agent such as zinc dust, any „low-valent" titanium species [Ti] which might emerge must immediately lead to product formation because it is generated in a „site selective" manner within the coordination sphere of the substrate.

This new *"instant procedure"*[13] turned out to exhibit a very favorable profile:

1. It may be applied both to conventional McMurry couplings as well as to reductive heterocycle syntheses.
2. The formation of the active species and its subsequent reaction with the carbonyl compound are combined into a single preparative step. The reactions are simple to perform just by mixing and heating all ingredients in an inert solvent.
3. The aggressive and potentially hazardous reducing agents previously employed are avoided without any loss in performance. This also greatly facilitates the up-scaling of the process[18].
4. The reaction can even be carried out in non-ethereal solvents such as MeCN, ethyl acetate or DMF which were precluded so far.

5. Carbonyl coupling reactions under „instant" conditions exhibit the same selectivity profile as those using pre-formed [Ti]. Limits are set only in the case of very acid sensitive starting materials which may suffer from the exposure to the Lewis-acidic $TiCl_x$.

6. It should be mentioned that our rational which led to the development of the „instant method" has been fully confirmed by recent studies of *Bogdanovic* et al.[19]. These authors did not only provide additional evidence that one has to consider different „McMurry agents" according to the way of their preparation; more importantly, they have also shown beyond doubt that complexation of the carbonyl compound to $TiCl_3$ is an essential prerequisite for successful coupling reactions using Zn as the reductant.

The driving force of any carbonyl coupling stems from the formation of titanium oxides as the inorganic by-products of the reaction. Since they are thermodynamically most stable and kinetically very resistant, all McMurry-type reactions have been notoriously (over)stoichiometric in titanium so far. We soon discerned that our new „instant procedure" may also bear the chance to solve this inherent problem. Because the active species is obtained from $TiCl_x$ in the presence of the substrate, a catalytic process might emerge if the titanium oxides or -oxyhalides initially formed can be recycled into $TiCl_x$ by ligand exchange. Chlorosilanes turned out to be suitable oxophilic additives for such a purpose. In fact, the addition of TMSCl or other simple chlorosilanes allowed for the first time to perform *intramolecular type II reactions catalytic in titanium* without loss in efficiency[14]. The turn-over number reached is dependent on the specific chlorosilane chosen. This concept of maintaining a catalytic cycle by means of an oxophilic additive is without precedence in the literature and by no means restricted to titanium chemistry. We have recently been able to show that it nicely applies to other relevant reactions induced by early transition metals in low oxidation states as well[20].

As a final preparative advancement, the use of *commercial titanium as an off-the-shelf reagent* deserves mentioning[14]. All previous attempts to use this material as a chemical have been in vain due to a very inert, non-porous, adherent and repairable oxide coating on its surface. However, since our catalytic protocol clearly showed that titanum oxides react reasonably well with chlorosilanes, we probed whether they might also degrade this passivating layer. In fact, heating of a mixture of a given substrate and commercial titanium dust in the presence of an excess of TMSCl results in a slow but very clean conversion. This exceedingly simple method nicely applies to inter- and intramolecular couplings of aromatic and α,ß-unsaturated aldehydes and ketones, and was successfully used for the synthesis of indole and benzofuran derivatives via oxo-amide or oxo-ester cyclizations, respectively.

These rather comprehensive studies mentioned above denote a considerable extension of the scope of titanium chemistry[3]. Not only are new types of products accessible, but the reactions can be carried out under a set of different experimental conditions which may be adopted to the specific needs of a given substrate. This flexibility gained both in structural and preparative terms may well serve to access biologically relevant target molecules in a straightforward manner. Our work has mainly focused on indoles as one of the ubiquitous motives of nature exhibiting a wide range of physiological properties. Some of our results are summarized below.

2 Applications to Target-Oriented Syntheses

A series of thiazolyl substituted indole derivatives has evoked much interest due to their favorable biological properties. Among them, the phytoalexin camalexin (7) produced by *Camelina sativa* in response to infection by *Alternaria brassicae* deserves mentioning because it displays an appreciable antifungal acitivity. This rather simple structure provided a good test for the applicability of our method. In fact, camalexin can be prepared in four simple steps in 50% *overall* yield from commercially available starting materials[8].

C-Acylation of *Dondoni*'s thiazole 3 with 2-nitrobenzoyl chloride, followed by hydrogenolysis of the nitro group of the resulting ketone 4 over Pd/C provided amine 5 in excellent yield. After its N-formylation, a final titanium-induced oxo-amide coupling reaction under „instant conditions" smoothly led to the desired target molecule. It should be noted that upon treatment of 5 with acid derivatives other than formic acid, this scheme provides ready access to a wealth of camalexin analogues for screening purposes.

Camalexin (7)

This aspect is also particularly relevant for applications of our method to the synthesis of pharmaceutically active compounds. Among the innumerable drugs bearing an indole moiety as the pharmacophore, we have selected the tumor growth inhibitor zindoxifene (8)[7], the endothelin receptor antagonist 9[9], as well as compound 10 which is a known precursor of diazepam® in order to demonstrate the advantages of the titanium-based approach. The latter compound is prepared in only two steps by acylation of the commercially available 5-chloro-2-aminobenzophenone with ethyl oxalyl chloride followed by a chemo- and regioselective reductive cyclization of the indole ring[6, 13, 14]. Preformed titanium-graphite, titanium on alumina[22], the particularly convenient „instant procedure", the catalytic version thereof, as well as commercial titanium powder in combination with

TMSCl are almost equally effective in this particular case (Table 2.1). We have also demonstrated that the reaction can be carried out on a multigram scale in 0.16 M solution without any intermolecular side reactions interfering[18].

Zindoxifene (**8**)   **9**   **10**

The preparation of indole **9** which was recently disclosed to be a potent endothelin receptor antagonist closely follows this line[9]. Again, the crucial titanium-mediated cyclization step was performed on a multigram scale with TiCl$_3$ and Zn under „instant conditions" and is distinguished by an excellent selectivity favoring the oxoamide- over any alternative coupling process. Notably, the chosen synthetic route allows to access numerous analogues of this potential drug. Such structural variations can be performed as a combinatorial step[21].

Table 2.1 Synthesis of indole **10** by means of various titanium samples

Entry	[Ti]	Solvent	Isolated Yield
1	TiCl$_3$ + 3 C$_8$K	DME	93%
2	TiCl$_3$ + 2 Na/Al$_2$O$_3$	THF	72%
3	TiCl$_3$, Zn („instant")	THF	87%
4	TiCl$_3$ (10 mol%), Zn, TMSCl	MeCN	79%
5	commercial Ti, TMSCl	THF	74%

A more elaborate example of a titanium-induced heterocycle synthesis is the trans-annular cyclization of oxo-amide **11** to (+)-aristoteline **12**[13]. Again the „instant procedure" worked nicely, providing the desired alkaloid in 75% isolated yield after a work-up with aqueous EDTA. This treatment is necessary in order to release products bearing basic nitrogen functions from the complexes formed with the Lewis-acidic titanium salts during the course of the reaction.

The class of indolo[2,3-a]quinolizine alkaloids comprising a whole range of derivatives isolated from various (medical) plants has recently found renewed interest. Some members of this family exhibit antiviral and antitumor activites most likely due to intercalation into DNA.

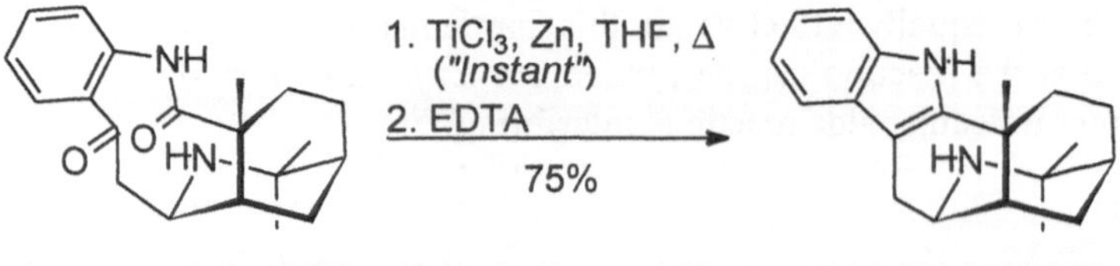

1. TiCl₃, Zn, THF, Δ ("Instant")
2. EDTA
75%

11   (+)-Aristoteline (**12**)

Since they essentially differ only in the substitution of their D-ring, the proven flexibility of the titanium-based indole synthesis might enable an advantageous entry into this class of alkaloids.

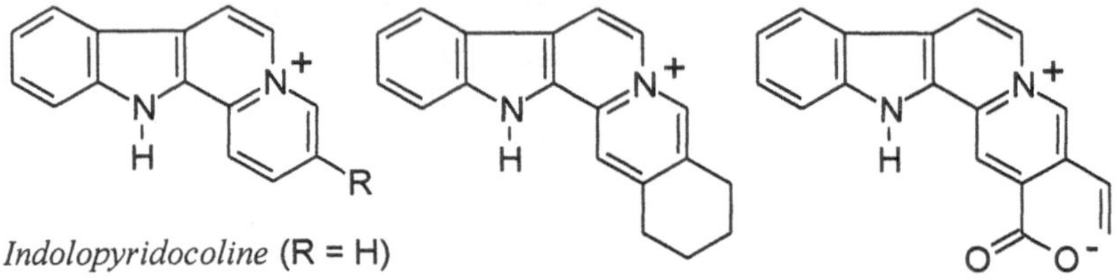

Indolopyridocoline (R = H)

Flavopereirine (R = Et)   *Sempervirine*   *Vincarpine*

Our first approach[8] starts from 2-iodoaniline **13**, which can be converted into oxo-amine **14** in two steps. N-Acylation with commercialy available pyridine-2-carboxylic acid chloride leads to compound **15**, which readily cyclizes to the corresponding indole derivative **16** on treatment with titanium-graphite as the preferred reagent. The „instant method" is less appropriate in this particular case, because of the rather acid sensitive aldol substructure of this substrate. Cleavage of the methyl ether with BBr₃ results in a spontaneous cyclization of the C-ring, which can be aromatized by means of DDQ. This affords indolopyridocolin **18** which constitutes the parent compound of this alkaloid family. Following the same synthetic scheme, flavopereirine **19** and 6,7-dihydroflavo-pereirine **17** are prepared just by choosing 5-ethylpyridine-2-carboxylic acid chloride for the acylation of the key-component **14**[8]. This underlines the flexibility of the titanium-based approach, particularly with regard to the substituent at C-2 of the newly formed indole nucleus.

316

An even more direct route makes use of amine **20** as the starting material[9]. This compound is very well accessible from 2-nitroacetophenone on a multigram scale. The same sequence of N-acylation with an appropriately substituted pyridine-2-carboxylic acid derivative followed by reductive ring closure by means of titanium-graphite affords indole **22**. Hydrolysis of its acetal group in the presence of a dehydrating agent leads to the fully aromatic back-bone of the natural product. This route can of course also be easily adapted to the synthesis of other members of this interesting family of alkaloids.

17 *6,7-Dihydroflavopereirine* (R = Et)

18 *Indolopyridocolin* (R = H)
19 *Flavopereirine* (R = Et)

Amine **20** also served as the starting material in our synthesis of secofascaplysin **24**[9]. This metabolite extracted from the sponge *Fascaplysinopsis reticulata* is biogenetically derived from the pentacyclic alkaloid fascaplysin **23**, which was found to inhibit i. a. the reverse transcriptase of the HIV virus. Biological data on secofascaplysin itself have not yet been reported.

Fascaplysin (**23**)   *Secofascaplysin* (**24**)

Its synthesis is highly straightforward: N-Acylation of **20** with the acid chloride **25**, which itself is readily available from anthranilic acid and oxalyl chloride, gives oxo-amide **26**. Cyclization of this compound by means of titanium-graphite in THF followed by hydrolysis of the acetal function of the crude indole formed leads to the desired target molecule in 60% isolated yield. Particular emphasis deserves the very regio- and chemoselective course of this transformation: the cyclization of the indole ring clearly overcomes any conceivable alternative pathway of substrate **26** which bears five different reducible sites. This example highlights best a very favorable feature of the titanium-based hetero-cycle synthesis.

Secofascaplysin (**24**)

As a final example, the first syntheses of the newly discovered pyrrole alkaloids lukianol A **34** and lamellarin O dimethylether **32** are briefly summarized[11]. While the former metabolite was isolated from an unidentified encrusting tunicate collected in the lagoon of the Palmyra atoll, the latter compound was extracted from the marine sponge *Dendrilla cactos* living near the South Australian coast.

Again our approach relies on a chemo- and regioselective cyclization of a polyfunctional precursor. Starting from commercially available 4,4'-dimethoxychalcone we have prepared isoxazole **28** in two steps, which masks an amino-carbonyl compound. Reductive cleavage of its N-O bond, acylation of the resulting amine **29** followed by titanium-mediated closure of the rather labile oxo-amide **30** affords the trisubstituted pyrrole core **31** of both targets. The unexpectedly low configurational stability of **30** under reaction conditions is mainly responsible for the somewhat lower yield obtained. N-Alkylation of **31** with 4-methoxyphenacyl bromide leads to lamellarin O dimethylether (**32**), which may be converted into the corresponding enollactone **33** by conventional means. Final cleavage of the methoxy groups with BBr$_3$ affords lukianol A (**34**) in almost quantitative yield. This 8-step synthesis highlights the close chemical relationship between these two natural products which, most curiously, are produced by taxonomically quite distinct organisms living at rather remote places in the Indian and Pacific ocean, respectively.

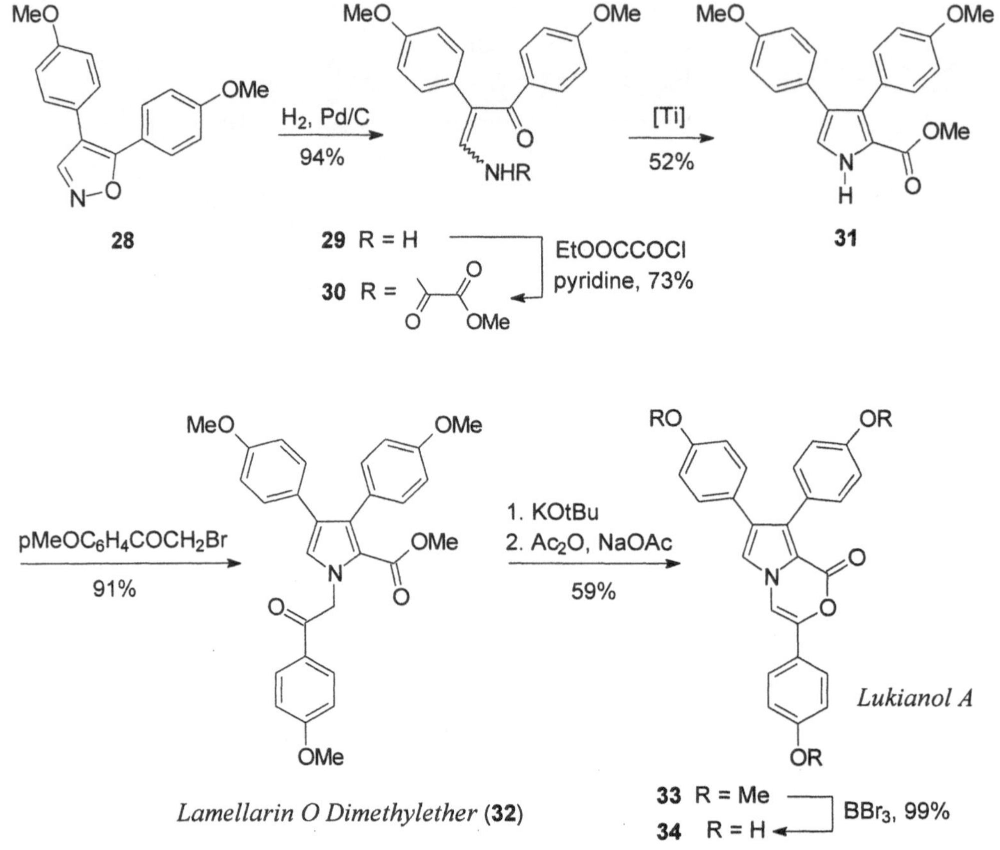

28

29 R = H

30 R =

EtOOCCOCl
pyridine, 73%

31

H$_2$, Pd/C
94%

[Ti]
52%

pMeOC$_6$H$_4$COCH$_2$Br
91%

1. KOtBu
2. Ac$_2$O, NaOAc
59%

Lamellarin O Dimethylether (**32**)

Lukianol A

33 R = Me

34 R = H

BBr$_3$, 99%

320

3 Macrocycle Synthesis: Scope and Limitations of Titanium Chemistry

The polar surface of low-valent titanium exerts a strong template effect on carbonyl groups. As a consequence intramolecular McMurry reactions are highly biased and provide ready access to medium- and large ring cycloalkenes. Several elegant applications to the synthesis of macrocyclic natural products have nicely confimed the validity of this concept[2, 3]. However, most of the targets chosen so far belong to the terpene series and are hardly functionalized.

As part of our work on new preparative procedures for reductive carbonyl coupling reactions we have also screened a representative set of 1,ω-dicarbonyl compounds as substrates. During these studies the intrinsic template effect of titanium became evident on several occasions. The efficient closure of the 36-membered ring 35[14, 22], the synthesis of several photoresponsive crownophanes such as 36[23], as well as the first successful cyclization of a bis-acylsilane to a cyclic 1,2-bissilylethene derivative 37[24] nicely feature this aspect. These cyclizations proceed readily even without applying high dilution conditions.

35   **36**   **37**

Encouraged by these results we were tempted to rely on a titanium-based macro-cyclization as the key step in an approach to the orsellinic acid macrolide lasiodiplodin, a plant growth regulating metabolite isolated from a culture broth of the fungus *Botrys-diplodia theobromae*[25].

Starting from cheap 3,5-dimethoxytoluene, a few steps lead to ester 40 in good overall yield. Its methyl group can be converted into an aromatic aldehyde function by dilithiation, trapping with diphenyldisulfide, followed by cleavage of the dithioacetal thus obtained by means of NBS in aqueous acetone. Deblocking of the terminal dimethylacetal under standard conditions cleanly affords the desired dialdehyde 42 as the cyclization precursor. Treatment of this compound with low-valent titanium under high dilution gave the desired cycloalkene 43 which after hydrogenation and deprotection leads to the natural product. However, this cyclization turned out to be rather tricky and hardly reproducible, with the yields varying from 25-82%[25]. Despite considerable efforts to improve on this key step, the results remained unsatisfactory from a preparative point of view. Therefore this example clearly denotes a limitation of the McMurry reaction, which can be troublesome when applied to highly oxygenated substrates. In order to succeed we were hence urged to explore alternative routes to this and related target molecules.

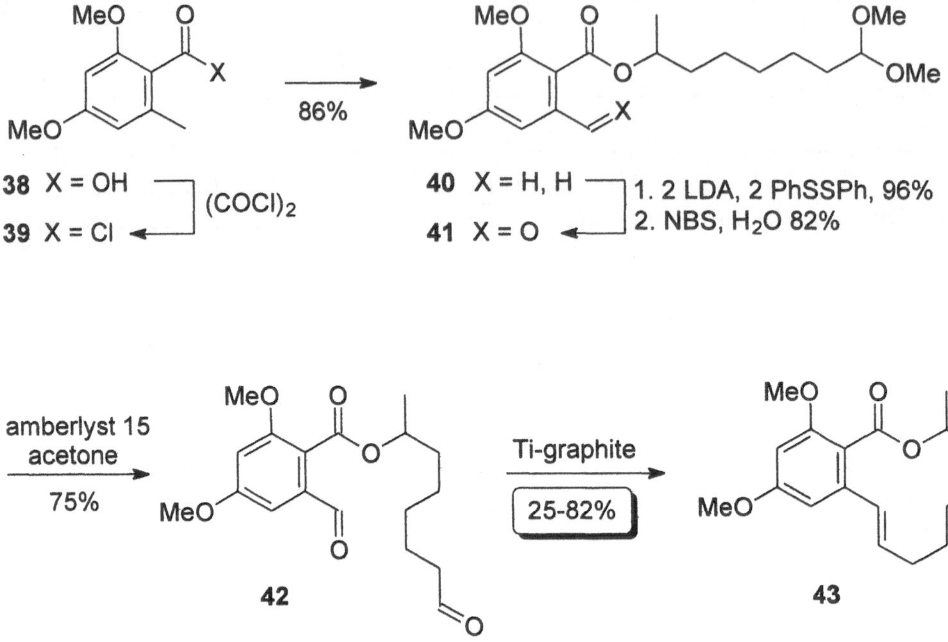

MeO
O
X
MeO
86%
OMe
OMe
38 X = OH
39 X = Cl
(COCl)2
40 X = H, H
41 X = O
1. 2 LDA, 2 PhSSPh, 96%
2. NBS, H2O 82%
amberlyst 15
acetone
75%
42
Ti-graphite
25-82%
43

4 Macrocyclization Reactions by Ring Closing Metathesis

Formally speaking, the cyclization of a dicarbonyl compound to a cycloalkene via a McMurry coupling closely resembles to ring closing metathesis (RCM) of a 1,ω-diene, although these reactions have nothing in common as to their mechanisms.

The development of a new generation of single component catalysts for olefin metathesis which combine a high performance with a reasonable tolerance towards polar functional groups has brought great impetus to this branch of research[26]. Among them, the ruthenium carbene **44** developed by *Grubbs* et al. (denoted in the Schemes as „**Ru**“)[27] and the molybdenum carbene **45** introduced by *Schrock* et al. (denoted as „**Mo**“)[28] deserve particular mentioning.

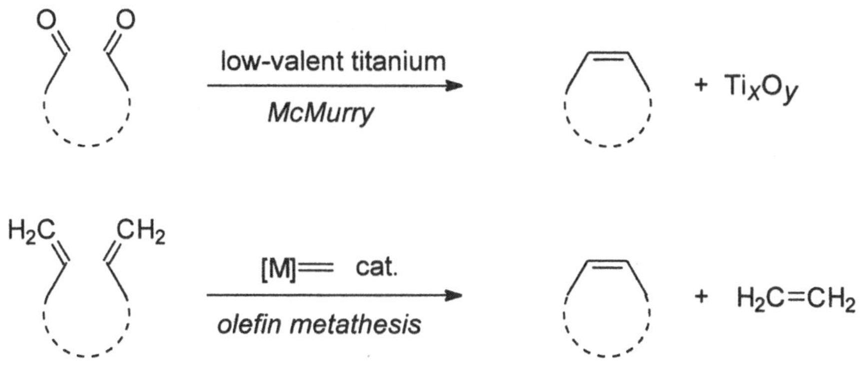

Both catalysts have been applied to RCM with considerable success, making a wealth of 5-, 6- and 7-membered carbo- and heterocycles readily available from simple precursors. However, it has been claimed that RCM does not well apply to the synthesis of medium- and large ring systems except for conformationally biased cases. In fact, very few successful macrocyclization reactions by RCM have been reported in the literature so far[26].

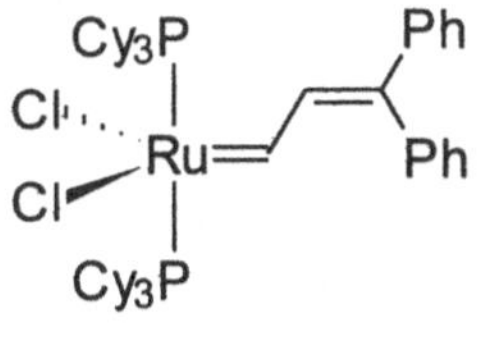

44

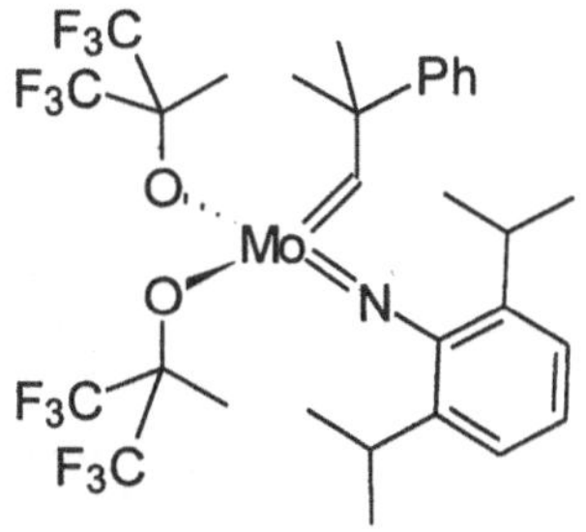

45

Keeping in mind that RCM cuts one molecule into two, we reasoned that the gain in entropy might provide sufficient driving force even for the cyclization of substrates devoid of any conformational constraints[29]. Two syntheses of Exaltolid® (**50**), a musk-odored perfumery ingredient from the root oil of *Archangelica officinalis*, have fully confirmed this hypothesis.

Thus, acylation of 5-hexen-1-ol with 10-undecenoyl chloride or of 10-undecen-1-ol with 5-hexenoic acid, followed by RCM of the resulting esters **46** and **48** in the presence of the Grubbs carbene **44** as the pre-catalyst, and finally hydrogenation of the unsaturated macro-cyles provided this valuable 16-membered lactone **50** in excellent overall yield. Other macrolactone syntheses proceeded equally well and have clearly shown that the ring size formed is of minor concern[29]. All of these applications favorably compare to established macrocyclization strategies in terms of efficiency, accessibility of the substrates, number of steps, flexibility, atom economy and overall yield.

Exaltolid (**50**)

However, during a synthesis of enantiomerically pure (12*R*)-(+)-12-methyl-13-tri-de-canolide **55**, a minor but very precious olfactory component of the *Angelica* root oil, we discerned that another parameter is essential for productive RCM[29]. Thus, all attempts to cyclize substrate **51** were unsuccessful, with the isolated yield never exceeding 10%. In contrast, compound **53** as the cyclization precursor reacted properly to the 14-membered lactone **54** in 72% yield. Since these two reactions differ only in the *site of ring closure* it is obvious that this parameter rather than ring size is a key for successful RCM.

With this information in hand we turned back to the synthesis of (R)-(+)-lasiodiplodin (63)[30]. Our RCM-based approach starts from commercially available (S)-propenoxide 56, which was ring-opened with 4-pentenylmagnesium bromide in the presence of catalytic amounts of Cu(I). The resulting enantiomerically pure alcohol 57 was esterified with the salicylic acid derivative 58 under Mitsunobu conditions resulting in complete inversion of the configuration. The latter compound is readily accessible on a multigram scale by means of a Kolbe-Schmitt reaction of cheap 3,5-dimethoxyphenol with pressurized CO_2. Treatment of the ester 59 with triflic anhydride in pyridine set the stage for a subsequent Stille cross coupling in order to attach an allylic side chain to the aromatic nucleus (60 → 61). Most gratifyingly, RCM of diene 61 proceeded smoothly giving the desired 12-membered macrolide 62 in a very well reproducible 94% isolated yield! Saturation of its double bond over Pd/C provided (R)-lasiodiplodin methyl ether which can be deprotected to the natural product according to literature procedures.

56 → (BrMg-pentenyl, CuCl(COD) cat., 81%) → **57** → (MeO, MeO, OH, OH salicylic acid **58**, 83%) →

59 R = H ┐ Tf₂O, pyridine
60 R = SO₂CF₃ ┘ 91%

→ (allyl-SnBu₃, Pd(0) cat., 93%) → **61**

→ "Ru" (6 mol%), 94% → **62** → 1. H₂, Pd/C (94%), 2. Lit. → *Lasiodiplodin* (**63**)

The preparative advantages of this synthesis compared to the McMurry approach outlined above are striking. Specific mention deserves the fact that the enantiomerically pure target is reached in only 7 synthetic operation in 40% *overall* yield starting from commercially available precursors. Furthermore, all C-C-bond formations - except for the initial Kolbe-Schmitt reaction - are metal-catalyzed. And finally, RCM has proved not only to be applicable but to be even among the most efficient entries into macrocyclic systems, provided that the site of ring closure is properly chosen. Several other syntheses from our laboratory fully confirm this conclusion. Thus, zearalenone (**64**) as another member of the orsellinic acid macrolide series can also be approached by a similar route[25]. The structurally unique family of „azamacrolides" including epilachnene (**65**) and several other closely related compounds (e. g. **66**) is also very well accessible by an RCM-based strategy[31].

326

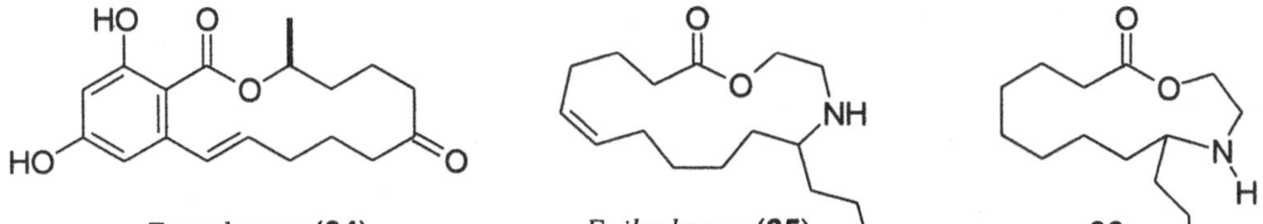
HO
O
HO
O
O
O
O
NH
O
O
N
H

Zearalenone (**64**) *Epilachnene* (**65**) **66**

The biological function of these metabolites, all of which bear a basic nitrogen atom within a lactone ring, is quite surprising: isolated from the secretion of pupae of the Mexican bean beetle *Epilachna varivestis*, these compounds are the first example reported in the literature showing that insects in the pupal state may defend themselves against ant attacks by chemical means.

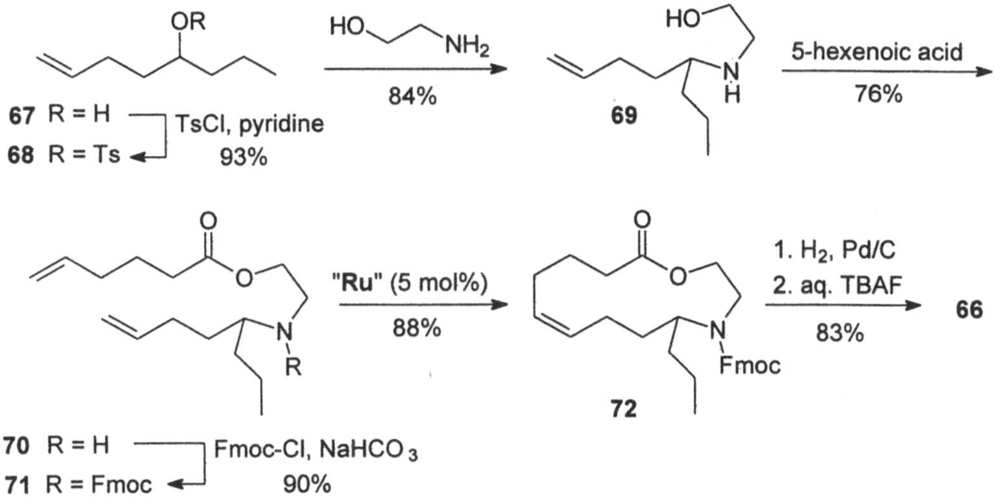
OR
HO
NH2
84%
67 R = H
68 R = Ts
TsCl, pyridine
93%
HO
N
H
69
5-hexenoic acid
76%
O
O
N
R
"Ru" (5 mol%)
88%
O
O
N
Fmoc
72
1. H2, Pd/C
2. aq. TBAF
83%
66
70 R = H
71 R = Fmoc
Fmoc-Cl, NaHCO3
90%

Our synthesis of product **66** as a minor component of these secretions is representative. Tosylation of 1-octen-5-ol **67** (obtained by reaction of 3-butenylmagnesium bromide with butanal), substitution of the tosylate **68** by ethanolamine, followed by esterification of the alcohol group of **69** with 5-hexenoic acid and N-protection readily gave the cyclization precursor **71**. This diene cyclized in high yield upon treatment with catalytic amounts of the ruthenium-carbene **44** under high dilution conditions. Two standard operations then convert the macrocyclic **72** into the desired natural product. Syntheses of other members of this interesting class of macrolides are presently underway[31].

A final example is meant to illustrate that medium sized rings are also within the realm of RCM[32]. This emerges clearly from our approach to the cyclooctenoid sesquiterpene dactylol (**79**). The known reluctance of 8-membered carbocycles to ring closure and their propensity to transannular reactions severely hampered all previous syntheses of this target, with the overall yields of these rather lengthy sequences never exceeding 1%. We have developed an unprecedently short and efficient entry into this demanding bicyclo[6.3.0] skeleton.

73 → MeCu·PBu$_3$ → [OM] → 1. H (aldehyde) / 2. MsCl, DMAP / 65% → **74** → Bu$_3$SnH, ZnCl$_2$, Pd(0) cat. / 83%

75 → 1. (methallyl)CeCl$_2$ / 83% → **76** (Me$_3$SiO) → 1. "Mo" (3 mol%) / 2. aq. TBAF / 92% → *Dactylol* (**79**), HO

2. (Me$_3$Si)$_2$NH, AcCl, DMAP / 93% → **78** (Me$_3$SiO) → 1. "Mo" (3 mol%) / 2. aq. TBAF / 85% → **80**, HO

A three component coupling process enabled us to assemble the *trans*-disubstituted cyclopentanone derivative **75** which was subsequently treated with methallyl cerium to afford a mixture of tertiary alcohols which can be separated by flash chromatography. While attempted cyclization of these O-unprotected dienes failed due to their likely coordination to the Lewis-acidic metal center of the catalyst, RCM of the O-silylated derivatives **76** and **78** thereof occurred smoothly. Dactylol (**79**) and its even more highly strained *cis*-fused analogue **80** were thus obtained in excellent yields each after *in situ* desilylation of the crude products with TBAF. This 6-step synthesis not only affords dactylol in 17% overall yield but also clearly surpasses previous ones in all respects. It provides encouraging precedence for further applications of ring closing metathesis to the synthesis of relevant targets. Some projects along these lines are actively pursued in our laboratory.

Acknowledgement

It is a great privilege for me to thank all my coworkers for their invaluable intellectual and experimental contributions. Their names appear in the references. I would also like to thank Prof. B. Bogdanovic for embarking in a joint project on low-valent titanium chemistry. Generous financial support of our work by the *Volkswagen Stiftung*, Hannover, the *Fonds der Chemischen Industrie*, Frankfurt, and the *Max-Planck-Institut für Kohlen-forschung* is gratefully acknowledged.

References

1. For this pioneering work see: (a) T. Mukaiyama, T. Sato, J. Hanna, *Chem. Lett.* **1973**, 1041. (b) S. Tyrlik, I. Wolochowicz, *Bull. Soc. Chim. Fr.* **1973**, 2147. (c) J. E. McMurry, M. P. Fleming, *J. Am. Chem. Soc.* **1974**, *96*, 4708.
2. Reviews: (a) J. E. McMurry, *Chem. Rev.* **1989**, *89*, 1513. (b) T. Lectka in *Active Metals. Preparation, Characterization, Applications* (A. Fürstner, Ed.), VCH, Weinheim, **1996**, 85.
3. For a timely review see: A. Fürstner, B. Bogdanovic, *Angew. Chem.* **1996,** *108*, 2583.
4. J. E. McMurry, D. D. Miller, *J. Am. Chem. Soc.* **1983**, *105*, 1660.
5. a) A. Fürstner, D. N. Jumbam, *Tetrahedron* **1992**, *48*, 5991; b) A. Fürstner, D. N. Jumbam, H. Weidmann, *Tetrahedron Lett.* **1991**, 6695.
6. A. Fürstner, D. N. Jumbam, *J. Chem. Soc. Chem. Commun.* **1993**, 211.
7. A. Fürstner, D. N. Jumbam, G. Seidel, *Chem. Ber.* **1994**, 1125.
8. A. Fürstner, A. Ernst, *Tetrahedron* **1995**, *51*, 773.
9. A. Fürstner, A. Ernst, H. Krause, A. Ptock, *Tetrahedron* **1996**, *52*, 7329.
10. A. Fürstner, A. Ptock, H. Weintritt, R. Goddard, C. Krüger, *Angew. Chem.* **1995**, *107*, 725; *Angew. Chem. Int. Ed. Engl.* **1995**, *34*, 678.
11. A. Fürstner, H. Weintritt, A. Hupperts, *J. Org. Chem.* **1995**, *60*, 6637.
12. A. Fürstner, D. N. Jumbam, N. Shi, *Z. Naturforsch.* **1995**, *50B*, 326.
13. A. Fürstner, A. Hupperts, A. Ptock, E. Janssen, *J. Org. Chem.* **1994**, *59*, 5215.
14. A. Fürstner, A. Hupperts, *J. Am. Chem. Soc.* **1995**, *117*, 4468.
15. (a) A. Fürstner, H. Weidmann, *Synthesis* **1987**, 1071. (b) A. Fürstner, R. Csuk, C. Rohrer, H. Weidmann, *J. Chem. Soc. Perkin Trans. 1* **1988**, 1729.
16. (a) D. L. J. Clive, K. S. K. Murthy, A. G. H. Wee, J. S. Prasad, G. V. J. da Silva, M. Majewski, P. C. Anderson, C. F. Evans, R. D. Haugen, L. D. Heerze, J. R. Barrie, *J. Am. Chem. Soc.* **1990**, *112*, 3018. (b) D. L. J. Clive, K. S. K. Murthy, C. Zhang, W. D. Hayward, S. Daigneault, *J. Chem. Soc. Chem. Commun.* **1990**, *509.* (c) D. L. J. Clive, C. Zhang, K. S. K. Murthy, W. D. Hayward, S. Daigneault, *J. Org. Chem.* **1991**, *56*, 6447. (d) G. P. Boldrini, D. Savoia, E. Tagliavini, C. Trombini, A. Umani-Ronchi, *J. Organomet. Chem.* **1985**, *280*, 307. (e) P. Burger, H. H. Brintzinger, *J. Organomet. Chem.* **1991**, *407*, 207. (f) S. Pitter, G. Huttner, O. Walter, L. Zsolnai, *J. Organomet. Chem.* **1993**, *454*, 183.
17. A. Fürstner, *Angew. Chem.* **1993**, *105*, 171; *Angew. Chem. Int. Ed. Engl.* **1993**, *32*, 164.

18. A. Fürstner, A. Hupperts, G. Seidel, *Org. Synth.*, submitted for publication.
19. (a) B. Bogdanovic, A. Bolte, *J. Organomet. Chem.* **1995**, *502*, 109. See also: (b) L. E. Aleandri, S. Becke, B. Bogdanovic, D. J. Jones, J. Rozière, *J. Organomet. Chem.* **1994**, *472*, 97. (c) L. E. Aleandri, B. Bogdanovic, A. Gaidies, D. J. Jones, S. Liao, A. Michalowicz, J. Rozière, A. Schott, *J. Organomet. Chem.* **1993**, *459*, 87. (d) Review: L. E. Aleandri, B. Bogdanovic in *Active Metals. Preparation, Characterization, Applications* (A. Fürstner, Ed.), VCH, Weinheim, **1996**, 299.
20. A. Fürstner, N. Shi, *J. Am. Chem. Soc.* **1996**, *118*, 2533; *J. Am. Chem. Soc.*, in press.
21. A. Fürstner, A. Ernst, unpublished results.
22. A. Fürstner, G. Seidel, *Synthesis* **1995**, 63.
23. A. Fürstner, G. Seidel, C. Kopiske, C. Krüger, R. Mynott, *Liebigs Ann.*, **1996**, 655.
24. A. Fürstner, G. Seidel, B. Gabor, C. Kopiske, C. Krüger, R. Mynott, *Tetrahedron* **1995**, *51*, 8875.
25. A. Fürstner, N. Kindler, unpublished results.
26. Reviews: (a) R. H. Grubbs, S. J. Miller, G. C. Fu, *Acc. Chem. Res.* **1995**, *28*, 446. (b) H.-G. Schmalz, *Angew. Chem.* **1995**, *107*, 1981.
27. S. T. Nguyen, R. H. Grubbs, J. W. Ziller, *J. Am. Chem. Soc.* **1993**, *115*, 9858.
28. R. R. Schrock, J. S. Murdzek, G. C. Bazan, J. Robbins, M. DiMare, M. O'Regan, *J. Am. Chem. Soc.* **1990**, *112*, 3875.
29. A. Fürstner, K. Langemann, *J. Org. Chem.* **1996**, *61*, 3942.
30. A. Fürstner, N. Kindler, *Tetrahedron Lett.*, **1996**, 7005.
31. A. Fürstner, K. Langemann, unpublished results.
32. A. Fürstner, K. Langemann, *J. Org. Chem.*, in press.

Pd(II)-Catalyzed Carbonylation of Unsaturated Polyols and Aminopolyols

Volker Jäger, Tibor Gracza, Eric Dubois, Thomas Hasenöhrl, Walter Hümmer, Ulrich Kautz, Bettina Kirschbaum, Albrecht Lieberknecht, Lubos Remen, Duncan Shaw, Ulrich Stahl, Oliver Stephan

Institut für Organische Chemie der Universität Stuttgart, Pfaffenwaldring 55, D-70569 Stuttgart, Germany, and Department of Organic Chemistry, Slovak Technical University, Radlinskeho 9, 812 37 Bratislava, Slovakia

Summary

Unsaturated polyols (enitols) undergo Pd(II)-catalyzed intramolecular *oxycarbonylation* with high chemo-, regio-, and diastereoselectivity. From a variety of carbohydrate-derived substrates with up to five free OH groups bicyclic lactone products of the [3.3.0] type prevail. With partially protected or missing OH ("deoxy") functions, formation of lactones with [4.3.0] and [3.2.1] structure or of [3.3.0] bis(lactones) is preferred. *Amidocarbonylation* of *N*-Z-protected γ-aminoalkenols to yield pyrrolidino-γ-lactones proceeds equally well. *Aminocarbonylation*, with *N*-benzylamino-enepolyols, is feasible also, although with unsatisfactory yields yet. - Concerning the *mechanism* of such Pd(II)-mediated multi-step processes, the essential role of the Pd(0) → Pd(II) re-oxidant copper(II)chloride in the catalytic carbonylation is suggested from experiments following the CO consumption. Some *applications* of these new oxy- and amidocarbonylations are presented, concerning syntheses of iminopolyols (potential glycosidase inhibitors), bicyclic anti-tumor active lactones ("gonio-lactones"), di- and tripeptides incorporating the unnatural β-amino acid dihydroxy-homoproline, and novel nucleoside analogues.

1 Introduction

Palladium(II)-promoted carbonylation of alkenes in the presence of alcohols may lead to alkyl enoates ("hydrocarboxylation"), diesters ["bis(alkoxy-carbonylation)"] or β-alkoxy esters ("alkoxy-carboxylation" or "alkoxy-carbonylation")[1]. Related reactions employing amides are known[1,2]. Particulary attractive features of these processes are that

(i)   they may be catalytic with regard to palladium(II), with excess copper(II) chloride mostly serving as re-oxidant;

(ii) carbon monoxide in general can be introduced at atmospheric pressure, and

(iii) that *intra*molecular versions proceed well, giving access to a variety of heterocyclic products with diverse functional groups. Thus, facile introduction of two substituents onto unactivated CC double (and triple) bonds is achieved by nucleophilic addition to the Pd^{2+}-activated alkene, see Scheme 1.

Scheme 1

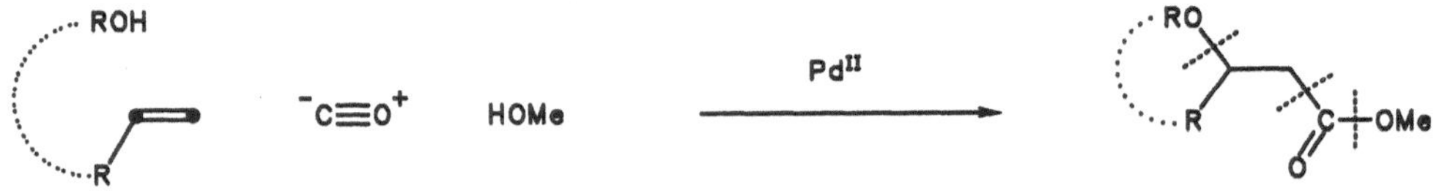

Within a long-term programme on syntheses of glycosidase inhibitors[3] (see Schemes 2, 3), we noticed reports on oxy- and amido-carbonylations leading to tetrahydrofuran- and pyrrolidine-annulated lactones[4-6].

Scheme 2. Glycosidase inhibitors. Lead compounds and structural variations prepared

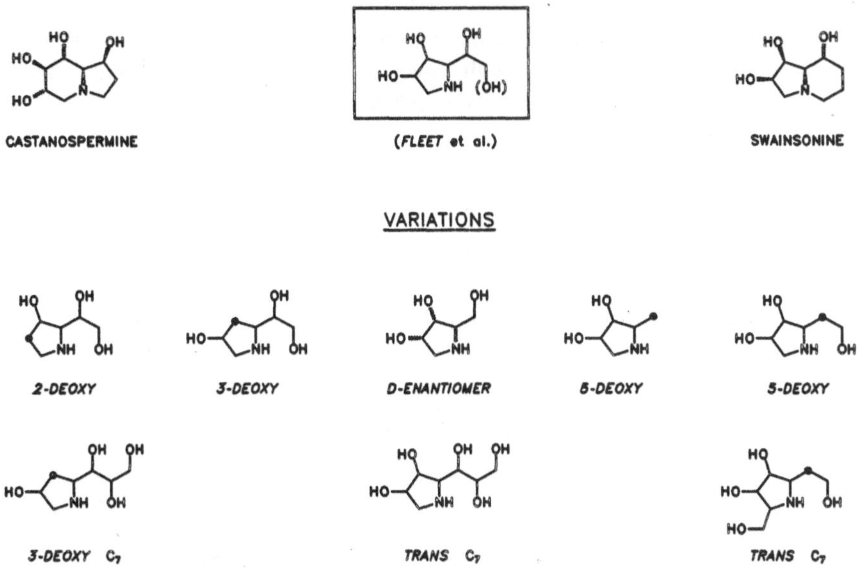

CASTANOSPERMINE (*FLEET et al.*) SWAINSONINE

VARIATIONS

2-DEOXY *3-DEOXY* *D-ENANTIOMER* *B-DEOXY* *5-DEOXY*

3-DEOXY C_7 *TRANS* C_7 *TRANS* C_7

Scheme 3. Dihydroxypyrrolidines. Retrosynthesis

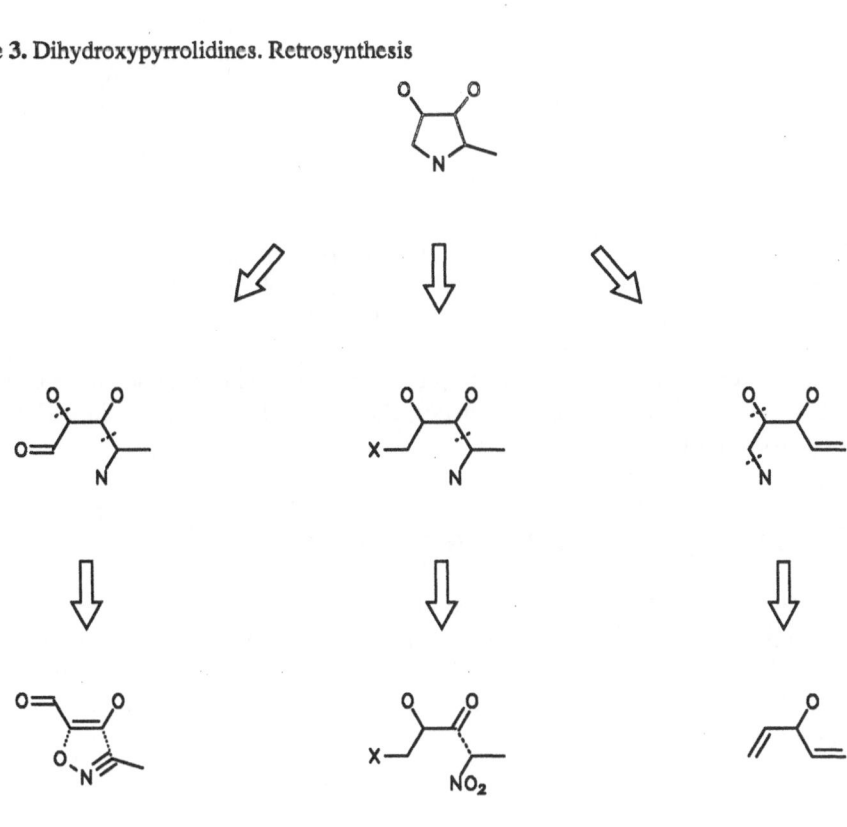

In these studies (see Scheme 4), notably effected by the groups of Tamaru and Yoshida[5,6] and by Semmelhack et al.[4], relatively simple, racemic alkenediols and amino-alkenols were used. However, applications of these elegant transformations to optically active, more complex substrates with several polar groups, f. e. carbohydrate-derived, unsaturated polyols or aminopolyols obviously were not intended.

Scheme 4. Pd(II)-catalyzed oxycarbonylation. Literature precedents with ω-alkene-1,3- and -1,4-diols

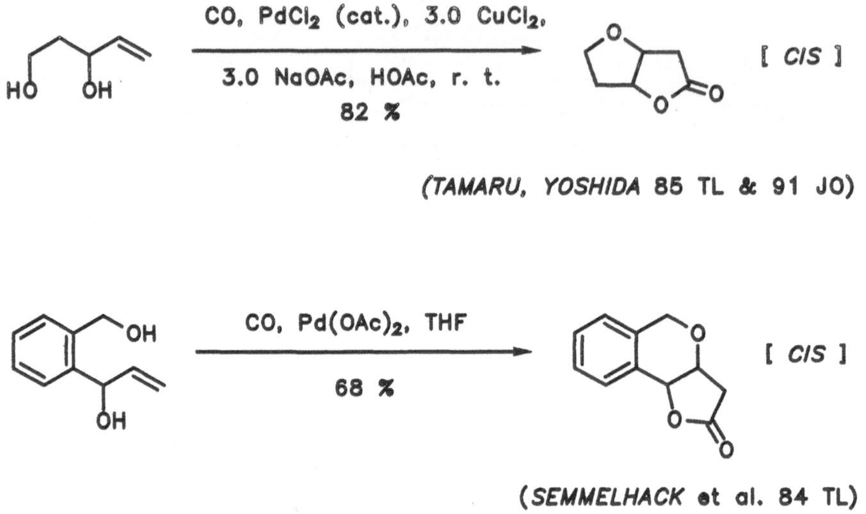

CO, PdCl$_2$ (cat.), 3.0 CuCl$_2$,
3.0 NaOAc, HOAc, r. t.
82 %

[*CIS*]

(*TAMARU, YOSHIDA* 85 TL & 91 JO)

CO, Pd(OAc)$_2$, THF
68 %

[*CIS*]

(*SEMMELHACK* et al. 84 TL)

With a first test using a pentenetriol substrate proving successful[7,8], a systematic study of the Pd(II)-catalyzed carbonylation with unsaturated polyols and aminopolyols was started. This was extended to examine the scope and selectivities, exploitations for syntheses in different areas, practical and mechanistic problems as well as variations of the participating internal nucleophiles, the subject of the present account. These studies have benefited from the large variety of optically active enepolyols and the like produced in earlier or parallel projects in our group[4,9-12].

2 Oxycarbonylation of Alkenepolyols. Scope

On Pd(II)-catalysis, allylic alcohols and 3-butenols under conditions of *hydrocarbonylation* form γ-butyrolactones[13,14]. With the latter type of alkenols, depending on the use of either Cu(II) or Cu(I) chloride oxidant, bis(carbonylation) can also be achieved in high yield[14].

The reaction of alkenols with carbon monoxide/Pd(II) salts proceeds with *intramolecular oxycarbonylation* when 5- and 6-membered lactones can be formed (see Scheme 5[8]). In these cases, palladium(II) chloride is best used as the catalyst (0.1 eq) and copper(II) chloride (3.0 eq) as re-oxidant, in acetic acid with acetate (3.0 eq) as a buffer.

With alkene*polyols* such as hexene-tetrols, two other, new cyclization modes are possible, considering only the formation of 5- or 6-membered rings, see Scheme 6. The question then is, if such enepolyol substrates would react with sufficient chemoselectivity [mono- vs. bis(carbonylation], regioselectivity, and diastereoselectivity to be useful in synthesis, or just produce mixtures?

Scheme 5. Known types of Pd(II)-catalyzed carbonylation of alkenols

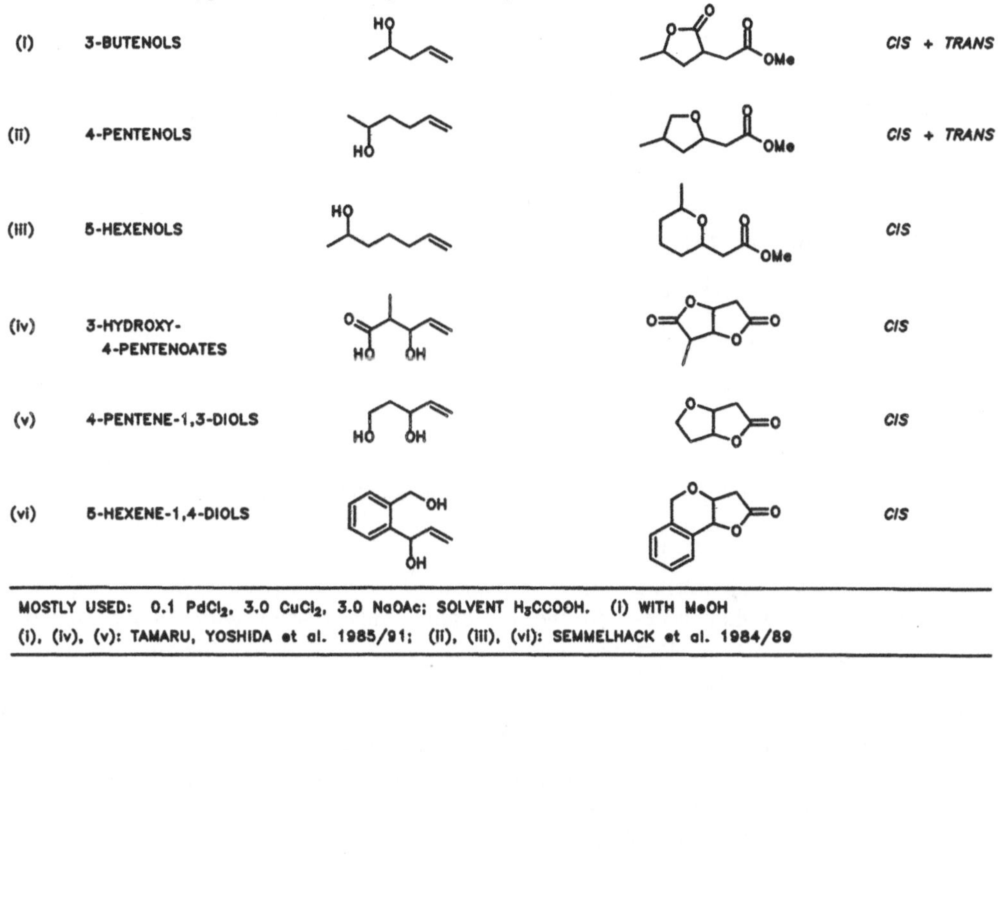

(I)	3-BUTENOLS			*CIS + TRANS*
(II)	4-PENTENOLS			*CIS + TRANS*
(III)	5-HEXENOLS			*CIS*
(iv)	3-HYDROXY- 4-PENTENOATES			*CIS*
(v)	4-PENTENE-1,3-DIOLS			*CIS*
(vi)	5-HEXENE-1,4-DIOLS			*CIS*

MOSTLY USED: 0.1 PdCl₂, 3.0 CuCl₂, 3.0 NaOAc; SOLVENT H₃CCOOH. (I) WITH MeOH
(I), (iv), (v): TAMARU, YOSHIDA et al. 1985/91; (II), (III), (vi): SEMMELHACK et al. 1984/89

335

Scheme 6. Carbonylation of alkenepolyols. Cyclization modes

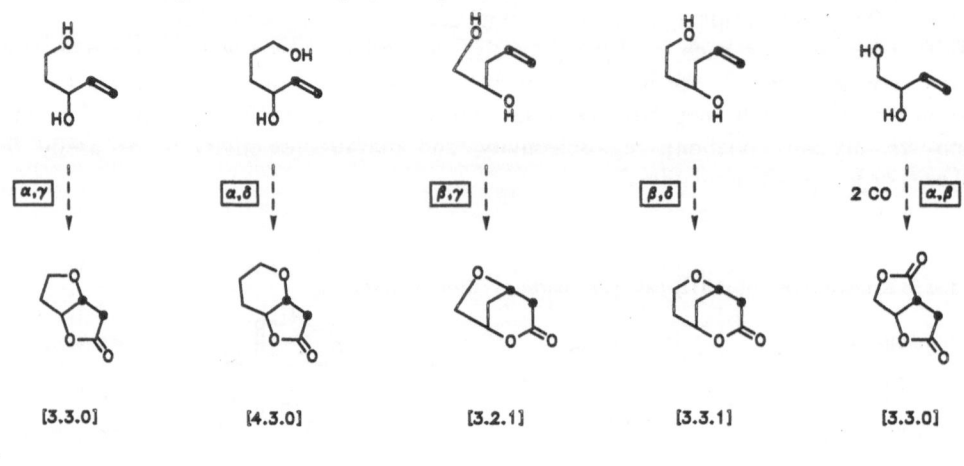

The optically active enepolyols (enitols) needed as substrates (cf. Schemes 7 - 12) were prepared relying mostly on work from our group. Thus, by asymmetric Sharpless epoxidation of divinylcarbinol, a quartet of epoxy-4-pentenol stereoisomers was ob‾tained[10,11,15], and transformed into 4-pentenetriols[11], 4-pentenediols[9,16-18], and 5-exenetriols[17,18]. Another access to the C_5 compounds started with D-ribonolactone acetonide[19], introducing the terminal double bond by reductive halo-lactone opening, a variant of the Boord elimination[20,21].

The C_6 and C_7 enitol diastereomers were prepared from carbohydrate precursors: L-*ribo* by Wittig methylenation of D-ribose acetonide; L-*arabino* from D-galactose, *vide infra* (Scheme 9)[8,18,22]; D-*xylo* and D-*lyxo* from D-glucose and D-mannitol, respectively (relying on the old, but efficient di-mesylate elimination)[8]; D-*manno*-heptenitols (Scheme 10) from D-mannose and Wittig reaction with the bis(acetonide)[18].

Oxycarbonylation of the 4-pentenetriols showed the typical course of these bicyclizations: a rather slow reaction at room temperature with CO introduced from a balloon, the green starting mixture turning ochre at the end. The bicyclic[3.3.0] products were isolated in good yield, with small amounts of the [3.2.1] type compounds (Scheme 7)[8]. This showed for the first time that - in principle - the β-hydroxy group may also be involved in the first step, the accommodation of the Pd(II) spezies.

Scheme 7. Oxycarbonylation of 4-pentenetriols

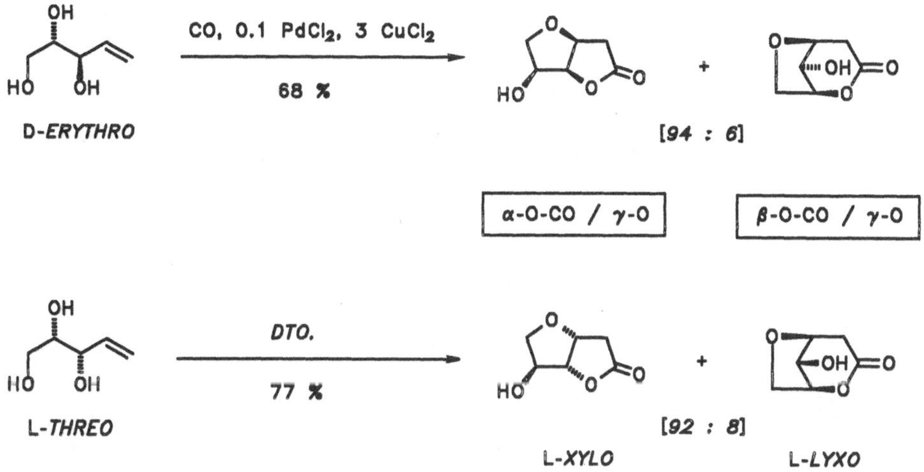

The two cyclization modes may be termed "α-O-CO/γ-O" and "β-O-CO/γ-O" or, shortly, "α,γ" and "β,γ" types (cf. Scheme 6). The ratio of lactones formed is *not* due to product equilibration, since L-*xylo* and L-*lyxo* isomers were identified in the L-*threo* pentenitol case[8]. The product ratio may reflect, however, the enthalpy difference of γ-butyro- *vs.* δ-valerolactone as known from hydrolytic equilibra ($\Delta\Delta G°$ 1.96 kcal/mol, 96.5:3.5)[23].

As shown by the results with the four hexenitol diastereomers (Schemes 8, 9), in these cases formation of the tetrahydrofuran-γ-lactones ("α,γ" type) prevails likewise. In one case, with the D-*lyxo* substrate, the 6/5-membered lactone was isolated as a minor product (14 %; Scheme 8)[8]. Again, this may be taken as a hint that the δ-OH group may be involved as the nucleophile trapping the Pd(II)/alkene (or allyl alcohol) complex formed initially.

337

Scheme 8. Oxycarbonylation of 5-hexenetriols I. *lyxo, xylo, ribo*

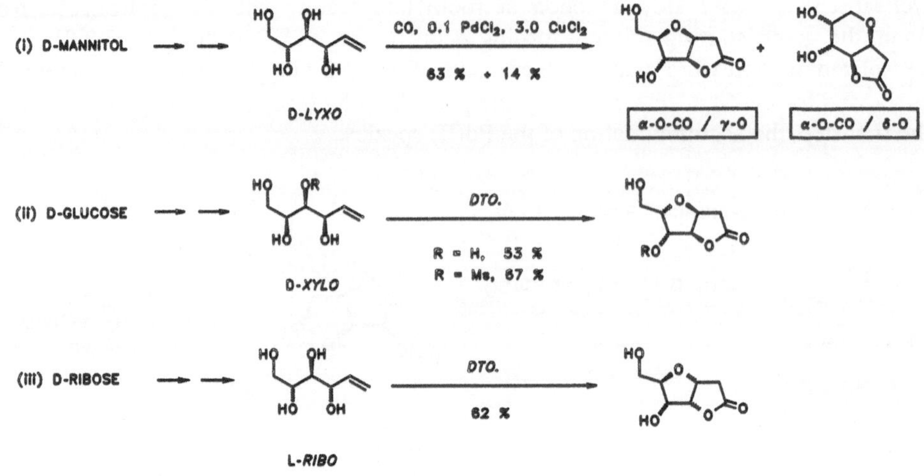

Scheme 9. Oxycarbonylation of 5-hexenetetrols II. L-*arabino*

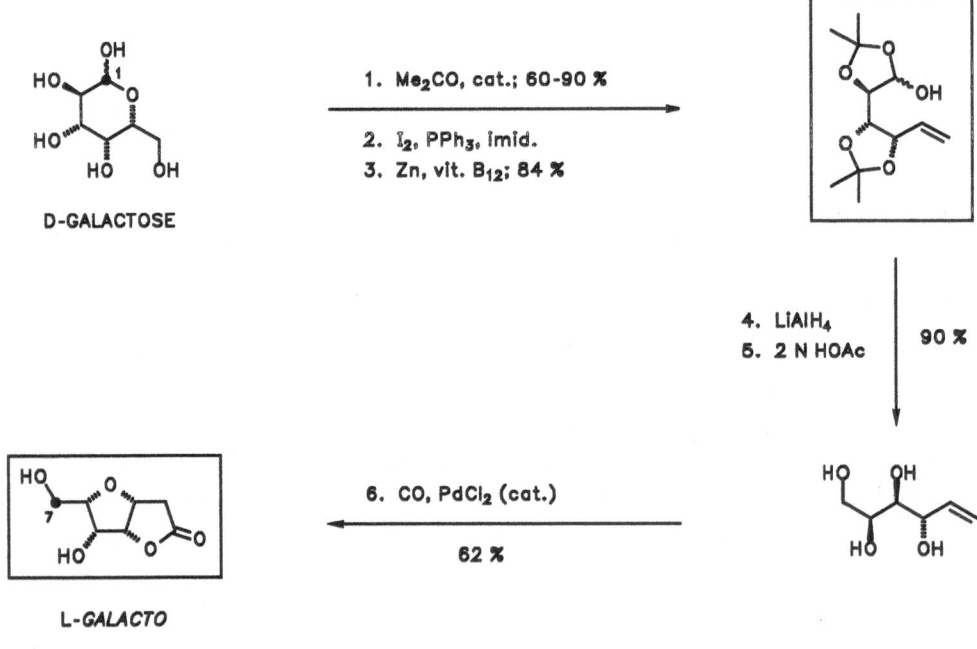

In Scheme 9 details of the access to the carbonylation substrate are also given[18,22], since this involves a new application of the vitamin B$_{12}$-catalyzed Boord elimination as advanced by Scheffold et al.[24]. This actually permitted to isolate the hexenose product, a diacetonide hemi-acetal of remarkable stability, in high yield and purity (after distillation!). Due to its availability in multi-gram quantity and to the unusual pattern of protection, this hexenose constitutes a new, very versatile building block for organic synthesis[22].

In the heptenitol series, so far only D-*manno* substrates were submitted to the Pd(II)-catalyzed oxycarbonylation[18]. The "free" enitol did accept carbon monoxide readily, however, two products - the "usual" [3.3.0] as well as the [4.3.0] bicycle[18a] - resulted ("α,γ" and "α,δ" type, cf. Scheme 6), as shown in Scheme 10.

Scheme 10. Oxycarbonylation of D-*manno*-6-heptenitols

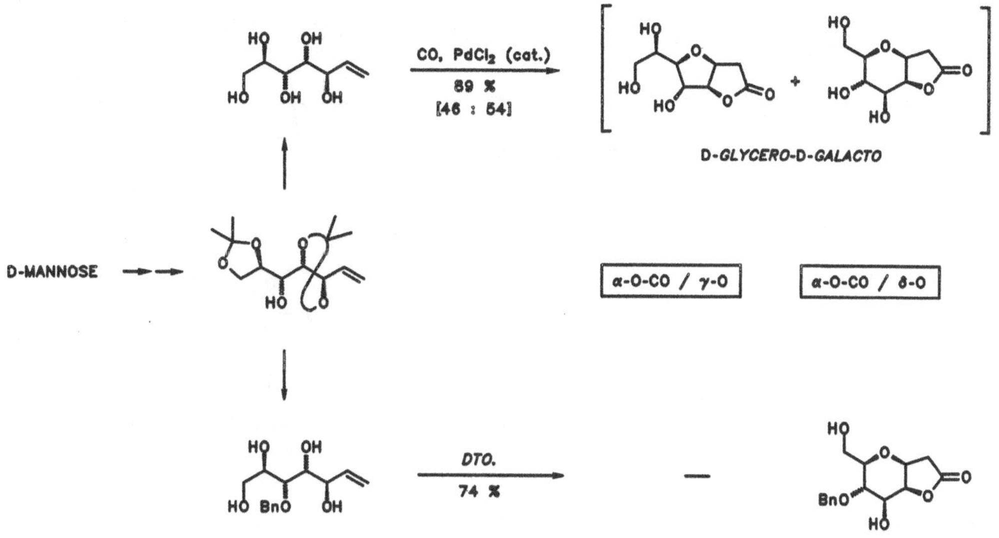

In the case of the heptenitol, a substrate with *five* free OH groups, both the secondary γ- and δ-OH groups compete for the cyclizing addition of the nucleophile to the Pd(II)-activated C=C moiety. Removal or blocking of either of these should then, in principle, permit to control the course of the oxycarbonylation. Indeed, when the γ-OH group was put out of action by O-benzylation, the tetrahydropyran-γ-lactone was isolated as the sole product[18].

The results with enitols given above suggested that various modes of bi-cyclization were within reach, other than the "α,γ" type predominating when *all* modes would compete. One way to allow for this, by blocking the most amenable γ-OH group, was already put to practice. To remove any action of this or of the initiating α-hydroxy group, the corresponding deoxy compounds were chosen as substrates. For access to these pentenediols, the epoxy-4-pentenols (at hand from Sharpless epoxidation of divinylcarbinol) were drawn upon once again. As noted earlier[10], optically active 4-pentene-1,2-, -1,3- and

-2,3-diols were prepared from these epoxyalcohols by regio-controlled reduction[16,18], and then submitted to the usual procedure (Scheme 11). The 1,3-diol gave the "α,γ" product, as expected, since in this case the optically active educt was submitted, a case reported before with the racemic mixture by Tamaru and Yoshida[5]. The 1,2- and the 2,3-diol were transformed equally well, the former leading to the [3.2.1] product, with involvement of the β- and γ-OH groups acting as CO acceptor/ director and nucleophilic addend, respectively[16,18].

Scheme 11. Oxycarbonylation of 4-pentenediols

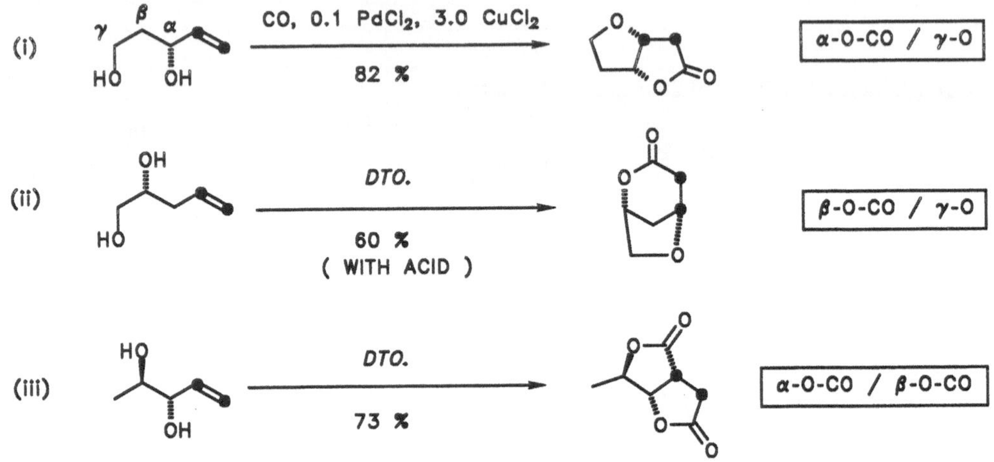

The α,β-diol [eq (iii) in Scheme 11] proved a viable substrate for carbonylation likewise, leading to the bis(lactone) with incorporation of *two* equivalents of carbon monoxide. This constitutes a new mode of the bis(carbonylation) path of alkenols that has only been observed to lead to monocyclic lactone esters so far[5,14].

5-Hexene-1,3,4-triol, with D-*erythro* configuration, and the 3-O-benzyl derivative were also prepared from divinylcarbinol, via the Sharpless product, i. e. the terminal epoxy-4-pentenol: Introduction of cyanide (cf. Scheme 24), hydrolysis, and LiAlH₄ reduction - directly or after O-benzylation of the intermediate β-hydroxy-γ-vinyl-γ-butyrolactone - gave good access to this pair of enepolyol substrates[18,25,26]. In both cases, the "unusual" carbonylations proceeded well. The free triol gave a mixture of the "α,δ"-oxycarbonylation and the "α,β"-bis(carbonylation) products. Blockade of 3-OH, in the form of its benzyl ether, made the oxycarbonylation the favoured process, to afford the tetrahydropyran-γ-lactone as the major product (Scheme 12)[18].

Scheme 12. Oxycarbonylation of 5-hexenetriols

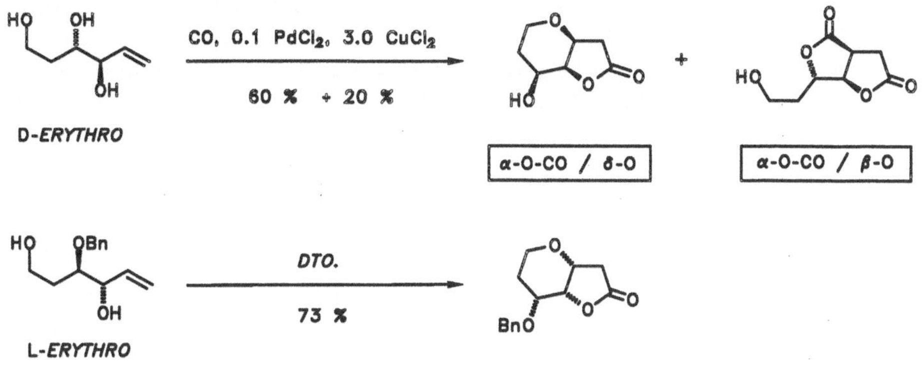

The conclusion from these results with unsaturated polyols is that various reaction modes, i. e. four of the five given in Scheme 6, may be followed in good yield. With the choice of α-, β-, γ- and δ-OH groups all present and competing, the tetrahydro-furan-γ-lactone products of the [3.3.0] type ("α,γ" mode) predominate, but blocking/removal of OH groups may well lead to either variation of the ring size of bicyclic lactone products or to bis(lactone) formation.

There is a wide range of substrates amenable to this process that also constitutes an indirect way of homologating aldoses at the *terminal* carbon to the corresponding deoxy-glyconolactones[8]. This should be a very useful option in organic synthesis, complementing the known, equally convenient procedures that deal with homologation at C1, for example by Wittig reaction with alkoxycarbonylmethylene phosphoranes[27] or Reformatsky-type additions[28]. The products from enitol oxycarbonylation provide new C-glycoside derivatives in the furanoside and pyranoside series[29], and may thus serve to extend the manifold uses already derived therefrom.

3 Oxycarbonylation of Alkenepolyols. Applications

The oxycarbonylation of enitols provides good access to new anhydro-deoxy-glyconolactones, as shown above. While there are many potential applications where these lactones might serve well, we had envisaged specific targets, i. e. new nucleoside analogues with potential anti-viral activity[30,31], and syntheses of recently discovered anti-tumor-active lactones, notably goniofufurone[32] and goniopypyrone[32-34].

The 2-deoxy-D-*gluco*-heptonolactone obtained from the mannitol-derived enitol [eq (i) in Scheme 8] represents a β-D-*arabino*-C-glycoside derivative[8]. On replacement of one or the other (or both) primary OH groups by nucleobases like adenine, access to new nucleoside analogues[30,31,35] might be given. The first part of this concept to elaborate new nucleoside analogues is outlined in Scheme 13. The β-D-*arabino* products with nucleobases adenine, uracil, thymine, and cytosine were all obtained in good yield by alkylation of respective sodium derivatives with the 5-O-mesyl-C-furanoside[36]. Unfortunately, neither results from broad or specific screening of these compounds have yet become available.

Another effort concerning application of enitol oxycarbonylation was directed towards the synthesis of bioactive styryl-lactones: From ethanol extracts of the stem bark of several Thai plants belonging to *Goniothalamus* species, a variety of bicyclic lactones, some of which showed significant cytotoxic activity towards human tumor cells, was isolated and identified, see Scheme 14[32-34].

Scheme 13. Inverse nucleoside analogues from 2-deoxy-D-*gluco*-1,4-heptonolactone

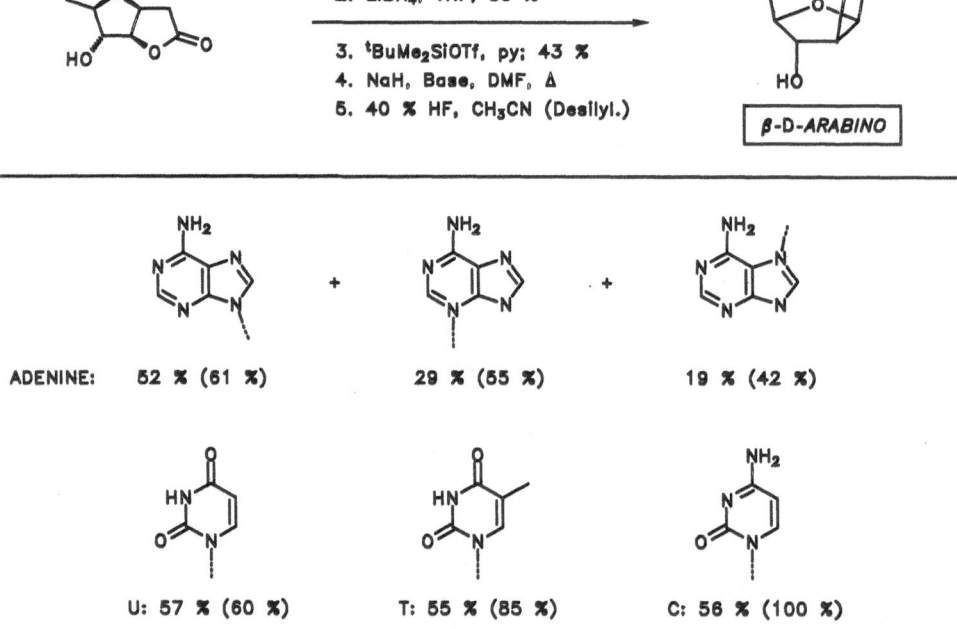

Scheme 14. Mono- and bicyclic lactones from Thai *Goniothalamus* Species

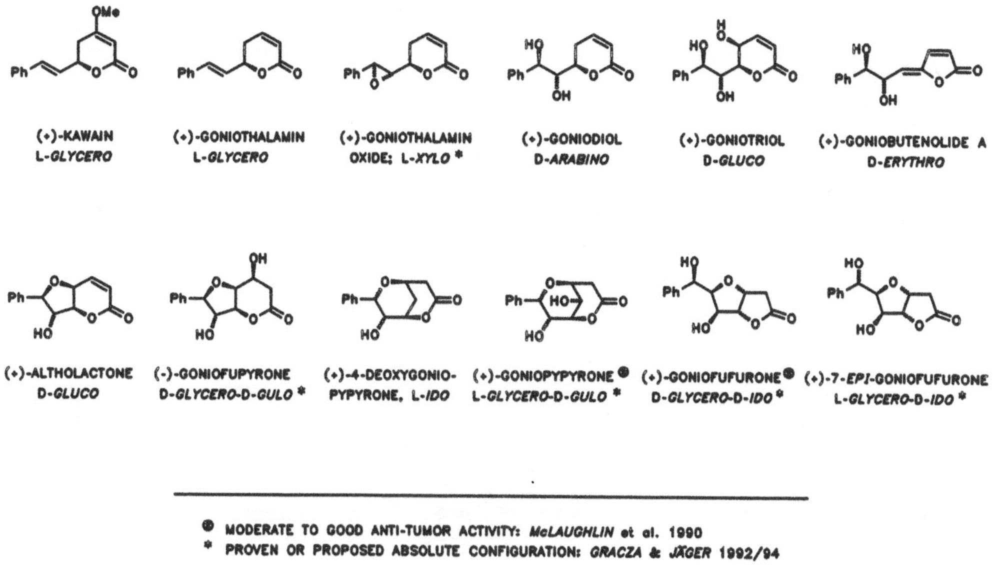

(+)-KAWAIN	(+)-GONIOTHALAMIN	(+)-GONIOTHALAMIN	(+)-GONIODIOL	(+)-GONIOTRIOL	(+)-GONIOBUTENOLIDE A
L-*GLYCERO*	L-*GLYCERO*	OXIDE; L-*XYLO* *	D-*ARABINO*	D-*GLUCO*	D-*ERYTHRO*

(+)-ALTHOLACTONE	(-)-GONIOFUPYRONE	(+)-4-DEOXYGONIO-	(+)-GONIOPYPYRONE ⊕	(+)-GONIOFUFURONE ⊕	(+)-7-*EPI*-GONIOFUFURONE
D-*GLUCO*	D-*GLYCERO*-D-*GULO* *	PYPYRONE, L-*IDO*	L-*GLYCERO*-D-*GULO* *	D-*GLYCERO*-D-*IDO* *	L-*GLYCERO*-D-*IDO* *

⊕ MODERATE TO GOOD ANTI-TUMOR ACTIVITY: *McLAUGHLIN* et al. 1990
* PROVEN OR PROPOSED ABSOLUTE CONFIGURATION: *GRACZA* & *JÄGER* 1992/94

Some of these lactones, notably goniofufurone and goniopypyrone, did represent structures that might readily be accessible from corresponding enepolyols by use of the oxycarbonylation process, see Scheme 15, and we have embarked on this right off, although only the relative configurations of these lactones were known at that time[37].

Scheme 15. (+)- and (-)-Goniofufurone from D-glucose. Strategy

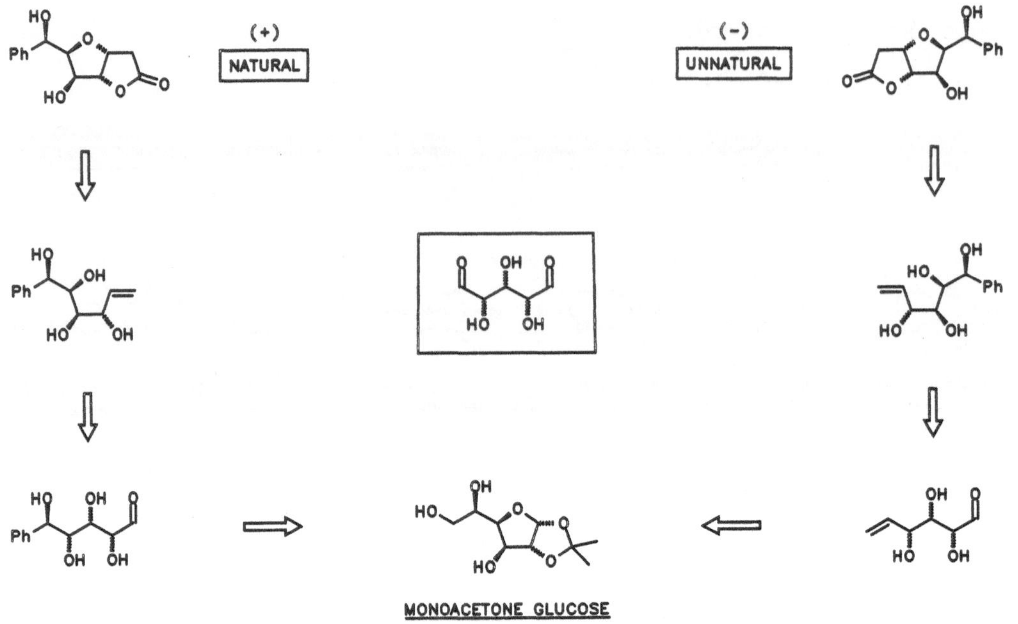

MONOACETONE GLUCOSE

The approach taken from D-glucose, with phenyl addition at C1 and C=C elaboration concerning C5/C6, was successful and led to a bicyclic lactone with all data identical to those reported, but with a different sign (negative) of specific optical rotation[37]. Of course, this established the absolute configuration of the natural, active compound (to be D-*glycero*-D-*ido*), to facilitate a number of syntheses of the correct enantiomer reported by other groups shortly afterwards[34,38-40]. With the absolute configuration of goniofufurone determined, we were also able to predict the configurations of several other "goniolactones", notably that of the most active compound goniopypyrone, along with a proposal as to its biogenetic origin by Michael cyclization of a monocyclic precursor, 7-*epi*-goniotriol[37] (not observed so far; however, only compounds with sufficient anti-tumor activity usually were followed)[32,34].

Consequently, access to the (+)-enantiomer, *i. e.* the natural, anti-tumor active goniofufurone of D-*glycero*-D-*ido* configuration was sought, and D-glucose was again identified as the most convenient starting material. The analysis of the target structure showed that a *chiral* equivalent of the *achiral* dialdopentose would permit the step-wise elaboration of either terminus, by phenyl addition first, followed by methylenation (see Scheme 15).

The synthesis of (+)-goniofufurone, the natural enantiomer, was effected along this line (Scheme 16)[38]. The Grignard reaction with the protected *xylo*-dialdose gave a (1:3) mixture of D-*gluco* and L-*ido* isomers that were di-O-benzylated (for better yields in the subsequent Wittig reaction). The two olefin diastereomers, now of L-*gulo* and L-*ido* configuration, were

344

separated by chromatography and submitted to the oxycarbonylation procedure that went smoothly in each case. The critical step, removal of the two O-benzyl groups without touching O-7, also in a benzylic position, required some experimentation in order to suppress the formation of the 7-deoxy compound. Actually, with catalytic hydrogenation carried out in ethyl acetate, 22 % of this 7-deoxy material was obtained, in the form of crystals suitable for X-ray analysis[38b]. With optimized conditions of the catalytic hydrogenation in methanol or ethanol, (+)-goniofufurone and its 7-epimer were obtained in 7 steps over-all from monoacetone glucose, with 6 and 18 % total yield, respectively[38a,c].

Scheme 16. Synthesis of (+)-goniofufurone and 7-epimer

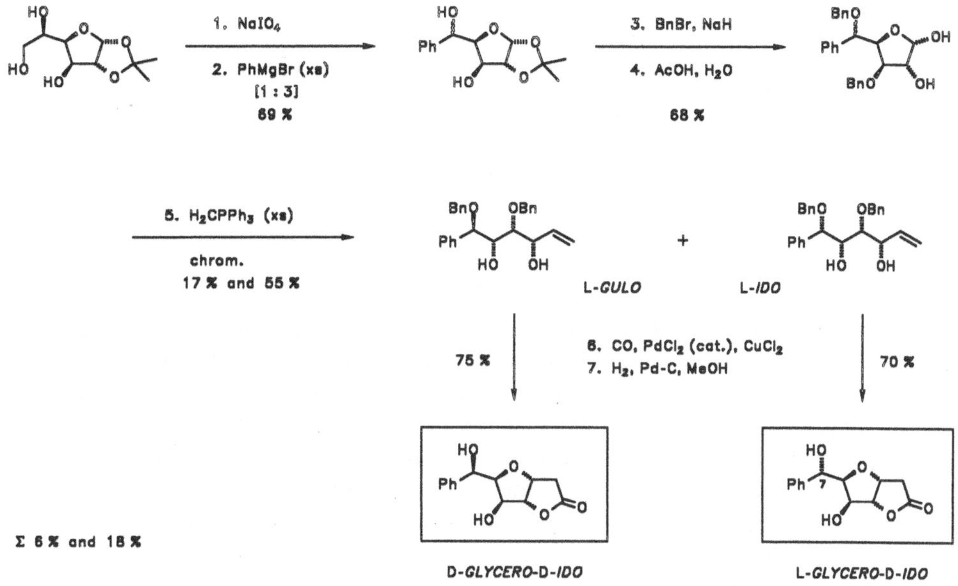

Taking advantage of the biogenetic relationship outlined above and the absolute configuration proposed[37] (*vide supra*), syntheses of natural (+)-goniopypyrone were rapidly achieved in three other groups[34]. The approach pursued in our group again relies on oxycarbonylation of a suitable enepolyol which in fact is the L-*ido*-1-phenyl-5-hexenitol met earlier in the goniofufurone synthesis (see Scheme 16). For goniopypyrone, however, bicyclization according to the "β-O-CO/δ-O" type is required (see Scheme 17), a cyclization mode not yet observed during the model studies.

Scheme 17. (+)-Goniopypyrone. Analysis

(I) STRUCTURE, RELATIVE CONFIGURATION, ANTI-TUMOR ACTIVITY: *McLAUGHLIN* et al., 1990/91

(II) PROPOSED ABSOLUTE CONFIGURATION AND BIOGENESIS: *GRACZA & JÄGER,* 1992

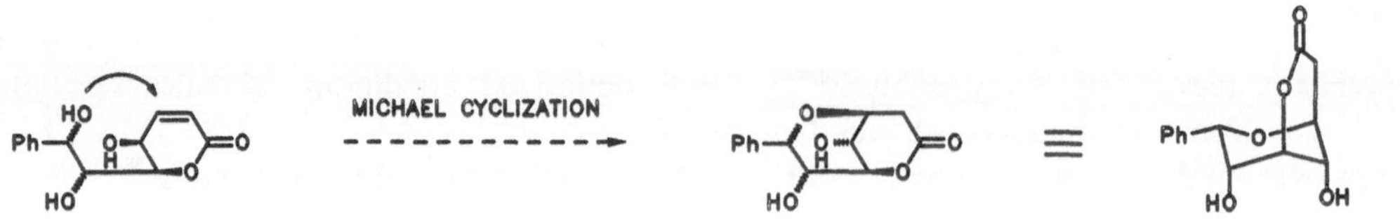

MICHAEL CYCLIZATION

7-*EPI*-GONIOTRIOL

(+)-L-*GLYCERO*-D-*GULO*

(III) SYNTHESES: *SHING* et al. 1993; *TSURUKI* et al. 1993; *ZHOU* et al. 1993.

(iv) RETRO OXYCARBONYLATION:

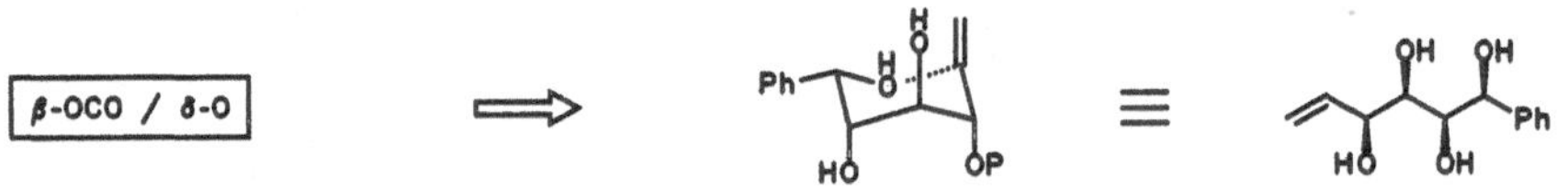

β-OCO / δ-O

L-*IDO*

In order to make formation of the β,δ-product the more favourable pathway, blocking of the α- and γ-OH groups seemed mandatory. The synthesis of such a precursor was carried out employing "glucoheptonolactone", a cheap starting material[40], see Scheme 18. Again, a differentially-protected, chiral derivative of an *achiral*, pseudo-asymmetric structure was drawn upon where the two termini can be elaborated by choice, similar to the use of D-glucose in the synthesis of both enantiomers of goniofufurone. The oxycarbonylation step with the benzylidene tetrol so far proved troublesome, and other substrates are being studied to this purpose.

Scheme 18. (+)-Goniopypyrone. Route to an oxycarbonylation substrate

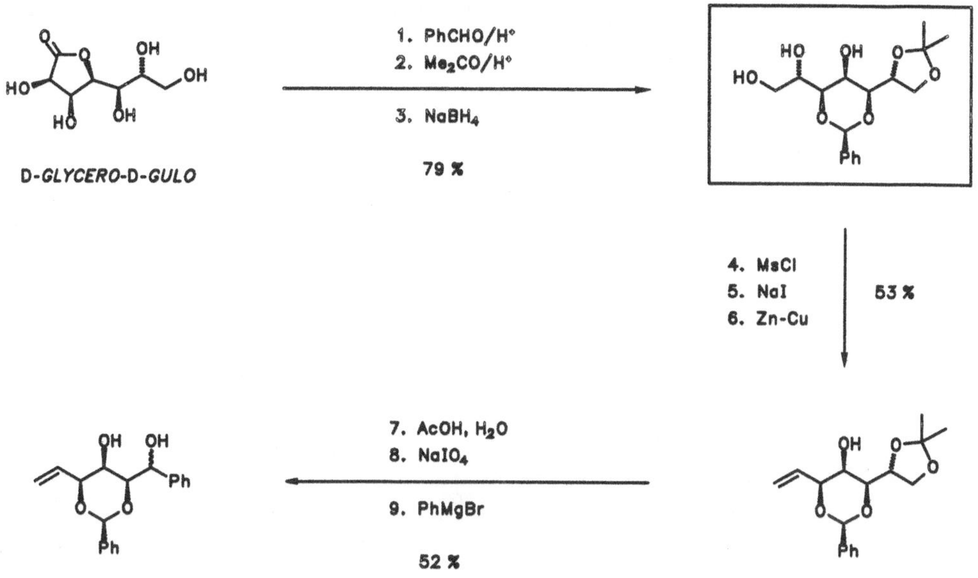

4 Carbonylation of Derivatives of Unsaturated Amino Polyols

In contrast to oxygen nucleophiles that coordinate only weakly to palladium(II), amines bind well and rather displace an olefin from the metal than attack it as a nucleophile[2]. For this reason, aliphatic amines are not suitable and derivatives with compatible, deactivating ("protecting") groups are needed, f. e. N-SO$_2$Ar, N-Ar, N-COOR, NCONR$_2$[2,41]. The first examples of intramolecular *amidocarbonylation* plus lactonization were also reported by Tamaru, Yoshida et al.[6] N-Methoxycarbonyl, -tosyl and -carbamoyl derivatives of racemic 1-amino-4-pentene-3-ols were found suitable for optimal adjustment of the N-nucleophilicity, concerning intramolecular addition to the C=C bond[6].

This reaction looked promising with regard to the synthesis of polyhydroxypyrrolidines that often had proven effective glycosidase inhibitors[3]. We have therefore prepared a series of optically active amino enepolyols as new, more complex substrates for amidocarbonylation. The C$_5$ compounds were again produced from divinylcarbinol, via its epoxides, cf. Scheme 19[42]. A homologous aminohexenetriol was prepared from D-glucosamine, recurring to the reductive elimination of 6-halopyranosides, see Scheme 20 (i)[43]. Entry to the 4-aminohexenetriol series was found by the highly *threo* selective addition of vinylmagnesium bromide to the N,O-dibenzyl derivative of threose imine, see Scheme 20 (ii)[44]. This sequence came as an off-spring of a parallel project in our group concerned with the preparation of amino polyols from optically active α-benzyloxy imines

347

by addition of organometallic reagents[12,45]. Objectives achieved there comprise the family of statins, *i. e.* amino hydroxy acids[45], and antitumor-active *trans*-dihydroxypyrrolidines such as deacetylanisomycin[44].

Scheme 19. Amino-4-pentenediols from divinylcarbinol

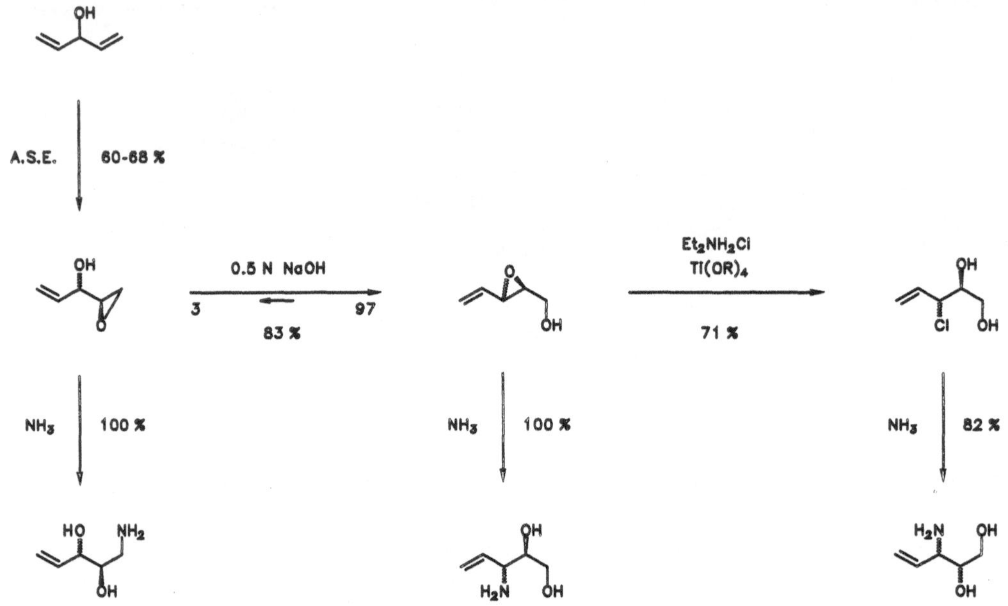

Scheme 20. Preparation of aminodeoxy-D-xylo-5-hexenitols

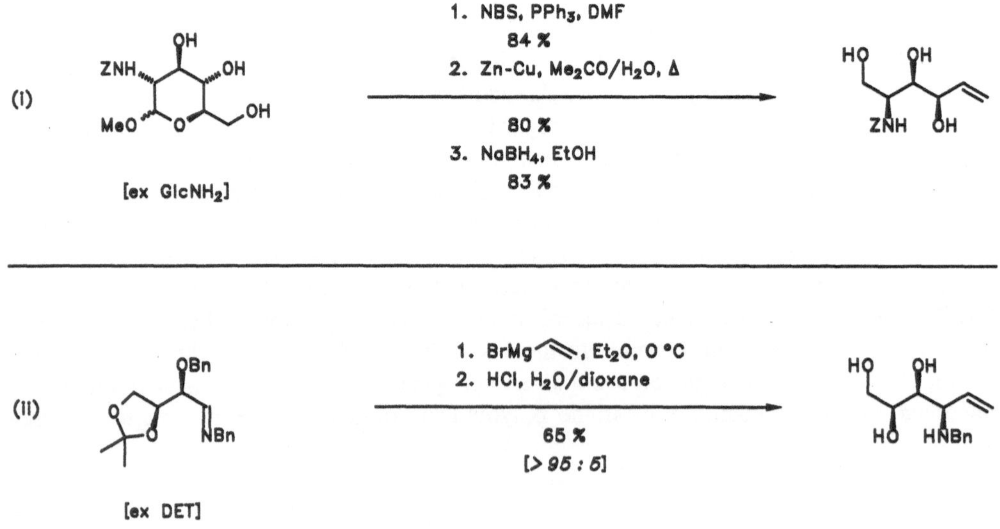

(I)

1. NBS, PPh₃, DMF
   84 %
2. Zn-Cu, Me₂CO/H₂O, Δ
   80 %
3. NaBH₄, EtOH
   83 %

(II)

1. BrMg⟍, Et₂O, 0 °C
2. HCl, H₂O/dioxane

65 %
[> 95 : 5]

348

For attempts at amidocarbonylation of aminoalkenenpolyols, we introduced the N-benzyloxycarbonyl (Z) and the N-benzylaminocarbonyl groups[46]. This extends the findings of Tamaru and Yoshida to *benzyl* derivatives that we hoped would present advantages, such as simplified monitoring of the reaction (by TLC/UV detection) and, in particular, facile *removal* of the N-moderating group afterwards. Indeed, the Z-version with the aminopentenediol proceeded with high yield [eq (i) in Scheme 19], making available the versatile dihydroxy-homoproline derivative (*vide infra*) in multi-gram quantity[46]. Similarly, the C_7 pyrrolidino-lactone was obtained, along with 10 % of the 7-O-acetyl derivative (carbohydrate numbering) and 4 % of the tetrahydropyran product[43]. This is taken as a hint that the NHZ and OH groups show similar nucleophilicity, and that this might be exploited by differentiating protection in other substrates.

Scheme 21. Pd(II)-catalyzed amido- and related carbonylations. I

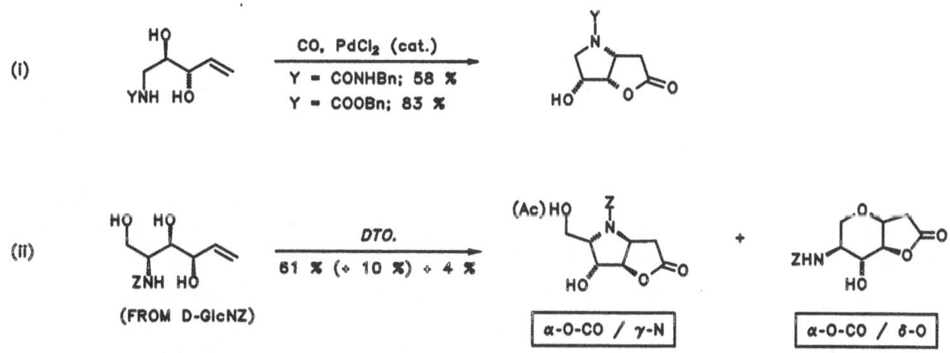

USES: (I) POTENTIAL MANNOSIDASE INHIBITION (-), 90 AC 1171
DI- & TRIPEPTIDES INCORPORATING DIHYDROXY-HOMOPROLINE
(II) REDUCTION GIVES 2,5,6-TRIDEOXY-2,5-IMINO-L-*IDO*-HEPTITOL,
A WEAK INHIBITOR OF α-L-FUCOSIDASE, β-GLUCOSIDASE (ALMOND)

The pyrrolidino-lactones were both transformed into iminopolyols, first by reduction of the lactone part with lithium borohydride, and then by removal of the benzyloxycarbonyl group on catalytic hydrogenation[43,46]. The *N*-Z-iminotriol was also transformed into the *N*-methyl compound by means of lithium aluminium hydride (Scheme 22)[46].

Scheme 22. 1,4-Iminopolyols from dihydroxypyrrolidine-lactones

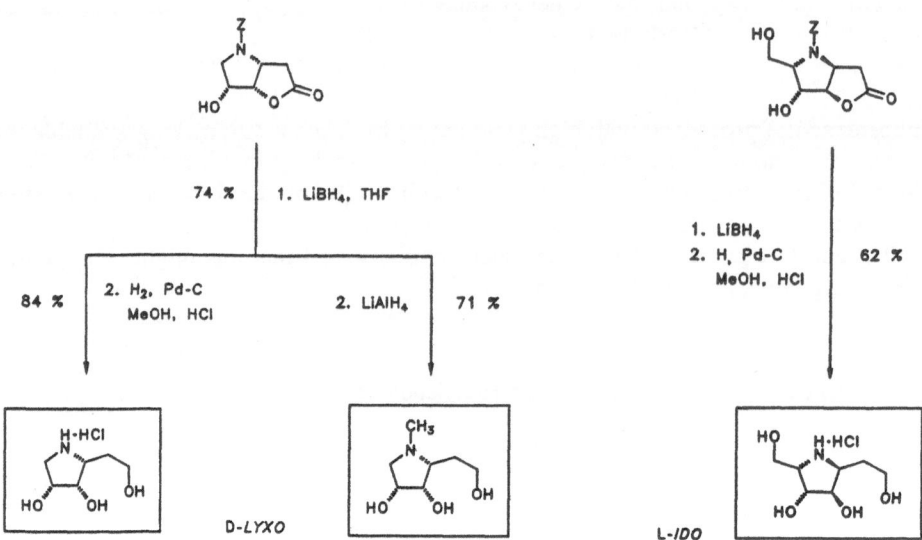

The three iminopolyols shown were chosen and prepared in view of potential glycosidase inhibition, as structural analogues of the lead compound 1,4-dideoxy-1,4-imino-D-mannitol from Fleet's group (cf. Scheme 2)[3]. The D-*lyxo* compounds (*manno* analogues) were inactive towards α-mannosidase from jack beans, however, and screening of the L-*ido*-iminoheptitol did reveal weak inhibition of α-fucosidase (from bovine epididymis) and β-glucosidase (from almonds) in the range of ca. 1 mM for IC_{50} only[47]. The conclusion from this part concerning the minimum requirements for the structure of iminopolyols for high activity is that 6-OH may be lacking (Fleet), but that neither of the other three OH groups can be omitted[3].

The C_6 pyrrolidino-lactone constitutes an unnatural, novel analogue of (hydroxy)proline and we thought it worthwhile to incorporate this β-amino acid into simple di- and tripeptides. These syntheses are summarized in Scheme 23[48].

Dipeptide formation at the C-terminus could be effected by heating the lactone with *t*-butyl esters of various amino acids [eq (a) in Scheme 23], a rarely used, but very convenient method (if it works)[49]. The free dipeptides were obtained in very good yield by a four-step deprotection sequence starting with acetylation of the free hydroxy groups [eq (b)-(e)]. Otherwise, the γ-hydroxy group would interfere and reform the lactone with amide cleavage as seen when acid-catalyzed cleavage of the t-butyl ester group was attempted[48].

On removal of the *N-Z* group, the amino lactone is open to coupling at its N-terminus. After tedious experimentation with a variety of coupling agents - here we profited very much from the experience with peptide chemistry gathered in the group of U. Schmidt[50] - the EDC method gave good results except for the proline case where TPTU proved best [EDC: 1-ethyl-3-(3'-dimethylaminopropyl)-carbodiimide hydrochloride; TPTU: 2-[2-oxo-1(2H)-pyridyl]-1,1,3,3-tetramethyluronium tetrafluoroborate]. Deprotection and purification then gave the lactone dipeptides in a straight-forward manner. Finally, both approaches were combined and the protected tripeptides, with dihydroxyhomoproline as the central part, could be obtained[48].

Scheme 23. Di- and tripeptides incorporating dihydroxy-D-homoproline (DHHoPro)

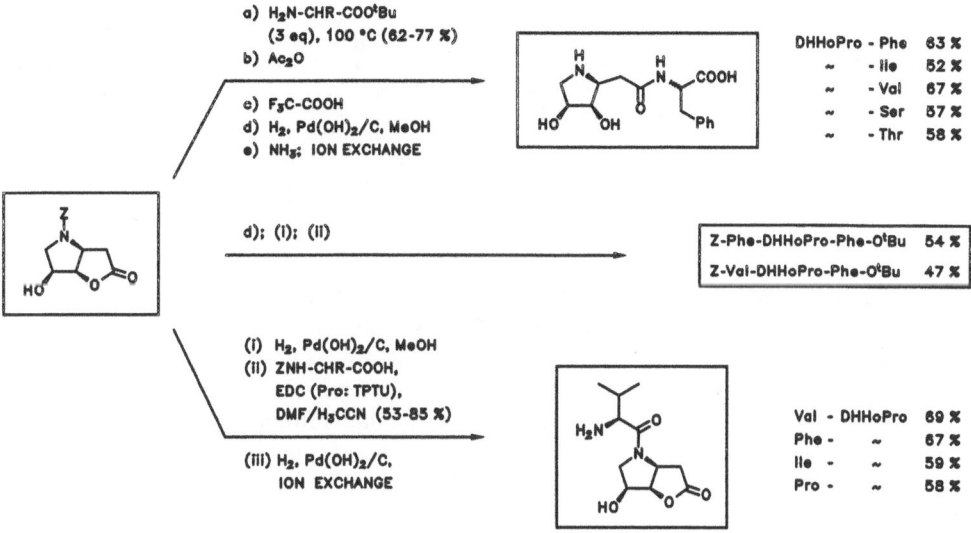

In the pioneering, systematic studies by Tamaru, Yoshida et al.[5,6] *6-aminohexene-3-ols* were also included[6]. Low yields of piperidino-lactones (ca. 35 %) and formation of other products were stated with N-urea derivatives as the best substrates; the *N-Z-aminohexenol* and the sulfonamide proved unreactive[6]. We have carried out some related experiments with the optically active *erythro*-aminohexenediol, available from Sharpless epoxidation of divinylcarbinol, cyanide addition and reduction, as outlined in Scheme 24[25,26].

Scheme 24. Pd(II)-catalyzed amido- and related carbonylations. II

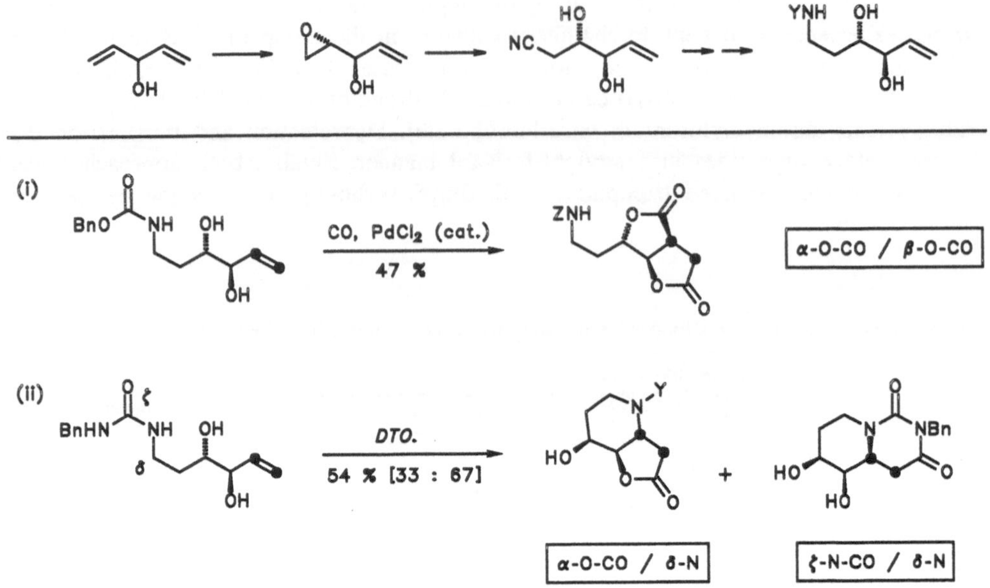

(I)

CO, PdCl$_2$ (cat.)
47 %

α-O-CO / β-O-CO

(II)

DTO.
54 % [33 : 67]

α-O-CO / δ-N ζ-N-CO / δ-N

The *N*-Z-derivative reacted solely with the butenediol part and the bis(carbonylation) product was obtained in moderate yield, just as seen with the corresponding polyols (Scheme 11, 12). On the other hand, the N-benzylurea diol led to a mixture of the lactone and of the dihydroxypiperidino-acetamide derivative, see eq (ii) in Scheme 24[25,26].

It was of interest therefore, and still is, to find other nitrogen functions suitable as internal nucleophiles, to add to the Pd(II)-complexed olefin part, and also for accommodation of carbon monoxide which would result in formation of a lactam ring. As pointed out above, Pd(II)-catalyzed addition of *aliphatic amines* are notoriously hampered. It was very surprising therefore to see that some benzylamino alkenepolyols *did undergo* carbonylation (see Scheme 25)[18,26,51].

In the first case, the benzylamino group acts as the nucleophile [eq (i)]; with the β- and α-benzylamino substrates it takes the role of CO acceptor, to produce lactams [eq (ii), (iii)][18,51]. The second case also shows that the benzylamino moiety is more "active" than the α-OH group since the usually preferred tetrahydrofuran-lactone ("α,γ'-type) was *not* isolated from the reaction mixture[26]. On the other hand, the benzyloxycarbonyl-substituted amino group is less active than a suitably placed diol group: the "β,γ'-oxycarbonylation product prevails [eq (iv) in Scheme 25][26].

Scheme 25. Pd(II)-catalyzed aminocarbonylation

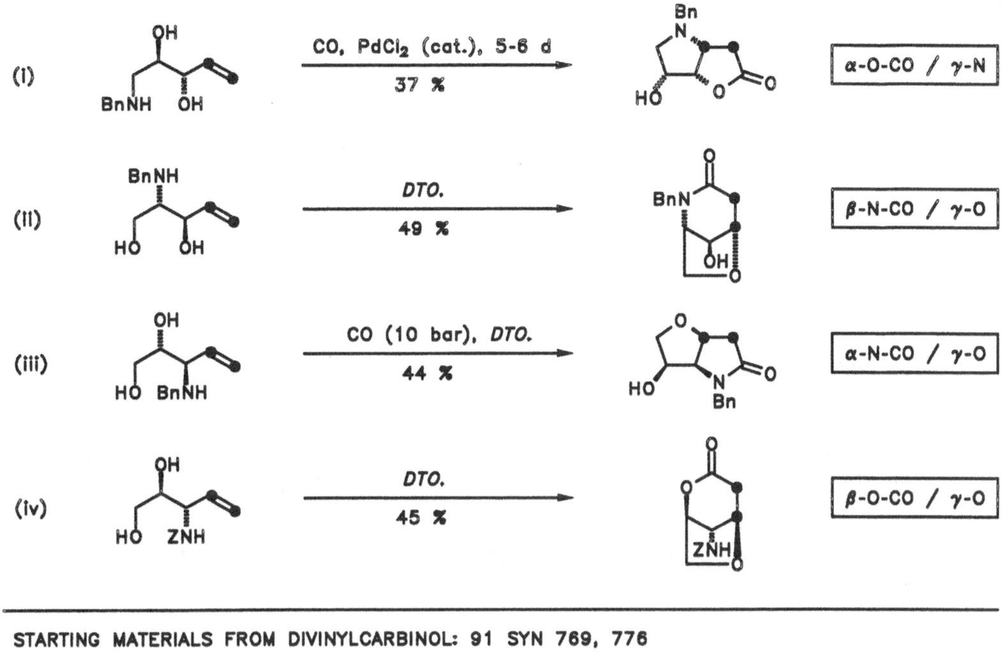

(I) ... CO, PdCl₂ (cat.), 5-6 d / 37 % ... α-O-CO / γ-N

(II) ... DTO. / 49 % ... β-N-CO / γ-O

(iii) ... CO (10 bar), DTO. / 44 % ... α-N-CO / γ-O

(iv) ... DTO. / 45 % ... β-O-CO / γ-O

STARTING MATERIALS FROM DIVINYLCARBINOL: 91 SYN 769, 776

At present, these reactions present new and interesting cases of benzylamino participation, but the yields so-far have not reached the range of preparative usefulness. Improvements may be found by varying each reagent and reaction parameter, but more detailed knowledge on the course of these reactions is warranted likewise.

In looking for other nucleophilic partners for the Pd(II)-catalyzed carbonylation of alkenepolyols, the oxime group was chosen next. A cyclization of this kind was recently described by Grigg et al., leading to a six-membered nitrone[52]. Closure to the five-membered pyrroline N-oxide "got stuck", however, forming a PdCl₂ complex with the oxime nitrogen and the C=C bond acting as chelating ligands[52]. We recently started experiments with oximes derived from enoses to this respect. Carbonylation was not observed - instead, new, mild procedures for (standard) conversions of oximes to the nitrile and the dimethyl acetal were discovered (Scheme 26)[53]. While these results were rather unexciting, in parallel experiments - without the use of palladium - a new entry to several classes of highly interesting dipoles was found, to be described elsewhere[54].

Scheme 26. Quest for new nucleophilic terminators. Results I

(i) ANSWER #1 : A NEW & MILD PROCEDURE FOR ALDOXIME CONVERSION INTO NITRILES:

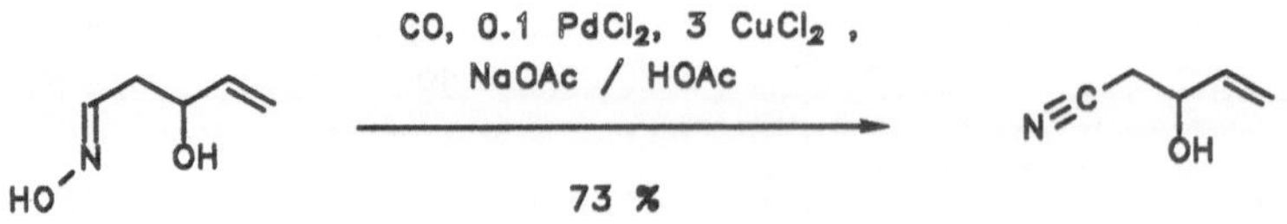

CO, 0.1 PdCl$_2$, 3 CuCl$_2$,
NaOAc / HOAc

73 %

(ii) ANSWER #2 : A NEW & MILD PROCEDURE FOR ALDOXIME CONVERSION INTO ACETALS:

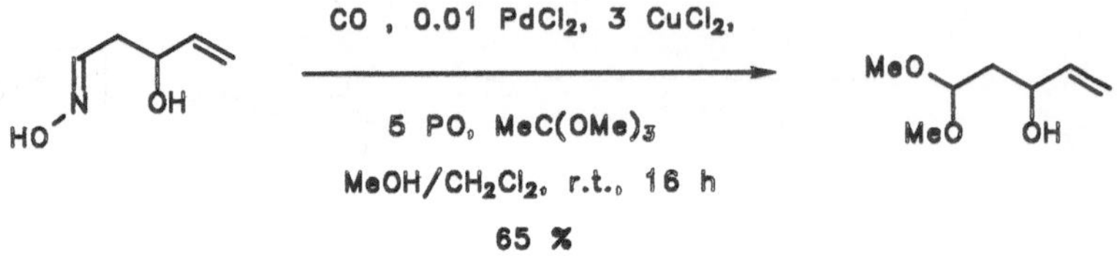

CO , 0.01 PdCl$_2$, 3 CuCl$_2$,

5 PO, MeC(OMe)$_3$
MeOH/CH$_2$Cl$_2$, r.t., 16 h

65 %

5 Mechanistic Course of Pd(II)-Catalyzed Oxy- and Amido-carbonylations

The standard view on the palladium(II)-catalyzed oxy- and amidocarbonylations is expressed in Scheme 27[1,4-6]. The various intermediates and the situation concerning the different ligands on the palladium centre have been the subject of intense discussions[1,4-6,55-57]. The role of the re-oxidant copper(II) chloride is not clear: In some cases it has been substituted by other oxidants, e. g. p-benzoquinone, and even CO/oxygen mixtures have been employed successfully[57]. The latter system seems the most promising one to develop this elegant transformation into a truly catalytic process - which is not the case with the standard reagent mix including three equivalents of copper chloride!

We have started to explore these problems, first by monitoring the CO consumption during the reaction. The stoichiometric equation given in Scheme 28 does account for one equivalent of CO, and for two of CuCl$_2$. As the test substrate we chose the *N*-Z-protected aminopentenediol mentioned earlier, because of easy monitoring and excellent yield under standard conditions (see Scheme 21). Some of these experiments are summarized in the diagramms[51]. In the first, it is seen that the reaction consumes a total of almost *four* equivalents of carbon monoxide before turning ochre at the end [experiment (i)], an unexpected and very surprising result.

In other experiments, one of the major ingredients PdCl$_2$/CuCl$_2$/substrate was left out in each case. There was no reaction taking place without PdCl$_2$, and when CuCl$_2$ was not added, the CO volume consumed after 3 days matched about half the quantitiy of the PdCl$_2$ "catalyst" employed (1.0 eq substrate, 1.0 PdCl$_2$, 3.0 NaOAc), but none of the lactone product was found[51].

Scheme 27. Assumed course of Pd(II)-catalyzed oxycarbonylation

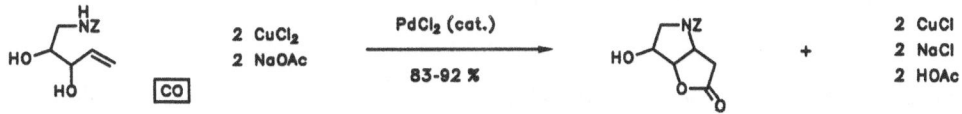

Scheme 28. Stoichiometry of an amidocarbonylation: Theory and experiment

EXPERIMENTS:

(i)

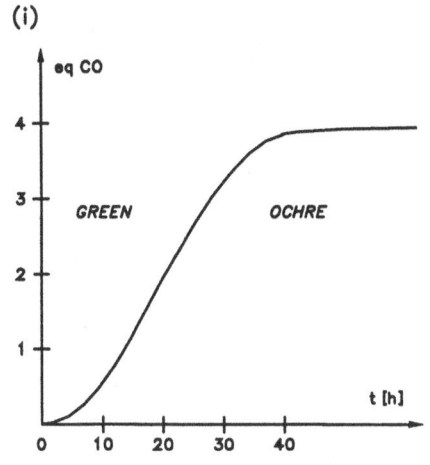

(ii)

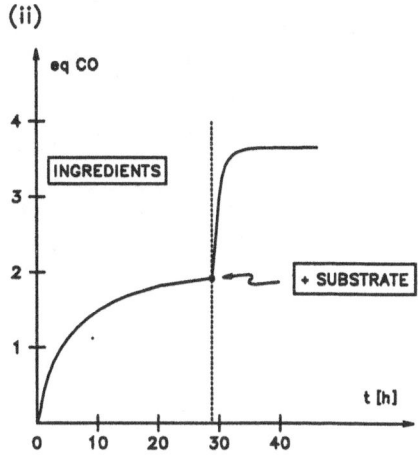

355

When the *substrate* was omitted, the mixture of PdCl$_2$/CuCl$_2$/NaOAc (0.1/3/3 eq) in acetic acid took up carbon monoxide nevertheless, see Experiment (ii) in Scheme 28. Two equivalents of CO were absorbed, and this "saturated" system exhibited *much higher catalytic activity*: On addition of the substrate, the reaction took only 1.5 h (usually ca. 30 h), with a conversion of ca. two thirds of the substrate[51]. In related experiments concerned with the dicarbonylation of 1-alkynes, Alper et al. had observed a yellow solid formed by the action of CO on a suspension of PdCl$_2$/CuCl$_2$ (1:2) in tetrahydrofuran; this also proved catalytically active producing ca. 50 % of dicarbonylated material on alkyne addition with CO/O$_2$ introduced[57]. Except for an IR analysis showing strong CO absorption, this solid could not be characterized further due to its instability in the absence of CO[57]. In Scheme 29 two complexes of the kind that may be involved here are depicted, one from a structure analysis[58], and the other as proposed by Alper et al.[57].

Scheme 29. Copper(I) carbonyl complexes

The experiments shown in Scheme 28 give support to the formation of an intermediate carbonyl complex involving copper *and* palladium. In the presence of a polyol substrate, which may act as a chelating ligand, the reaction is retarded and perhaps auto-catalytic as seen from the sigmoid shape of the consumption curve. It also seems clear that preformation of the catalyst system may lead to the development of more active, reliable catalyst mixtures, a task where cooperation with and advice from experts in transition metal chemistry will be needed.

Acknowledgements

This work was made possible by a generous grant from the Volkswagen-Stiftung, in context with the Schwerpunktprogramm "Metallorganische Reaktionen für die Organische Synthese". Thanks are further due to BMFT/BGA (AIDS-Forschungsförderung) and to Fonds der Chemischen Industrie for financial support, to the Alexander von Humboldt-Stiftung for awarding post-doctoral fellowships to Dr. T. Gracza, Dr. E. Dubois, and Dr. D. Shaw, and to Bayer AG and Hoechst AG for donation of chemicals. We also thank Prof. Dr. J.-C. Fiaud, Université de Paris-Orsay/ICMO, Prof. Dr. D. Sinou, Université Claude Bernard, Lyon, for discussions and Dr. S. Picasso and Prof. P. Vogel, Université de Lausanne, for comprehensive tests concerning glycosidase inhibition. For practical help and advice we particularly thank Dipl.-Ing.(FH) H. Griesser, also Mr. T. Gräther, Ms. S. Bucovarova, and many student participants of the undergraduate research programme at the University of Stuttgart. - Finally, last minute production of this manuscript by Mrs. G. Kraschewski, F.-M. Kiess, O. Schwardt, and J. Greul is gratefully acknowledged.

References

1. For reviews on Pd(II)-mediated chemistry and carbonylations see:
    H. M. Colquhoun, D. J. Thompson, M. V. Twigg, Carbonylation, Plenum Press, New York, **1991**; J. P. Collman, L. S. Hegedus, J. R. Norton, R. G. Finke, Principles and Applications of Organotransition Metal Chemistry, University Science Books/Oxford University Press, Mill Valley, CA, **1987**; R. F. Heck, Palladium Reagents in Organic Synthesis, Academic Press, London **1990**; J. Falbe in: Houben-Weyl, Methoden der Organischen Chemie, J. Falbe (ed.), 4. Aufl., Vol. E 18/2, Thieme, Stuttgart, **1986**; J. Falbe, Carbon Monoxide in Organic Synthesis, Springer, New York, **1970**; J. Tsuji, Palladium Reagents and Catalysts, Wiley, Chichester **1995**; P. Eilbracht in: Houben-Weyl, Methoden der Organischen Chemie, G. Helmchen, R. W. Hofmann, J. Mulzer, E. Schaumann (eds.), Vol. E 21c, Thieme, Stuttgart, **1996**, p. 2488-2734; C. Elschenbroich, A. Salzer, Organometallchemie, 3rd ed., Teubner, Stuttgart, **1993**; L. S. Hegedus, Organische Synthese mit Übergangsmetallen, VCH, Weinheim **1995**.
2. L. S. Hegedus, Tetrahedron **1984**, *40*, 2415; J.-E. Bäckvall in OSM 4, D. Enders, H.-J. Gais, W. Keim (eds.), Vieweg, Braunschweig, **1993**, p. 165; H. Hiemstra, R. A. T. M. van Benthem, Bull. Soc. Chim. Belg. **1994**, *103*, 559.
3. V. Wehner, V. Jäger, Angew. Chem. **1990**, *102*, 1181; Angew. Chem. Int. Ed. Engl. **1990**, *29*, 1169; V. Jäger, R. Müller, T. Leibold, M. Hein, M. Schwarz, M. Fengler, L. Jaroskova, M. Pätzel, P.-Y. LeRoy, Bull. Chem. Soc. Belg. **1994**, *103*, 491.; cf. B. Winchester, G. W. J. Fleet, Glycobiology **1992**, *2*, 199.
4. M. F. Semmelhack, C. Bodurow, J. Am. Chem. Soc. **1984**, *106*, 1496; M. F. Semmelhack, N. Zhang, J. Org. Chem. **1989**, *54*, 4483; M. F. Semmelhack, C. Bodurow, M. Baum, Tetrahedron Lett. **1984**, *25*, 3171; M. F. Semmelhack, C. Kim, N. J. Zhang, C. Bodurow, M. Sanner, W. Dobler, M. Meier, Pure Appl. Chem. **1990**, *62*, 2035; cf. M. McCormick, R. Monahan III, J. Soria, D. Goldsmith, D. Liotta, J. Org. Chem. **1989**, *54*, 4485.

5. Y. Tamaru, T. Kobayashi, S. Kawamura, H. Ochiai, M. Hojo, Z. Yoshida, Tetrahedron Lett. **1985**, *26*, 3207; Y. Tamaru, M. Hojo, Z. Yoshida, J. Org. Chem. **1991**, *56*, 1099.

6. Y. Tamaru, T. Kobayashi, S. Kawamura, H. Ochiai, Z. Yoshida, Tetrahedron Lett. **1985**, *26*, 4479; Y. Tamaru, M. Hojo, Z. Yoshida, J. Org. Chem. **1988**, *53*, 5731.

7. Post-doctoral work by T. Gracza, Würzburg, **1988**.

8. T. Gracza, T. Hasenöhrl, U. Stahl, V. Jäger, Synthesis **1991**, 1108.

9. V. Jäger, T. Franz, W. Schwab, B. Häfele, D. Schröter, D. Schäfer, W. Hümmer, E. Guntrum, B. Seidel in Chemistry of Heterocyclic Compounds, J. Kovac, P. Zalusky (eds.), Vol. 35, Elsevier, Amsterdam, **1988**, p. 58.

10. V. Jäger, I. Müller, R. Schohe, M. Frey, R. Ehrler, B. Häfele, D. Schröter, Lect. Heterocycl. Chem. **1985**, *8*, 79.

11. V. Jäger, D. Schröter, B. Koppenhoefer, Tetrahedron **1991**, *47*, 2195.

12. V. Jäger, V. Wehner, Angew. Chem. **1989**, *101*, 512; Angew. Chem. Int. Ed. Engl. **1989**, *28*, 469; T. Franz, M. Hein, U. Veith, V. Jäger, E.-M. Peters, K. Peters, Angew. Chem. **1994**, *106*, 1308; Angew. Chem. Int. Ed. Engl. **1994**, *33*, 1531.

13. H. Alper, D. Leonard, Tetrahedron Lett. **1985**, *26*, 5639; H. Alper, D. Leonard, J. Chem. Soc., Chem. Commun. **1985**, 511; H. Alper, Aldrichimica Acta **1991**, *24*, 3; E. A. Bassam, H. Alper, J. Org. Chem. **1991**, *56*, 5357.

14. S. Toda, M. Miyamoto, H. Kinoshita, K. Inomata, Bull. Chem. Soc. Jpn. **1991**, *64*, 3600.

15. B. Häfele, D. Schröter, V. Jäger, Angew. Chem. **1986**, *98*, 89; Angew. Chem. Int. Ed. Engl. **1986**, *25*, 87.

16. D. Schröter, T. Hasenöhrl, V. Jäger, manuscript in preparation.

17. D. Schröter, Dissertation, Universität Würzburg, 1989.

18. (a) T. Hasenöhrl, Dissertation, Universität Stuttgart, **1995**; (b) Crystal structure of the 4.3.0. lactone: S. Henkel, T. Hasenöhrl, A. Lieberknecht, V. Jäger, Z. Kristallogr., submitted.

19. V. Jäger, B. Häfele, Synthesis **1987**, 801.

20. M. Schlosser in Houben-Weyl, 4. Aufl., Vol. V/1 b; E. Müller (ed.); Thieme, Stuttgart, **1972**, p. 213.

21. For related examples see:
    V. Jäger, H. Grund, W. Schwab, Angew. Chem. **1979**, *91*, 91; Angew. Chem. Int. Ed. Engl. **1979**, *18*, 78; B. Bernet, A. Vasella, Helv. Chim. Acta **1979**, *62*, 1990; F. Heinzer, D. Bellus, Helv. Chim. Acta **1981**, *64*, 2279; R. Csuk, A. Fürstner, B. I. Glänzer, H. Weidmann, J. Chem. Soc., Chem. Commun. **1986**, 1149; A. Fürstner, D. Jumbam, J. Teslic, H. Weidmann, J. Org. Chem. **1991**, *56*, 2213.

22. M. Kleban, Dissertation, Stuttgart, **1996**.

23. A. Streitwieser, C. H. Heathcock, E. M. Kosower, Introduction to Organic Chemistry, 4th ed., MacMillan, New York, **1992**, p. 876; J. Lehmann, Chemie der Kohlenhydrate, Thieme, Stuttgart, **1976**, p. 78.

24. R. Scheffold, G. Rytz, L. Walder, Mod. Synth. Meth. **1983**, *3*, 355.

25. W. Hümmer, Dissertation, Würzburg, **1990**.

26. U. Stahl, Dissertation, Würzburg, **1993**.

27. B. E. Maryanoff, A. B. Reitz, Chem. Rev. **1989**, *89*, 863; cf. ref. 39, 40.

28. R. Csuk, B. I. Glänzer, J. Carbohydr. Chem. **1990**, *9*, 797, 802.

29. S. Hanessian, A. G. Pernet, Adv. Carbohydr. Chem. Biochem. **1976**, *33*, 111; P. DeShong, G. A. Slough, V. Elango, G. I. Trainor, J. Am. Chem. Soc. **1985**, *107*, 7788; A. Boschetti, F. Nicotra, L. Panza, G. Russo, J. Org. Chem. **1988**, *53*, 4181; D. E. Levy, C. Tang, The Chemistry of C-Glycosides, Pergamon, Oxford, **1995**.

30. A. Holy, E. De Clerq, I. Votruba in: Nucleotide Analogues as Antiviral Agents, J. C. Martin (ed.), ACS Symposium Series No. 401, ACS, Washington, **1989**, p. 51; V. E. Marquez, ibid., p. 140.

31. V. E. Marquez, C. K. Tseng, S. P. Treanor, J. S. Driscoll, Nucleosides Nucleotides **1987**, *6*, 244.

32. X.-P. Fang, J. E. Anderson, C.-J. Chang, P. E. Fanwick, J. L. McLaughlin, J. Chem. Soc., Perkin Trans 1 **1990**, 1655.

33. Cf. isolation and structure analyses of (+)-7-*epi*-goniofufurone, (+)-4-deoxygonio-pypyrone, (+)-goniodiol: X.-P. Fang, J. E. Anderson, C.-J. Chang, J. L. McLaughlin, J. Nat. Prod. **1991**, *54*, 1034; the absolute configuration of (+)-goniobutenolide A, proposed as being L-*threo* in Ref. 38a has meanwhile been established as D-*erythro*: Ref. 39 and D. Xu, K. B. Sharpless, Tetrahedron Lett. **1994**, *27*, 4685.

34. For reviews and literature compilations until **1994** see Ref. 38a, 39.

35. Cf. U. Hacksell, G. D. Daves, Jr., J. Org. Chem. **1983**, *48*, 2870; H. Ohrui, G. H. Jones, J. G. Moffatt, M. L. Maddox, A. T. Christensen, S. K. Byram, J. Am. Chem. Soc. **1975**, *97*, 4602; H. Machida, S. Sakata in: Nucleosides and Nucleotides as Antitumor and Antiviral Agents, C. K. Chu, D. C. Baker (eds.), Plenum Press, New York, **1993**, p. 1; J. Engels, Nachr. Chem. Techn. Lab. **1991**, *39*, 543.

36. O. Stephan, Dissertation, Stuttgart, **1996.**

37. T. Gracza, V. Jäger, Synlett **1992**, 191.

38. (a) T. Gracza, V. Jäger, Synthesis **1994**, 1359; (b) S. Henkel, T. Gracza, V. Jäger, Z. Kristallogr., submitted; (c) the respective statement in Ref. 38a concerns non-selective reduction in *ethyl acetate*; in MeOH and EtOH selective O-debenzylation occurs, see Ref. 38a, cf. K. R. C. Prakash, S. R. Rao, Tetrahedron **1993**, *49*, 1505.

39. Syntheses of goniofufurone, -pypyrone and several related natural "goniolactones": T. K. M. Shing, H.-C. Tsui, Z.-H. Zhou, J. Org. Chem. **1995**, *60*, 3121.

40. The syntheses of Shing's group, Ref. 39 and in earlier papers, rely on gluco-heptonolactone, with frequent use of Z-selective Wittig reactions.

41. L. S. Hegedus, G. F. Allen, D. J. Olsen, J. Am. Chem. Soc. **1980**, *102*, 3583; L. S. Hegedus, J. M. McKearin, J. Am. Chem. Soc. **1982**, *104*, 2444; Y. Tamaru, M. Hojo, H. Higashimura, Z. Yoshida, J. Am. Chem. Soc. **1988**, *110*, 3994; R. C. Larock, T. R. Hightower, L. A. Hasvold, K. P. Peterson, J. Org. Chem. **1996**, *61*, 3584; cf. D. Lathbury, P. Vernon, T. Gallagher, Tetrahedron Lett. **1986**, *27*, 6009.

42. W. Hümmer, T. Gracza, V. Jäger, Tetrahedron Lett. **1989**, *30*, 1517; V. Jäger, W. Hümmer, U. Stahl, T. Gracza, Synthesis **1991**, 769; V. Jäger, U. Stahl, W. Hümmer, Synthesis **1991**, 776.

43. W. Hümmer, E. Dubois, T. Gracza, V. Jäger, submitted for publication.

44. U. Veith, O. Schwardt, V. Jäger, Synlett, in press.

45. U. Veith, S. Leurs, V. Jäger, J. Chem. Soc., Chem. Commun. **1996**, 329; N. Meunier, U. Veith, V. Jäger, J. Chem. Soc., Chem. Commun. **1996**, 331.

46. V. Jäger, W. Hümmer, Angew. Chem. **1990**, *102*, 1182; Angew. Chem. Int. Ed. Engl. **1990**, *29*, 1171.

47. Essays done by Dr. S. Picasso, Prof. P. Vogel, Institut de Chimie Organique, Université de Lausanne; cf. A. Brandi, S. Cicchi, F. M. Cordero, R. Frignoli, A. Goti, S. Picasso, P. Vogel, J. Org. Chem. **1995**, *60*, 6806; M. Kleban, S. Picasso, P. Vogel, V. Jäger, submitted.

48. B. Kirschbaum, Dissertation, Stuttgart, **1996**.

49. F.-W. Holly, R. A. Barnes, F. R. Koniuszy, K. Folkers, J. Am. Chem. Soc. **1948**, *70*, 3088; T. Wieland, Chem. Ber. **1948**, *81*, 323; A. A. Patchett, B. Witkop, J. Am. Chem. Soc. **1957**, *79*, 185; R. Neidlein, P. Greulich, Helv. Chim. Acta **1992**, *75*, 2545.

50. For example: U. Schmidt, R. Meyer, V. Leitenberger, F. Stäbler, A. Lieberknecht, Synthesis **1991**, 409; U. Schmidt, R. Meyer, V. Leitenberger, H. Griesser, A. Lieberknecht, Synthesis **1992**, 1025.

51. U. Kautz, Diplomarbeit, Stuttgart, **1996**.

52. M. Frederickson, R. Grigg, J. Markandu, J. Redpath, J. Chem. Soc., Chem. Commun. **1994**, 2225.

53. D. E. Shaw, V. Jäger, unpublished results **1994**.

54. V. Jäger, D. E. Shaw, to be submitted.

55. D. E. James, J. K. Stille, J. Am. Chem. Soc. **1976**, *98*, 1810; J. K. Stille, R. Divakaruni, J. Org. Chem. **1979**, *44*, 3474.

56. Cf. carbonylation of acetylenic alcohols: E. G. Samsel, J. R. Norton, J. Am. Chem. Soc. **1984**, *106*, 5505, and earlier papers cited.

57. Dicarbonylation of 1-alkynes and use of oxygen reoxidant: D. Zargarian, H. Alper, Organometallics **1991**, *10*, 2914 and references given.

58. M. Pasquali, C. Floriani, A. Gaetani-Manfredotti, Inorg. Chem. **1981**, *20*, 3382; G. Rucci, M. P. Zanzottera, M. P. Lachi, M. Camia, J. Chem. Soc., Chem. Commun. **1971**, 652.

New Organometallic Solutions to Problems in Polyene Natural and Unnatural Products Synthesis

Bruce H. Lipshutz,* D.J. Buzard, Isaac Carrico,[1] David Dickson, Brian James, Craig Lindsley, Sabina Pecchi, Shelly Vance,[1] and Brett R. Ullman

Department of Chemistry, University of California, Santa Barbara, CA 93106 USA
[Fax: 805-893-8265; E Mail: lipshutz@sbmm1.ucsb.edu]

1 Introduction

By most yardsticks nowadays, polyenes are 'hot'. Of course, the term 'polyenes' is broad and seemingly all-encompassing, where any substance containing two or more carbon-carbon double bonds technically falls within this domain. But the field of polyene chemistry today is far more focused, with accents on such imposing targets as (a) the polyene macrolides,[2] typified by the clinically useful antifungal agent nystatin; (b) leukotriene C,[3] a member of the arachidonic acid cascade responsible in part for inducing human ailments such as asthma; and (c) the ubiquitous polyprenoidal coenzyme Q,[4] the *para*-quinone ultimately responsible for electron transport in most mammalian cells (Figure 1). While these are examples of naturally occurring conjugated, skipped, and doubly skipped polyenes, respectively, there are other classes of both natural and unnatural polyenes of great current interest. One of these among the latter group, which is admittedly a bit of a 'stretch' of the polyene concept, includes the biaryl structural motif,[5] most notably the binaphthyl array.[6] The immense importance of nonracemic BINOL[7] and BINAP[8] in asymmetric synthesis is hardly open to debate. And then there is the entire field of "polyacetylenes",[9] which is of considerable interest in materials science due to their non-linear optical (NLO) properties.

Figure 1

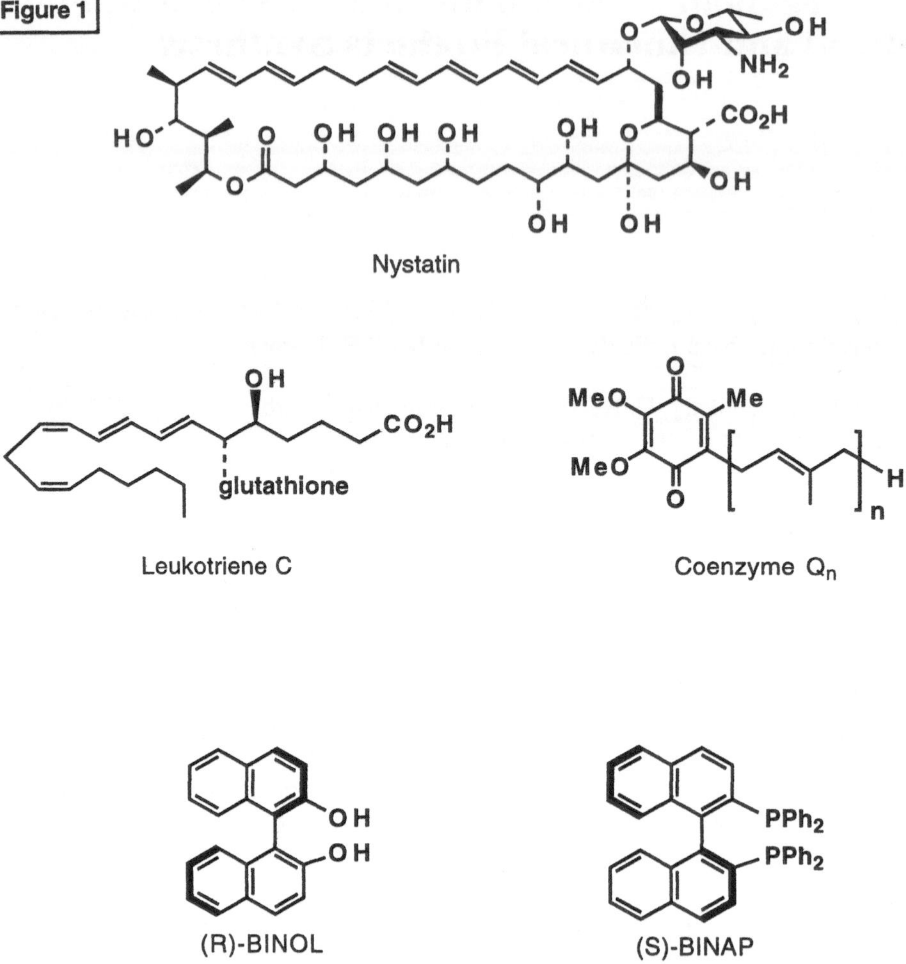

Nystatin

Leukotriene C

Coenzyme Q$_n$

(R)-BINOL

(S)-BINAP

Our programs in polyene chemistry over the past two years have now begun to bear fruit; two contributions have recently appeared describing the development of Ni^0-mediated couplings of benzylic chlorides and vinylalanes to arrive at the immediate precursors to coenzyme Q$_{3-5}$, as illustrated below.[10] Ongoing work is addressing the higher homologs, CoQ$_{6-9}$, valued at between \$10,000 and \$40,000+ per gram![11]

MeO
MeO, Me
MeO
MeO Cl

+ AlMe₂ — ... H   (n = 1-3)

$\xrightarrow{\text{cat Ni}^{\circ}, \text{rt}}$ (82-89%)

MeO
MeO, Me
MeO
MeO

$\xrightarrow{\text{CAN}}$ CoQ₃₋₅

In this presentation, however, we discuss our efforts on two other fronts: the first, in the biaryl area, involves the preparation of a new nonracemic BINOL equivalent, the approach to which is designed to be applicable to a range of substituted analogs, including the BINAP system. Secondly, we disclose a new linchpin approach to all *E*-conjugated polyenes, key subunits of many polyene macrolides, and the potential for this strategy to apply to other targets such as retinoids.

2 A New '*cyclo*-BINOL'

By far, most of the routes to optically active BINOL rely on a resolution of racemic material. Classical[12] as well as enzymatic[13] protocols have been developed in response to the increasing use of this ligand, as well as its relatively high commercial cost. Few chemical approaches exist, however, and while we described in 1994 one based on an intramolecular cyanocuprate-induced oxidative coupling (Scheme 1),[14] we endeavored to find an alternative that did not require moisture and air sensitive organometallic intermediates. Moreover, as the substitution pattern changes with each modified BINOL, it is not obvious that the resolution procedures developed for the parent system will be applicable.

We began our quest with two concepts in mind: (1) 2-naphthol is easily oxidized with FeCl₃ in H₂O to racemic BINOL,[15] suggesting an inexpensive, even environmentally friendly process might be used to advantage; (2) axial chirality could be derived from a nonracemic tether (*e.g.*, as in Eq. 1),[16] making the coupling of an intramolecular nature.

These guidelines translated to our selection of 2,7-dihydroxynaphthalene as educt, where the "back end" of the eventual 2,2'-binaphthol skeleton would be bridged by an appropriate nonracemic tether. In time, using model studies (*i.e.*, achiral tethers) and molecular modeling,[17] it was established that a 6-carbon unit afforded the greatest likelihood of controlled intramolecular biaryl coupling. The essential requirement in selecting a tether was that it possess, or allow for introduction of, asymmetry. Ultimately, after consideration of several alternatives (*e.g.*, from the 'chiral pool'), *E*-β-hydromucconic acid, in the form of its diester **1**, was selected in anticipation of asymmetry to be derived from a Sharpless asymmetric dihydroxylation (AD).[18]

363

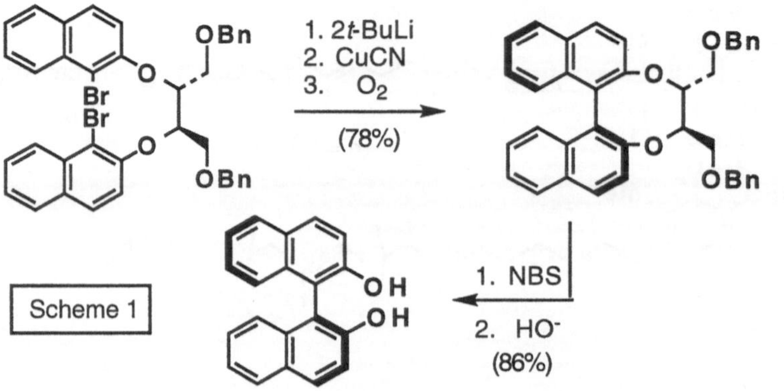

Scheme 1

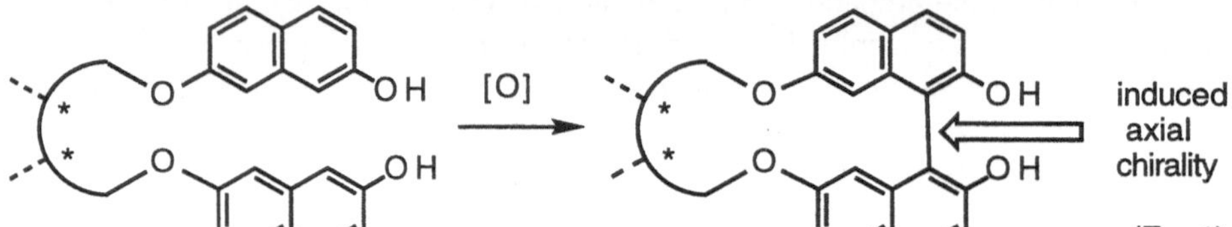

1. 2*t*-BuLi
2. CuCN
3. O₂
(78%)

1. NBS
2. HO⁻
(86%)

induced axial chirality

(Eq. 1)

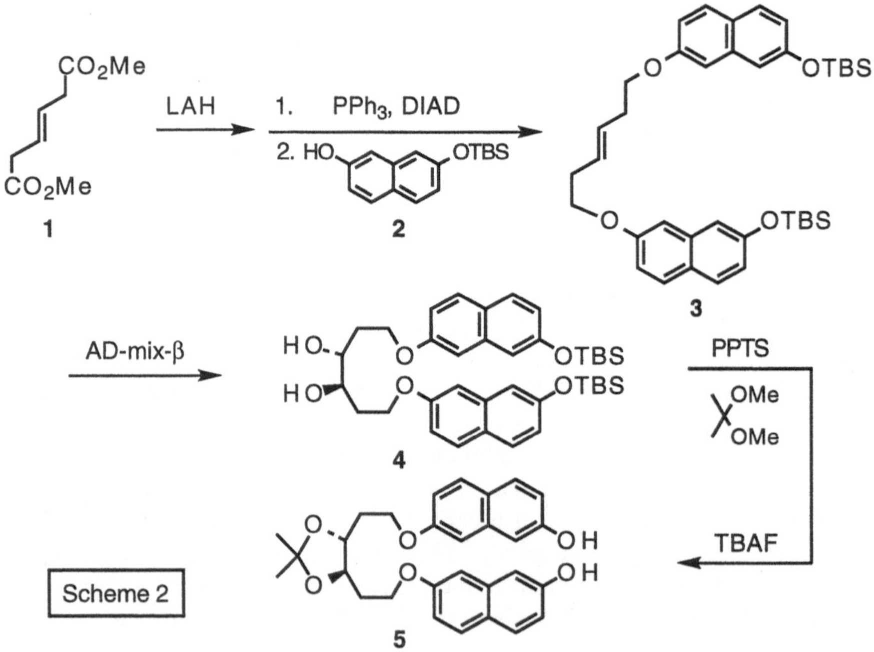

Scheme 2

LAH

1. PPh₃, DIAD
2. HO—naphthalene—OTBS

1

2

3

AD-mix-β

4

PPTS

$\overset{\text{OMe}}{\underset{\text{OMe}}{\times}}$

TBAF

5

364

Thus, dimethyl diester **1** (diacid + MeOH, cat H$^+$) was reduced with LAH to the diol, which was subjected to a double Mitsunobu reaction using the monosilylated[20] derivative of 2,7-naphthalenediol, **2** (Scheme 2). Treatment of adduct **3** with AD-mix-β[18] afforded diol **4**, the ee of which was assessed at essentially 100% using ^{19}F NMR analysis of the *bis*-MTPA derivative.[21] Standard conversion of **4** to its acetonide and then fluoride directed removal of the TBS moieties provided the pre-BINOL system **5**.

Oxidation of **5** could be effected by several different salts, including FeCl$_3$,[22] Mn(acac)$_3$,[23] and CuCl$_2$/*t*-BuNH$_2$.[24] Yields, however, were not satisfactory under a variety of conditions (aqueous, or otherwise). Best results were obtained using catalytic quantities (8 mol %) of CuCl(OH)•TMEDA complex[25] through which was continuously bubbled O$_2$ (Equation 2). Surprisingly, concentration turned out to be a critical factor, as at 0.01 M or higher, polymeric materials were evident. But at lower concentrations, yields in the 90-95% range could be realized reproducibly. The cyclization gave a ratio of axial diastereomers, about 12:1 favoring the predicted isomer.[26] Molecular modeling[27] had suggested that none of the minor isomer would form; however, these calculations do not necessarily reflect relative transition state energies, and in this case, even at ambient temperatures or below, both forms were produced (although less product is formed at lower temperatures).

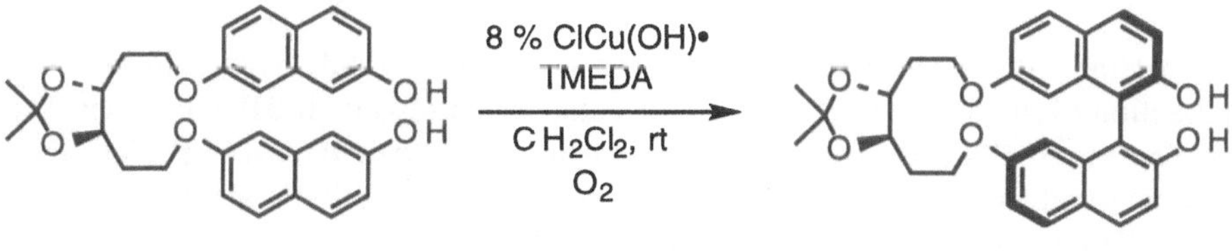

Nonetheless, pure materials are available either by standard chromatographic separation, or preparative HPLC. The sequence of six steps produces crystalline **6** in an overall yield of 41%. A Chem3D view of **6** is shown in Figure 2, as are alternative views looking down the axis of chirality in both (R)-BINOL and (R,R,R)-*cyclo*-BINOL.

Figure 2

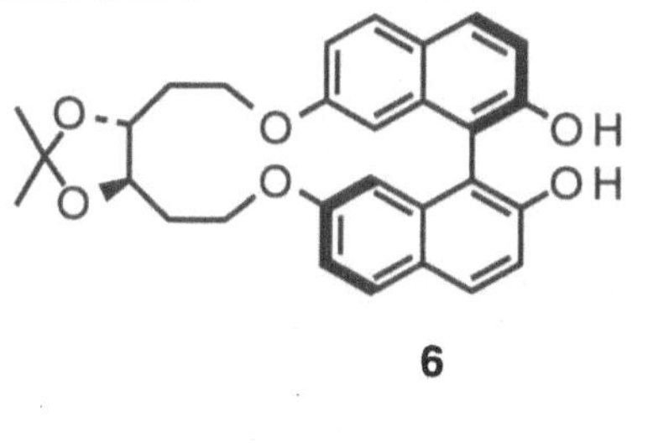

6

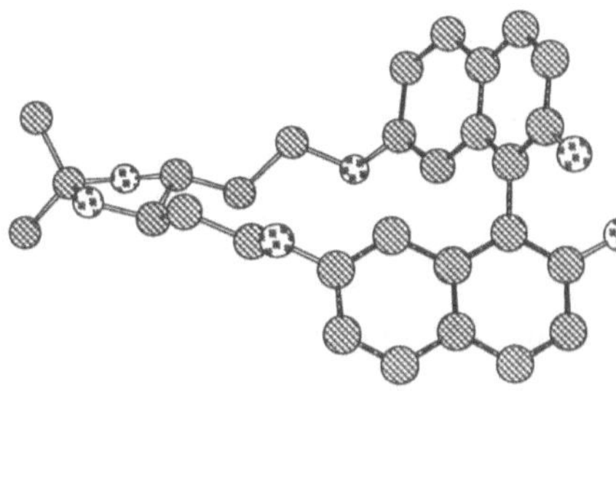

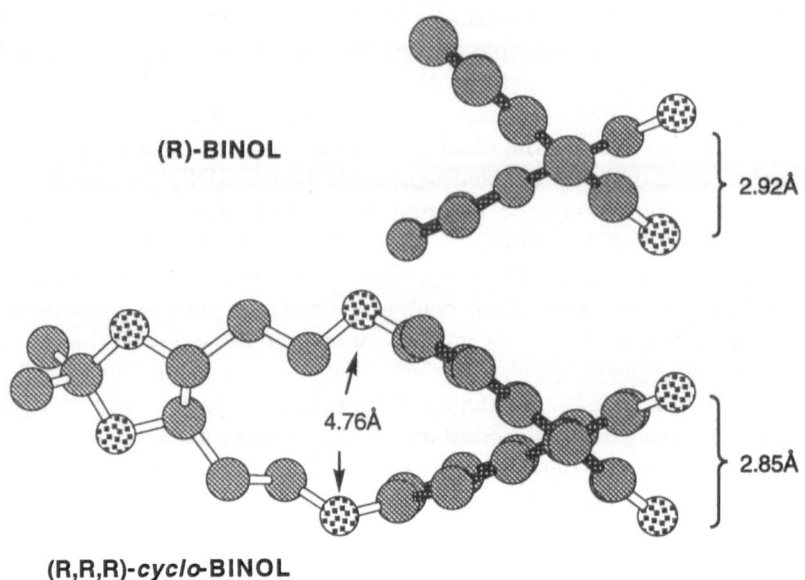

(R)-BINOL

2.92Å

4.76Å

2.85Å

(R,R,R)-*cyclo*-BINOL

With **6** in hand, its suitability as a ligand for asymmetric catalysis was tested. The allylation reaction of aldehydes was selected for comparison of **6** with BINOL under Keck conditions (Scheme 3).[28] In side-by-side reactions using benzaldehyde, *cyclo*-BINOL led to both an identical yield for, and level of chiral induction in, homoallylic alcohol **7** as seen with BINOL. Catalyst **6** was also readily recovered by standard silica gel chromatography to the extent of 70-75%.

RCHO → [cat (R)-BINOL / cat (R)-*cyclo*-BINOL]

SnBu₃ ... Ti(O-*i*-Pr)₄, CH₂Cl₂, -20°C, 48 h

HO H, R → **7**

R	BINOL	*cyclo*-BINOL
C$_6$H$_5$	84%, 95%ee	84%, 95%ee
C$_6$H$_{11}$	56%, 87%ee	58%, 85%ee

Scheme 3

The ability of **6** to function as a simple BINOL equivalent establishes its viability in asymmetric syntheses and provides a strong endorsement of the original proposal for further planning of substituted analogs. Along these lines, bromo diol derivative **8** is likely to play a valuable role. Using the Merck procedure,[29] it is also reasonable to anticipate conversion of **6** to the *cyclo*-BINAP analog **9**, which is currently under investigation. Finally, it has not gone unnoticed that the diol portion of **6** provides opportunities as well. That is, removal of the ketal unmasks the *vic* diol **10**, to which appendages could be attached, *e.g.,* to impart water solubility. Alternatively, **10** could be used to generate chiral, nonracemic ketals, *e.g.,* of enones, to direct nucleophilic additions[30] or cyclopropanations.[31]

| 8 | 9 | 10 |

3 New Linchpins for Conjugated Polyene Constructions

3.1 A Hexatriene Dianion Equivalent

The polyene macrolide antibiotics, which date back to the discovery of nystatin in 1950, are widely used antifungal agents.[2] Although well over 200 are known, synthetic efforts toward these macrocycles have only recently begun to accrue.[32]

Most of the advances can be traced to methodological developments in acyclic stereocontrol, where many novel and clever constructions of the polyol portions, which derive in nature from acetate or propionate residues, selectively give rise to the various stereo-relationships. Remarkably, however, there is little attention paid to the other side of these molecules! By far, the majority of reports on this topic rely on a Wittig or Horner-Emmons strategy,[33] building the polyene step-wise, with attention to hydrolyses, adjustments in oxidation levels, separations of *E/Z* mixes, and/or final isomerizations to the required all *E* form. We surmised that there must be a more direct way to accomplish all *E* polyene constructions, especially in light of the known tendency of stereodefined vinyl organometallics to undergo couplings, *e.g.* via Pd(0) catalysis, with retention of olefin geometry.[34] The upshot of this thinking was the design of a termini-differentiated 1,6-dimetallohexatriene **11**, in latent form, potentially capable of coupling with readily available partners to arrive at the target polyenes in very few steps. The *E*-stannylated hexadienyne **12** was envisioned as the linchpin equivalent to this hexatriene, since the vinylstannane portion should undergo Stille couplings[35] with vinylic halides after which protodesilylation would give a terminal alkyne susceptible to hydrozirconation. This intermediate would then be subject to transmetalation (to Pd, Cu, Zn, Al, etc.) followed by any of a variety of C-C bond-forming processes, including those involving vinylic leaving groups. Thus, in principle, a two-pot process might be realized for generating from four to seven conjugated, all *E* polyenes.

11   ≡   **12**

The linchpin, fortunately, turned out to be relatively straightforward in its preparation (Scheme 4). Using the Quintard procedure[36] for addition of our stannyl cuprate[37] to acetylene acetal **13** gave, upon hydrolysis enal **14**. Wittig coupling with ylide **15**[38] provided the trimethylsilyl-protected material, **16**, although the $E:Z$ ratio was highly dependent on conditions. Excellent results were obtained using either $KN(TMS)_2$ as base in toluene, or $NaN(TMS)_2$ in THF, thereby giving the linchpin in ratios ranging from $92E:8Z$ to $95E:5Z$. Although chromatographic separation was not possible, it was found that the Z isomer is reluctant to undergo coupling, and hence the first product of our proposed two-step route turned out to be virtually all E.

There are several options open for effecting couplings with the vinylstannane portion of **16**. For example, lithium-tin exchange with n-BuLi followed by addition of $ZnCl_2$ in THF gives the corresponding vinylzinc reagent which, in the presence of 5 mol % Pd(0), couples with vinylic iodides in high yields.[39] More convenient, however, is to use CuCN (1.5 eq) in a Stille-like process under modified conditions of Farina,[40] (either using $Pd_2(dba)_3$ / $AsPh_3$ or $Pd_2(dba)_3$ / $(furyl)_3P$) which go more slowly but in comparable yields at room temperature. Both proton and ^{13}C NMR analyses reveal the presence of trace amounts of the Z-coupled material, implying that the E form of the linchpin is far more prone toward coupling under the conditions employed. Some representative examples are illustrated in Scheme 5. Removal of the TMS moiety in these adducts is accomplished with K_2CO_3 in EtOH, which sets the stage for further manipulation of the terminal alkyne produced.

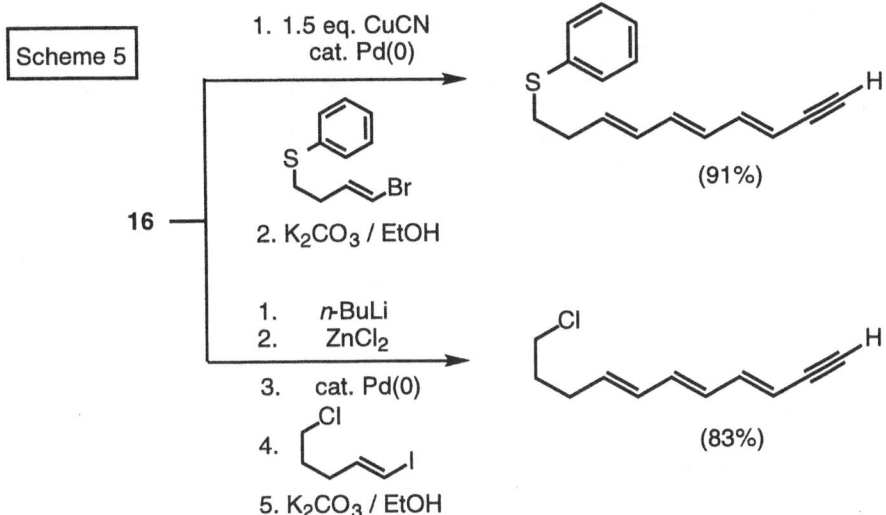

Scheme 5

16

1. 1.5 eq. CuCN
   cat. Pd(0)

2. K$_2$CO$_3$ / EtOH

(91%)

1. *n*-BuLi
2. ZnCl$_2$
3. cat. Pd(0)
4.
5. K$_2$CO$_3$ / EtOH

(83%)

Fortunately, Schwartz' reagent (Cp$_2$Zr(H)Cl)[41] adds cleanly across the alkyne to give an *E*-vinyl zirconocene. Although we have not been able to further transmetalate from Zr to Zn using Et$_2$Zn,[47] introduction of Me$_2$AlCl leads to a reactive vinylalane which couples with ethyl chloroformate to install the conjugated carboethoxy unit (*e.g.*, Eq. 3).[43] Carboalumination[44] of these 1-alkynes also proceeds readily at room temperature, and the vinylalanes produced likewise react with chloroformates (Eq. 4). Acid chlorides are also excellent electrophiles for these couplings, exemplified by the trivial synthesis of navenone B (Eq. 5). [45]

(Eq. 3)

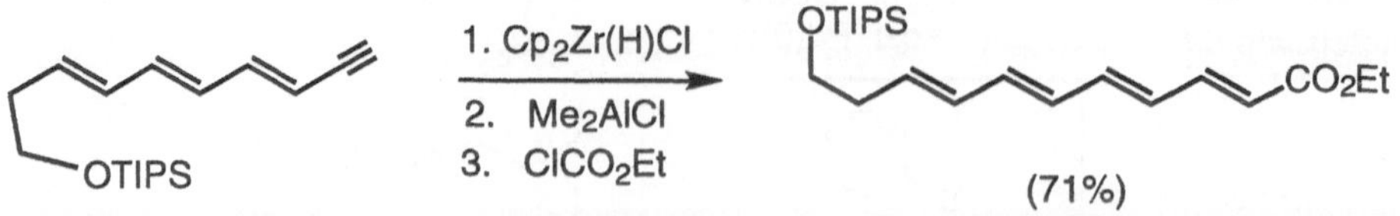

OTIPS
1. Cp2Zr(H)Cl
2. Me2AlCl
3. ClCO2Et
OTIPS
CO2Et
(71%)

(Eq. 4)

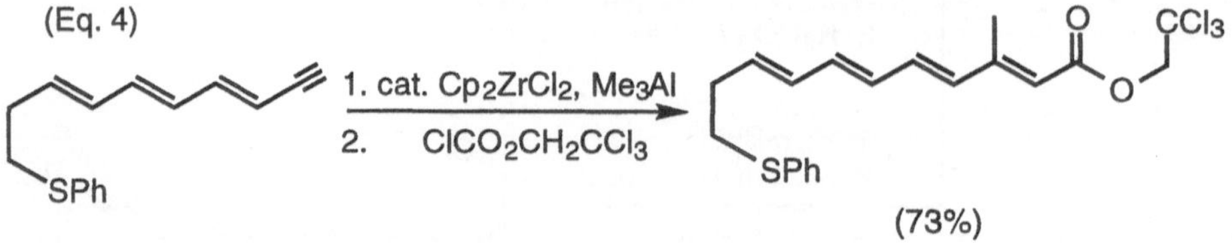

SPh
1. cat. Cp2ZrCl2, Me3Al
2. ClCO2CH2CCl3
O
CCl3
O
SPh
(73%)

(Eq. 5)

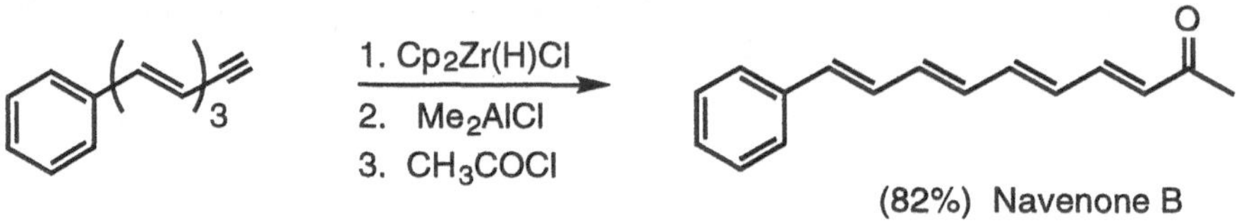

3
1. Cp2Zr(H)Cl
2. Me2AlCl
3. CH3COCl
O
(82%) Navenone B

A particularly noteworthy example is that which arrives at the polyene portion of the polyene macrolides mycoticin A and B, surgumycin, roseofungin, RK-397, etc.,[46] which are oxopentaenes containing an ester group on one end, and a homoallylic alcohol at the other terminus. An initial coupling of the linchpin with dienic iodide **17**[47] leads to tetraenyne **18**, which then undergoes the hydrozirconation-transmetallation-chloroformate quenching to arrive at all *E*-pentaenoate **19** in 52% overall yield (Scheme 6).[45]

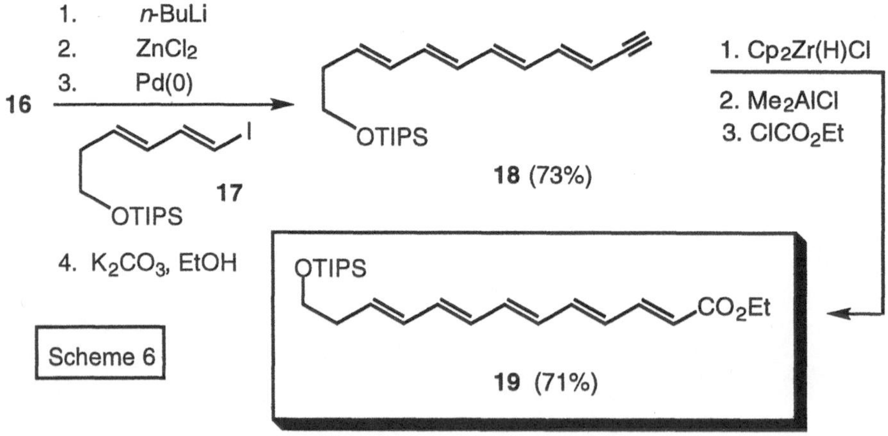

1. n-BuLi
2. ZnCl2
3. Pd(0)
16
I
17
OTIPS
4. K2CO3, EtOH
OTIPS
18 (73%)
1. Cp2Zr(H)Cl
2. Me2AlCl
3. ClCO2Et
OTIPS
CO2Et
19 (71%)

Scheme 6

Thus far, all attempts to further couple intermediates **20** with vinylic halides using Pd(0) catalysis, unfortunately, have met with failure. The initial vinylic zirconocenes themselves (**20**), or their presumed transmetallated derivatives **21** and **22**, all show no interest in undergoing ligand exchange of a tetra- or pentaenyl moiety for halide X in **23**, preferring instead to give only the proton-quenched product **24** (Figure 3), along with homocoupled product from the vinyl halide. Until a satisfactory coupling is found, this route solves the tetra- and pentaene problems in polyene synthesis, but leaves untouched the hexaene and heptaene systems. It was time, therefore, for a second generation linchpin.

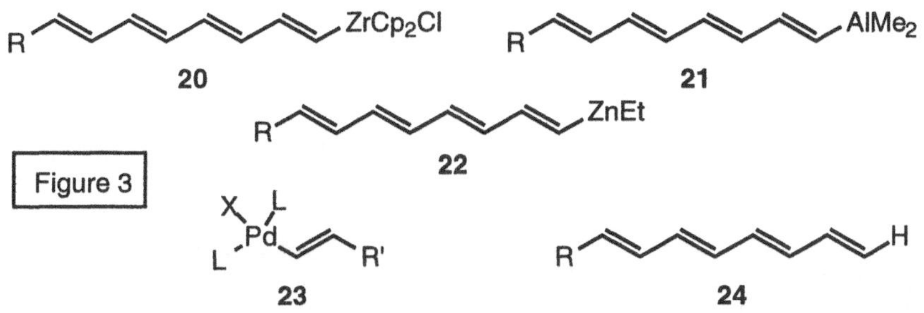

Figure 3

3.2 A New Bromotrienyne

In order to gain entry to the hexaene and, perhaps, the heptaene manifold, a linchpin that is further elongated was planned. Advantage was taken of the known conversion of pyridine sulfur trioxide to bromoenal **27** *via* isolable enolate salt **26**,[48] which is susceptible to the same Wittig extension used *en route* to **16** (Scheme 4), to now produce trienyne linchpin **28** in only three steps (Scheme 7). The initial *E* : *Z* ratio in our hands is only *ca.* 60 : 40 - 50 : 50,[49] however, the isomers are readily separated. Subsequent olefination proceeds to give an 85 : 15 mix favoring the desired all *E* product, and one pass through a standard silica gel column for purification purposes improves this ratio to ≥ 93 : 7.

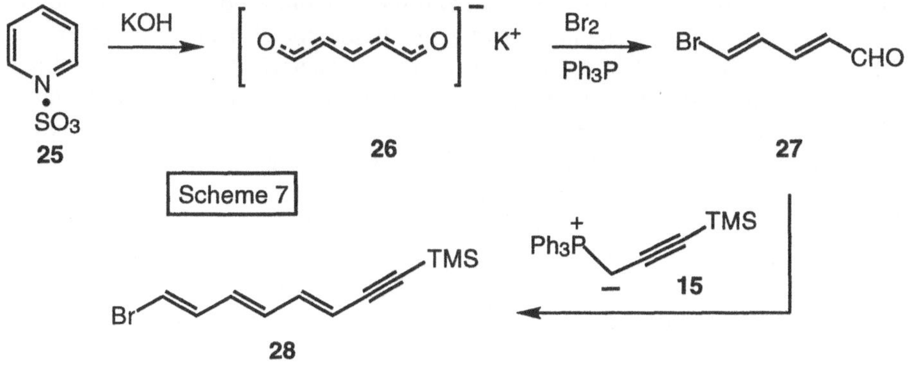

Scheme 7

Bromide **28** presents quite a different set of options for polyene synthesis relative to those using vinylstannane **16**. Perhaps the most obvious is that in **28** the trienyne is suitable as the electrophilic, rather than nucleophilic, partner in a Pd(0) coupling. Indeed, treatment of **28** with the zinc reagent **30** derived from iodide **29** leads to pentaenyne **31** after removal of the Me₃Si group. Use of the sequence developed for the tetraene series (*i.e.*, hydrozirconation, transmetallation to aluminum, and chloroformate quenching; Equations 3,4, and Scheme 8) ultimately affords hexaenoic ester **32** in 53% overall yield. Other examples using this technology are being examined to determine the scope of the method.

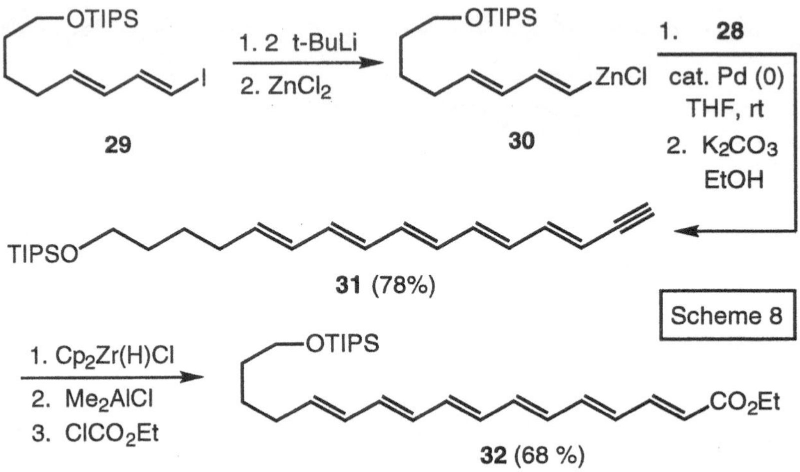

29

1. 2 t-BuLi
2. ZnCl₂

30

1. **28**
   cat. Pd (0)
   THF, rt
2. K₂CO₃
   EtOH

31 (78%)

Scheme 8

1. Cp₂Zr(H)Cl
2. Me₂AlCl
3. ClCO₂Et

32 (68 %)

3.3 Retinoids

In addition to the well-known service of retinoids in vision,[50] they are now becoming 'center stage' compounds for their roles in cell proliferation and differentiation, in particular with respect to their antitumor activities.[51] Indeed, only recently has the FDA approved Roche's Vesanoid for treatment of a rare form of leukemia. Extensive trials on other forms of cancer are currently ongoing worldwide, and the economic potential here, beyond the already used retinoids (Accutane and Retin-A for treatment of acne) has not escaped the watchful eye of Wall Street.[52] Analogs of the natural materials, in particular those with modified rings, and those with altered and/or conformationally restricted polyene chains, have become a means of probing retinoid receptor specificity.[53] We have begun to apply our linchpin approach to the rapid construction of these important compounds as well.

Starting with dienyne **33**, derivable in quantity from β-ionone *via* an *Organic Synthesis* procedure,[54] Sonogashira attachment[55] of linchpin **28** leads, after carbonate-based desilylation, to the 9,10-dehydro derivative **34** in 89% overall yield (Scheme 9). A second Sonogashira coupling of **34** with Z β-bromo ethyl acrylate, also performed in diethyl ether, ultimately arrives at product **35**, an unusual didesmethyl, C-21 'analog' of all *trans* retinoic acid.

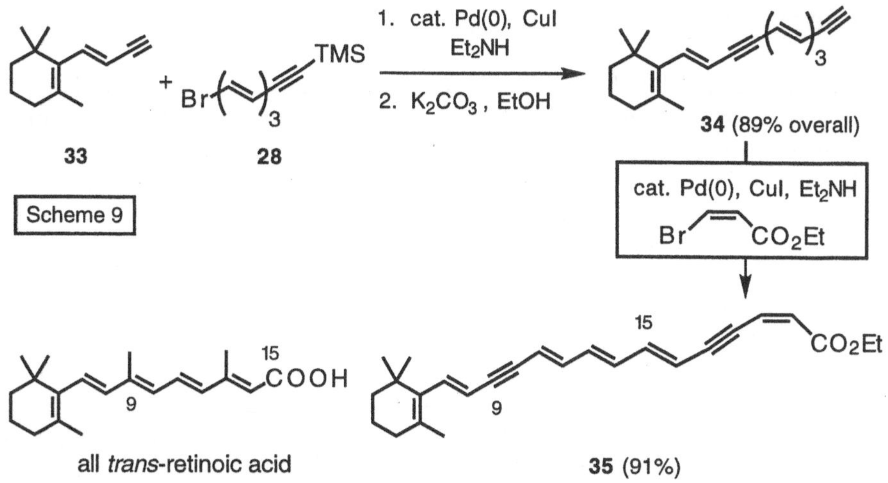

Scheme 9

34 (89% overall)

35 (91%)

all *trans*-retinoic acid

A C-9 desmethyl seventeen carbon analog could also be readily prepared using linchpin **16**. Thus, from the same educt (*i..e.*, β-ionone, Scheme 10), a palladium(0) coupling of vinyl iodide **39** proceeded smoothly to afford, after desilylation, the pentaenyne **37** in good isolated yield. Installation of the C-15 methyl moiety and C-17 carboethoxy group occurred without incident to produce the all *E* hexaenoic acid ester **38** to the extent of 65% overall.

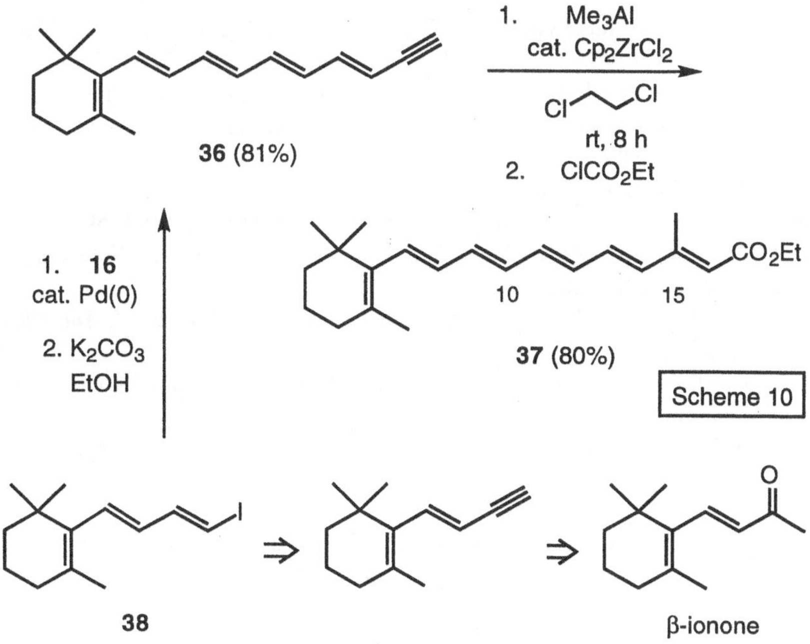

36 (81%)

37 (80%)

Scheme 10

38

β-ionone

373

4 Summary

The chemistry described highlights the increasingly important position which polyenes occupy as targets of modern organic synthesis. A novel *cyclo*-BINOL **6** was disclosed, which is readily available *via* a short sequence in optically pure form and potentially convertible to a BINAP analog. Linchpins **16** and **28** were also discussed as rapid entries to key subunits of clinically useful polyene macrolides, and to novel extended retinoids.

Acknowledgements

We are very pleased to acknowledge the continued financial support of our programs by the NIH (GM 40287) and the National Science Foundation (CHE 93-03883).

References

1. Undergraduate research associate supported by a grant to UCSB from the Howard Hughes Medical Institute.
2. Omura, S.; Tanaka, H. In *Macrolide Antibiotics: Chemistry, Biology and Practice*. Omura, S. Ed. Academic Press: New York, 1984; pp. 351–404; Rychnousky, S.D. *Chem. Rev.* **1995**, *95*, 2021.
3. Morris, H.R. *Nature* **1980**, *285*, 104; Smith, L.J.; Glass, M.,; Minkwitz, M.C. *Clin. Phar. Ther.*, **1993**, *54*, 430.
4. (a) Lenaz, G. Coenzyme Q. Biochemistry, Bioenergetics, and Clinical Applications of Ubiquinone, Wiley-Interscience, New York, **1985**. (b) Trumpower, B.L. Function of Ubiquinones in Energy Conserving Systems, Academic Press, New York, 1982. (c) Thomson, R.H. Naturally Occuring Quinones, 3rd Ed., Academic Press, New York, 1987. (d) Bliznakov, G.G.; Hunt, G.L. The Miracle Nutrient Coenzyme Q_{10}, Bantom Books, New York, **1987**.
5. Bringmann, G.; Walter, R.; Weirich, R. *Angew. Chem. Int. Ed., Engl.* **1990**, *29*, 977.
6. Rosini, C.; Franzini, L.; Raffaelli, A.; Salvadori, P. *Synthesis* **1992**, 503.
7. Cram, D.J.; Cram, J.M. *Acct. Chem. Res.* **1978**, *11*, 8. Jacques, J.; Fouquey, C. *Tetrahedron Lett.* **1971**, 4617.
8. Inoke, S.; Takaya, H.P.; Tam, K.; Otsuka, S.; Sato, T.; Noyori, R. *J. Am. Chem. Soc.* **1990**, *112*, 4897.
9. *Nonlinear* Optical *Materials: Theory and Modeling*, Karna, S.P.; Yeates, A.T. Eds., ACS Symposium Series 628, **1996**.
10. Lipshutz, B.H.; Bulow, G.; Lowe, R.F.; and Stevens, K.L. *J. Am. Chem. Soc.* **1996**, *118*, 5512.
11. Prices are quoted from the 1996 Sigma catalog.
12. Jacques, J.; Fouquay, C. *Org. Synth.* **1988**, *67*, 1. Truesdale, L. *Org Synth.* **1988**, *67*, 13. Mazalerat, J.–P.; Wakselman, M. *J. Org. Chem.* **1996**, *61*, 2695.

13. Fujimoto, Y.; Iwadate, H.; Ikekawa, N. *J. Chem. Soc. Chem. Commun.* **1985**, 1333. Kazlavkas, R.J. *J. Am. Chem. Soc.* **1989**, *111*, 4953.

14. Lipshutz, B.H.; Kayser, F.; Liu, Z.-P. *Angew. Chem. Int. Ed. Engl.* **1994**, *33*, 1842.

15. Pummerer, R.; Rieche, A.; Prell, E. *Chem. Ber.* **1926**, *59*, 2159. Toda, F.; Tanaka, K.; Iwata, S. *J. Org. Chem.* **1989**, *54*, 3007.

16. Bringman, G.; Keller, P.A.; Rolfing, K. *Synlett* **1994**, 423. Maitra, U.; Bandyopadhyay, A.K. *Tetrahedron Lett.* **1995**, *36*, 3749.

17. Molecular mechanics calculations were performed using the MM3 forcefield as applied by the Chem 3D modeling program, version 3.1.

18. Kolb, H.C.; VanNieuwenhze, M.S.; Sharpless, K.B. *Chem. Rev.* **1994**, *94*, 2483.

19. Hughes, D.L. *Org. React* **1992**, *42*, 335. Coleman, R.S.; Grant, E.B. *J. Am. Chem. Soc.* **1995**, *117*, 10889. Mitsunobu, O. *Synthesis* **1981**, 1.

20. As no general procedure exists for monosilylation of polyhydroxylated aromatics, a statistical distribution of products was obtained.

21. Dale, J.A.; Dull, D.L.; Mosher, H.S. *J. Org. Chem.* **1969**, *34*, 2543. Dale, J.A.; Mosher, H.S. *J. Am. Chem. Soc.* **1973**, *95*, 512.

22. Miller, L.L.; Jempty, T.C.; Mazur, Y. *J. Org. Chem.* **1980**, *45*, 751. Tanaka, M.; Mitsuhashi, H.; Wakamatsu, T. *Tetrahedron Lett.* **1992**, *33*, 4161.

23. Dewar, M.J.S.; Nakaya, T. *J. Am. Chem. Soc.* **1968**, *90*, 7134.

24. Smrcina, M.; Vyskocil, S.; Máca, B.; Polasek, M.; Claxton, T.A.; Abbott, A.P.; Kocovsky, P. *J. Org. Chem.* **1994**, *59*, 2156.

25. Noji, M.; Nakajima, M.; Koga, K. *Tetrahedron Lett.* **1994**, *35*, 7983.

26. The stereochemical outcome of the AD reaction was predicted based on the model of Sharpless.[18] Based on modeling studies the major diastereomer of **6** would have the R axial configuration. Based on the outcome of allylations the absolute axial stereochemistry of **6** is tentatively assigned as R.

27. Using the SYBYL molecular modeling package, version 6.2, an energy difference of 4.88 kcal/mol was found for the axial diastereomers of *cyclo*-BINOL **6**.

28. Keck, G.E.; Tarbet, K.H.; Geraci, L.S. *J. Am. Chem. Soc.* **1993**, *115*, 8467. Keck, G.E.; Geraci, L.S. *Tetrahedron Lett.* **1993**, *34*, 7827.

29. Cai, D.; Payack, J.F.; Bender, D.R.; Hughes, D.L.; Verhoeven, T.R.; Reider, P.J. *J. Org. Chem.* **1994**, *59*, 7180.

30. Alexakis, A.; Mangeney, P. *Tetrahedron Asymmetry* **1990**, *1*, 477, and references therein.

31. Yuan, T.-M.; Hsieh, Y.-T.; Yeh, S.-M.; Shyue, J.-J.; Luh, T.-Y. *Synlett* **1996**, 53. Yeh, S.M.; Huang, L.-H.; Luh, T.-Y. *J. Org. Chem.* **1996**, *61*, 3906.

32. For a few representative reports, see Bonini, C.; Giugliano, A.; Racioppi, R.; Righi, G. *Tetrahedron Lett.* **1996**, *37*, 2487. Mori, Y.; Asai, M.; Kawade, J.; Furukawa, H. *Tetrahedron*, **1995**, 51, 5315. Rychnovsky, S.C.; Griesgraber, G.; Jinsoo, K. *J. Am. Chem. Soc.* **1994**, *116*, 2621. Nicolaou, K.C.; Sorenson, E.J. *Classics in Total Synthesis* VCH Publishers, N.Y., 1996, Ch. 24.

33. Nicolaou, K.C. *Chemtracts-Organic Chemistry*, **1990**, *3*, 327. Hanessian, S., Bota, M., *Tetrahedron Lett.*, **1987**, *28*, 1151.

34. Hegedus, L.S., in *Transition Metals in the Synthesis of Complex Organic Molecules*, University Science Books, Mill Valley, CA, **1994**.

35. Stille, J.K.; Groh, B.L. *J. Am. Chem. Soc.* **1987**, *109*, 813.

36. Beaudet, I.; Parrain, J.-L.; Duchêne, A.; Quintard, J.-P. *Tetrahedron Lett.* **1991**, *32*, 6333.

37. Lipshutz, B.H.; Ellsworth, E.L.; Dimock, S.H.; Reuter, D.C. *Tetrahedron Lett.* **1989**, *30*, 2065.

38. Corey, E.J.; Ruden, R.A. *Tetrahedron Lett.* **1973**, 1495.

39. Knochel, P.; Singer, R.D. *Chem. Rev.* **1993**, *93*, 217. Erdik, E. *Organozinc Reagents in Organic Synthesis,* CRC Press, Boca Raton, FL, 1996.

40. Farina, V., presented at OMCOS 8, August 6-10, 1995, Santa Barbara, CA.

41. Hart, D.W.; Schwartz, J. *J. Am. Chem. Soc.* **1974**, *96*, 8115.

42. Wipf, P.; Xu, W. *Tetrahedron Lett.* **1994**, *35*, 5197.

43. Carr, D.; Schwartz, J. *J. Am. Chem. Soc.* **1979**, *101*, 3521.

44. Okukado, N.; Negishi, E. Tetrahedron *Lett.* **1978**, 2357.

45. Fenical, W.; Sleeper, H.L. *J. Am. Chem. Soc.* **1977**, *99*, 2367.

46. Isolation: Wasserman, H.H.; Van Verth, J.E.; McCaustland, D.J.; Borowitz, I.J.; Kamber, B.J. *J. Am. Chem. Soc.* **1967**, *89*, 1535. Synthesis: Poss, C.S.; Rychnovsky, S.D.; Schreiber, S.L. *J. Am. Chem. Soc.* **1993**, *115*, 3360.

47. Negishi, E. *Acc. Chem. Res.* **1982**, *15*, 340.

48. Soullez, D.; Ple, G.; Duhamel, L.; Duhamel, D. *J. Chem. Soc. Chem. Commun.* **1995**, 563.

49. The ratio quoted in the literature is 75E : 25Z.[48]

50. Yoshizawa, T., Kandori, H. *Progress in Retinal Research*; Osborne, N.N, Chader, G.J., Eds.; Pergamon Press: Oxford, 1992; Vol. 11, pp 33-55. Kandori, H., Sasabe, H., Nakanishi, K., Yoshizawa, T., Mizukami, T., Schichida, Y., *J. Am. Chem. Soc.* **1996**, *118*, 1002. Nakanishi, K., Crouch, R. *Isr. J. Chem.*, **1995**, *35*, 253.

51. Recent overviews: Spoin, M.B.; Roberts, A.B.; Goodman, D.S. eds., "The Retinoids: Biology, Chemistry, and Medicine," 2nd Ed; Raven: New York, 1993; Hashimoto, Y.; Shudo, K. *Cell. Biol. Rev.* **1991**, *25*, 209; Dawson, M.I.; Okamura, W.H. Eds., "Chemistry & Biology of Synthetic Retinoids", CRC Press: Boca Raton, FL, 1990.

52. Tanouye, E. The Wall Street Journal, Wednesday, November, 29, 1995. Tanouye, E. The Wall Street Journal, Wednesday, December 6, 1995.

53. Pfahl, M. in "From Molecular Biology to Therapeutics," Bernard, B.A.; Shroot, B. Eds; Karger: Basel, **1993**, pp. 83-93; Pfahl, M. in "Retinoids: From Basic Science to Clinical Applications," Livrea, M.A.; Vidali, G. eds; Birkhäuser: Basel, 1994, pp. 113-126.

54. Negeshi, E.; King, A.O.; Tour, J.M. *Org. Synth.*, **1985**, *64*, 44.

55. Sonogashira, K.; Tohda, Y.; Hagihara, N. *Tetrahedron Lett.*, **1975**, 4467.